Early Mathematics Learning and Development

Series Editor

Lyn D. English
Queensland University of Technology, School of STM Education
Brisbane, QLD, Australia

More information about this series at http://www.springer.com/series/11651

Aisling Leavy · Maria Meletiou-Mavrotheris
Efi Paparistodemou
Editors

Statistics in Early Childhood and Primary Education

Supporting Early Statistical and Probabilistic Thinking

Springer

Editors
Aisling Leavy
Department of STEM Education
Mary Immaculate College, University of Limerick
Limerick, Ireland

Efi Paparistodemou
Cyprus Pedagogical Institute
Latsia, Nicosia, Cyprus

Maria Meletiou-Mavrotheris
Department of Education Sciences
European University Cyprus
Nicosia, Cyprus

ISSN 2213-9273 ISSN 2213-9281 (electronic)
Early Mathematics Learning and Development
ISBN 978-981-13-1043-0 ISBN 978-981-13-1044-7 (eBook)
https://doi.org/10.1007/978-981-13-1044-7

Library of Congress Control Number: 2018945073

Printed on acid-free paper

This Springer imprint is published by the registered company Springer Nature Singapore Pte Ltd.
The registered company address is: 152 Beach Road, #21-01/04 Gateway East, Singapore 189721, Singapore

To Sercan, Róisín and Deren who will be disappointed to find this book is not about wizards, dragons or fictional characters
&
To Stathis, Nikolas, and Athanasia for giving me the power to embrace the uncertain future with curiosity and optimism
&
Panayiotis, Christoforos and Despina for creating chances

Foreword

Educate a child according to his way:
even as he grows old he will not depart from it.
Proverbs 22, 6

In the era of data deluge, people are no longer passive recipients of data-based reports. They are becoming active data explorers who can plan for, acquire, manage, analyse, and infer from data. The goal is to use data to understand and describe the world and answer puzzling questions with the help of data analysis tools and visualizations. Being able to provide good evidence-based arguments and critically evaluate data-based claims are important skills that all citizens should have and, therefore, that all students should learn as part of their formal education.

Statistics is therefore such a necessary and important area of study. Moore (1998) suggested that it should be viewed as one of the liberal arts and that it involves distinctive and powerful ways of thinking. He wrote: "Statistics is a general intellectual method that applies wherever data, variation, and chance appear. It is a fundamental method because data, variation, and chance are omnipresent in modern life" (p. 134). Understanding the powers and limitations of data is key to active citizenship and to the prosperity of democratic societies. It is not surprising therefore that statistics instruction at all educational levels is gaining more students and drawing more attention. Today's students need to learn to work and think with data and chance *from an early age*, so they begin to prepare for the data-driven society in which they live. This book is therefore a timely and important contribution in this direction.

This book provides a useful resource for members of the mathematics and statistics education community that facilitates the connections between research and practice. The research base for teaching and learning statistics and probability has been increasing in size and scope, but has not always been connected to teaching practice nor accessible to the many educators teaching statistics and probability in early childhood and primary education. Despite the recognized importance of developing young learners' early statistical and probabilistic reasoning and conceptual understanding, the evidence base to support such a development is rare.

By focusing on this important emerging area of research and practice in early childhood (ages 3–10), this publication fills a serious gap in the literature on the design of probability and statistics meaningful experiences into early mathematics teaching and learning practices. It informs best practices in research and teaching by providing a detailed account of comprehensive overview of up-to-date international research work on the development of young learners' reasoning with data and chance in formal, informal, and non-formal education contexts.

The book is also an important contribution to the growth of statistics education as a recognized discipline. Only recently, the first International Handbook of Research in Statistics Education has been published (Ben-Zvi, Makar, & Garfield, 2018), signifying that statistics education has matured to become a legitimate field of knowledge and study. This current book provides another brick in building the solid foundation of the emerging discipline by providing a comprehensive survey of state-of-the-art knowledge, and of opportunities and challenges associated with the early introduction of statistical and probabilistic concepts in educational settings.

By providing valuable insights into contemporary and future trends and issues related to the development of early thinking about data and chance, this publication will appeal to a broad audience that includes not only mathematics and statistics education researchers, but also teaching practitioners. It is not often that a book serves to synthesize an emerging field of study while at the same time meeting clear practical needs: educate a child during his early years with powerful ideas in statistics and probability even at an informal level, and even as he grows old he will not depart from it.

It is a deep pleasure to recommend this pioneering and inspiring volume to your attention.

Haifa, Israel

Dani Ben-Zvi
The university of Haifa

References

Ben-Zvi, D., Makar, K., & Garfield J. (Eds.) (2018). *International handbook of research in statistics education*. Springer international handbooks of education. Springer Cham.

Moore D. S. (1998). Statistics among the Liberal Arts. *Journal of the American Statistical Association*, *93*(444), 1253–1259.

Preface

Introduction

New values and competencies are necessary for survival and prosperity in the rapidly changing world where technological innovations have made redundant many skills of the past. The expanding use of data for prediction and decision-making in almost all domains of life has made it a priority for mathematics instruction to help all students develop their statistical and probabilistic reasoning (Franklin et al., 2007). Despite, however, the introduction of statistics in school and university curricula, the research literature suggests poor statistical thinking among most college-level students and adults, including those who have formally studied the subject (Rubin, 2002; Shaughnessy, 1992).

In order to counteract this and achieve the objective of a statistically literate citizenry, leaders in mathematics education have in recent years being advocating a much wider and deeper role for probability and statistics in primary school mathematics, but also prior to schooling (Shaughnessy, Ciancetta, Best, & Canada, 2004; Makar & Ben-Zvi, 2011). It is now widely recognized that the foundations for statistical and probabilistic reasoning should be laid in the very early years of life rather than being reserved for secondary school level or university studies (National Council of Teachers of Mathematics, 2000).

As the mathematics education literature indicates, young children possess an informal knowledge of mathematical concepts that is surprisingly broad and complex (Clements & Sarama, 2007). Although the amount of research on young learners' reasoning about data and chance is still relatively small, several studies conducted during the past decade have illustrated that when given the opportunity to participate in appropriate, technology-enhanced instructional settings that support active knowledge construction, even very young children can exhibit well-established intuitions for fundamental statistical concepts (e.g. Bakker, 2004; English, 2012; Leavy & Hourigan, 2018; Makar, 2014; Makar, Fielding-Wells & Allmond, 2011; Meletiou-Mavrotheris & Paparistodemou, 2015; Paparistodemou & Meletiou-Mavrotheris, 2008; Rubin, Hammerman, & Konold, 2006). Use of appropriate educational tools (e.g. dynamic statistics software), in combination with suitable curricula and other supporting material, can provide an inquiry-based learning environment through which genuine endeavours with data can start at a very young age (e.g. Ben-Zvi, 2006; Gil & Ben-Zvi, 2011; Hourigan & Leavy, 2016; Leavy, 2015; Leavy & Hourigan, 2015, 2018; Paparistodemou & Meletiou-Mavrotheris, 2010; Pratt, 2000). Through the use of meaningful contexts, data exploration, simulation, and dynamic visualization, young children can investigate and begin to comprehend abstract statistical concepts, developing a strong conceptual base on which to later build a more formal study of probability and statistics (Hall, 2011; Ireland & Watson, 2009; Konold & Lehrer, 2008; Leavy & Hourigan, 2016, 2018; Meletiou-Mavrotheris & Paparistodemou, 2015).

Edited Volume Objectives

The edited volume will contribute to the Early Mathematics Learning and Development Book Series, a volume focused on the development of young children's (ages 3–10) understanding of data and chance, an important yet neglected area of mathematics education research. The goal of this publication is to inform best practices in early statistics education research and instruction through the provision of a detailed account of current best practices, challenges, and issues, and of future trends and directions in early statistical and probabilistic learning worldwide. Specifically, the book has the following objectives:

1. Provide a comprehensive overview of up-to-date international research work on the development of young learners' reasoning about data and chance in formal, informal, and non-formal education contexts;
2. Identify and publish worldwide best practices in the design, development, and educational use of technologies (mobile apps, dynamic software, applets, etc.) in support of children's early statistical and probabilistic thinking processes and learning outcomes;
3. Provide early childhood educators with a wealth of illustrative examples, helpful suggestions, and practical strategies on how to address the challenges arising from the introduction of statistical and probabilistic concepts in preschool and school curricula;
4. Contribute to future research and theory building by addressing theoretical, epistemological, and methodological considerations regarding the design of probability and statistics learning environments targeting young children; and
5. Account for issues of equity and diversity in early statistical and probabilistic learning, so as to ensure increased participation of groups of children at special risk of exclusion from math-related fields of study and careers.

This timely publication approaches an audience that is broad enough to include all researchers and practitioners interested in the development of children's understanding of data and chance in the early years of life. Early childhood educators can access a compilation of best practices and recommended processes for optimizing the introduction of statistical and probabilistic concepts in the mathematics curriculum. Mathematics and statistics education researchers interested in exploring and advancing early probabilistic and statistical thinking can be informed about the latest developments in the field and about relevant research projects currently being implemented in various formal and informal educational settings worldwide. Academic experts, learning technologists, and educational software developers can become more sensitized to the needs of young learners of probability and statistics and their teachers, supporting the development of new methodologies and technological tools. National and transnational education authorities responsible for setting mathematics curricula and educational policies can get useful information regarding current developments and future trends in statistics education practices targeting young learners. Teacher education institutions can utilize the book for further improvement of their teacher preparation programmes. Finally, the book can also be useful to professionals and organizations offering parent training programmes in early mathematics education.

Edited Volume Contents

The edited volume has compiled a collection of knowledge on the latest developments and approaches to probability and statistics in early childhood and primary education (ages 3–10). It has collected incisive contributions from leading researchers and practitioners internationally, as well as from emerging scholars, on the development of young children's understanding of data and chance in the prior-to-school and early school years. The contributions address a variety of theoretical aspects underpinning the development of early statistical and probabilistic reasoning and their related pedagogical implications. The authors identify current best practices, place them within the overall context of current trends in statistics education research and practice, and consider the implications both theoretically and practically. The majority of the chapters report on original, cutting-edge empirical studies, which demonstrate validated practical experiences related to early statistical and probabilistic learning. Chapters presenting interim results from innovative, ongoing projects have also been included. The volume also contains conceptual essays which will contribute to future research and theory building by presenting reflective or theoretical analyses, epistemological studies, integrative and critical literature reviews, or forecasting of emerging learning technologies and tendencies.

The book includes 17 chapters that cover a broad range of topics on early learning of data and chance in a variety of both formal and informal education contexts. The chapters have been organized into three parts covering the following themes: (a) Part I: Theory and Conceptualization of Statistics and Probability in the Early Years; (b) Part II: Learning Statistics and Probability in the Early Years; (c) Part III: Teaching Statistics and Probability in the Early Years. Each section includes chapters that discuss the above from both research and innovative practice perspectives.

Part I: Theory and Conceptualization of Statistics and Probability in the Early Years

Chapters included in Part I focus on theoretical, epistemological, and methodological considerations related to early statistics education.

In Chap. 1, Katie Makar argues that conventional approaches to early statistics education tend to undervalue young children's capacity by adopting incremental approaches (from simple to complex) that isolate and disconnect statistical concepts from purposeful activity and their structural relations with other key statistical ideas, thus making them less coherent from students' perspective. The author theorizes how contextual experiences can be a powerful scaffold for young children to engage informally with powerful statistical ideas. She introduces the theoretical notion of *statistical context structures,* which characterize aspects of contexts that can expose children to key statistical ideas and structures (concepts with their related characteristics, representations, and processes). The author claims that use of statistical context structures to create repeated opportunities for children to experience informal statistical ideas has the potential to strengthen their understanding of core concepts when they are developed later. A classroom case study involving statistical inquiry by children in their first year of schooling (ages 4–5) is included in the chapter to illustrate characteristics of age-appropriate links between contexts and structures in statistics.

Chapter 2, authored by Zoi Nikiforidou, focuses on probabilistic thinking in preschool years. It provides a critical review of key theories and models on the early development of probabilistic thinking and highlights a number of pedagogical implications while introducing probabilistic concepts in the early years. The first part of the chapter contrasts findings from the first systematic explorations of the origins of probabilistic thinking conducted by Piaget and Inhelder (1975) that had indicated young children's difficulties in differentiating between certainty and uncertainty, to the findings of more recent studies which support pre-schoolers' capacity for sophisticated informal understanding of probability concepts. The second part reviews important curriculum-related aspects in embedding probabilities in the early childhood classroom so as to set foundations for probability literacy. The argument is made that early years practice should use young children's personal experiences with probabilistic situations and their initial understandings as stepping stones for a spiral curriculum that gradually builds probabilistic thinking and reasoning through meaningful tasks and collaborative learning environments.

Part II: Learning Statistics and Probability

Part II includes chapters which explore issues pertaining to learner and learning support in the early classroom, from both research and innovative practice perspectives.

In Chap. 3, Sibel Kazak and Aisling M. Leavy explore early primary school children's emergent reasoning about uncertainty from the three main perspectives on the quantification of uncertainty: classical, frequentist, and subjective. Their focus is on children's subjective notion of probability which, although being closely related to what people commonly use for everyday reasoning, is either neglected or has minimal mention in school curriculum materials. Combining a critical literature review with an analysis of empirical data arising from small group clinical interviews with children, the authors investigate the ways in which young children reason about the likelihood of outcomes of chance events using subjective probability evaluations before and after engaging in experiments and simulations, and the types of language they use to predict and describe stochastic outcomes.

Chapter 4 by Jane Watson describes a study which explored primitive understandings of variation and expectation by seven 6-year-old children in their beginning year of formal schooling. Children worked through four interview protocols which sought to present them with meaningful contexts that would allow them to display their naïve understandings. Across the contexts, students were asked to make predictions and to create or manipulate representations of data. At no time were the words "variation", "expectation", or "data" used with the children. Collected videos, transcripts, and written artefacts were analysed to document demonstration of understanding of the concepts of expectation and variation in relation to data. Findings support Moore's (1990) and Shaughnessy's (2003) view that appreciation of variation is the foundation of all statistical enquiry and the starting point for children's engagement with the practice of statistics. The 6-year-olds in the study had virtually no trouble recognizing and discussing variation in data, despite not always being able to explain its origin. Evidence of appreciation of variation in children occurred much more frequently than evidence of appreciation of expectation. This confirms Watson's (2005) claim that, in contrast to the traditional order of introduction of measures of centre and spread in the school curriculum, dealing with variation generally develops before the ability to express meaningful expectation related to that variation.

Chapter 5, by Celi Espasandin Lopes and Dana Cox, discusses the learning of probability and statistics by young children, centred on culturally relevant teaching and solving problems with themes derived from the children's culture and their daily life context. This chapter is part of a qualitative longitudinal research project that methodologically explores the temporal dimension of experience, in order to discern human action and take into account the social practices, the subjective experiences, identity, beliefs, emotions, values, contexts, and complexity of the participants. Using some of the data collected through the longitudinal study, Lopes and Cox identify structural elements and triggers of mathematical and statistical learning from activities, based on probabilistic and statistical content, prepared by the teachers who are responsible for the learners in the class. They also identify indicators of the development of different forms of combinatorial, probabilistic, and statistical reasoning that children acquire throughout their second and third year of primary school (ages 7–8).

The next chapter (Chap. 6), by Aisling M. Leavy and Mairéad Hourigan, builds on previously conducted research on young children's statistical reasoning when engaged in core components of data modelling. It describes a study which investigated young children's approaches to collecting and representing data in a data modelling environment. The investigation involved 26 primary school children aged 5–6 years in interpreting and investigating a context of interest and relevance to them. The children engaged in four 60-min lessons focusing on data generation and collection, identification of attributes, structuring and representation of data, and making informal inferences about the results. The authors focus on the outcomes of the first lesson which engaged children in generating and collecting data arising from a story context. They use the Worthington and Carruthers (2003) taxonomy of mathematical graphics to categorize the repertoire of inscriptions or marks used by children to track and record the appearance of their data values, and explore the justifications children provided for their invented inscriptions. They conclude that when the focus of statistical investigation is on reasoning about and understanding meaningful situations, the variety of marks young children make become both a record of and an abstraction for the real event and thereby serve an important communicative function in their efforts to make sense of and communicate statistical situations.

The aim of the design-based research study described in Chap. 7 by Jill Fielding-Wells was to investigate the ways in which a statistical inquiry could be facilitated in the early statistics classroom. The study insights emerged from observation and analysis of teacher–student interactions as an experienced teacher of inquiry scaffolded a class of 5–6-year-old students to engage with ill-structured statistical problems. The chapter details the framework employed in the study for introducing statistical inquiry to these young students and then provides an overview of the study findings. Sufficient detail of the classroom context is provided to enable the reader to envisage the learning. Implications and suggestions for educators are addressed.

Chapter 8, authored by Gilda Guimarães and Izabella Oliveira, examines young students' (aged 5–9) and their teachers' knowledge regarding activities involving classification, in the context of a statistical investigation. The chapter presents the results of three different studies conducted by the authors' research group, which involved students and/or teachers of the earliest school years. The first study involved 20 kindergarten children (aged 5), the second study 48 Grade 3 children (aged 8) and 16 early grade teachers, and the third study 72 Grade 4 children (aged 8–9). Findings of these studies demonstrate that people are able, from a very young age, to classify based on a previously defined criterion and to discover a classification criterion, but that they have difficulties in creating criteria to carry out a classification. The authors justify the reasons behind children's difficulties and make suggestions as to how instruction could utilize kindergarten children's ability to classify in different situations using pre-defined criteria to help them build skills in producing their own classification criteria.

Parts III–V: Teaching Statistics and Probability: Curriculum Issues, Tasks and Materials, and Modelling

Parts III–V focuses on issues related to statistics and probability teaching and on providing insights on how to support teachers and other educators in the adoption of the new pedagogical approaches that are needed for successful statistics instruction in the early years. The part is further divided into the following three subparts: (i) Curriculum Issues, (ii) Tasks and Materials, and (iii) Modelling.

Curriculum Issues

In Chap. 9, Randall E. Groth unpacks implicit disagreements among various early childhood standards for probability and statistics regarding the roles of student-posed statistical questions, probability language, and variability in young students' learning. He considers several different sources of disagreement including beliefs about students' abilities, beliefs about teachers' abilities, robustness and influence of the research literature, and priorities for early mathematics education in the early grades. The aim of the author is to define a space in which disagreements about curriculum standards for early childhood and primary statistics are made explicit and then respectfully analysed. In considering the different sources of disagreement, Groth makes suggestions for directions that could be taken by the field so as to provide high-quality statistics education for all young learners. Suggestions are made for ways to move towards a greater degree of consensus across standards documents. At the same time, steps that could be taken to support early statistics teaching and research in absence of consensus on curriculum standards are also highlighted. Specifically, Groth suggests the use of boundary objects, which allow related communities of practice to operate jointly despite the existence of disagreement.

In Chap. 10, Carmen Batanero, Pedro Arteaga, and María M. Gea argue that statistical graphs are complex semiotic tools requiring different interpretative processes of the graph components in addition to the entire graph itself. Based on this argument and on hierarchies proposed in previously conducted research, they analyse the content related to statistical graphs of the Spanish curricula, textbooks, and external compulsory examinations taken by 6–9-year-old children. Batanero et al. investigate the types of graphs introduced in the curriculum, the type of activity demanded, the reading levels required from children, as well as the graph semiotic complexity and the task context. This analysis leads the authors to the conclusion that the expected progression in young children's learning of statistical graphs as reflected in the Spanish current curricular guidelines, the textbooks, and the external assessment is in accord with contemporary research literature recommendations for the teaching of graphs. Curricular materials introduce a rich variety of different types of graphs, activities, tasks, and contexts, with reading levels being adequately ordered in progressive difficulty in the different grades as described by Curcio (1989) and Shaughnessy, Garfield and Greer (1996), and with the graph semiotic complexity (Batanero, Arteaga & Ruiz, 2010) being age-appropriate. Nonetheless, Batanero et al. caution that, in some of the textbooks, an excessive emphasis is being placed on computation with the graph data, resulting in a very high percentage of *reading between the data* (level 2) activities when compared to *reading beyond the data* (level 3) and *reading behind the data* (level 4) activities. Due to this and other important differences between textbooks observed, Batanero et al. highlight the responsibility of teachers when selecting the most adequate book for their students.

Tasks and Materials

Chapter 11, authored by Virginia Kinnear, explores the dual role that picture storybooks can play in contextualizing a statistical problem for investigation through the provision of both an engaging context for the task and of the context knowledge children can use to find a solution to the problem. The chapter presents the results of a small study conducted with fourteen 5-year-old children in a public school in Australia. The study's theoretical perspective, Models and Modeling (Lesh & Doerr, 2003), provided a theoretical framework for task design principles. Three picture storybooks were used to initiate three separate and consecutively implemented statistical problems (as data modelling activities). The study investigated the role of the picture storybooks in initiating children's interest in the statistical context of the problem and in handling the data to solve the statistical problem. The chapter identifies the characteristics of the books that interested children and discusses how knowledge of these characteristics could be used to inform educators' selection of picture storybooks, so as to stimulate students' interest in statistical problem-solving activities. The unique challenges in identifying books for contextualizing statistical problems are also discussed.

Chapter 12 by Efi Paparistodemou and Maria Meletiou-Mavrotheris presents a study which investigated early childhood teachers' planning, teaching, and reflection on stochastic activities targeting young children (4–6-year-olds). Five early childhood teachers (all females) participated in this research, which was organized in three stages. In Stage 1, the teachers were engaged in lesson planning. They selected a topic from the national mathematics curriculum on probability and statistics and developed a lesson plan and accompanying teaching material aligned with the learning objectives specified in the curriculum. In Stage 2, they implemented the lesson plans in their classroom, with the support of the researchers. Once the classroom implementation was completed, in Stage 3, teachers were interviewed and prepared and submitted a reflection paper, where they shared their observations on students' reactions during the lesson, noting what went well and what difficulties they faced and making suggestions for improvement. The researchers analysed the design of each lesson, observed teachers implementing their lesson, and interviewed them while they reflected on their instruction. The study has provided some useful insights into the varying levels of attention teachers paid to different kinds of activities during their lesson implementation, and into the different types of instructional material they used. Findings indicate that the early childhood teachers in this study appreciated the importance of using tools and real-life scenarios in their classrooms for teaching stochastics. They had rich ideas about the context, but needed extra effort to understand the stochastical ideas hidden in the tasks. Moreover, the findings also show that early childhood teachers' attention to different aspects of probability tasks can be developed through a reflective process on their teaching.

The next chapter, by Daniel Frischemeier, addresses the following two questions: in what manner is it possible to introduce early statistical reasoning elements (in regard to analysing large data sets) in German primary school? In what manner is it possible to lead Grade 4 students to fundamental statistical activities such as group comparisons? The first part of the chapter describes the design and implementation of a teaching unit on early statistical reasoning for German primary school students in Grade 4. The teaching unit was designed and developed using the design-based research approach (Cobb, Confrey, diSessa, Lehrer, & Schauble, 2003), and it incorporated key elements of the Statistical Reasoning Learning Environment (Garfield & Ben-Zvi, 2008): focus on central statistical ideas (group comparisons), use of real and motivating data sets (class and school data), use of engaging classroom activities (cooperative learning environments), employment of multiple representation levels (enactive, ikonic, symbolic), integration of appropriate technological tools (TinkerPlots) for analysing large and real data sets and comparing groups. The second part of the chapter presents results of an empirical study which investigated how a class of 11 ($n = 11$) Grade 4 students compared groups before and after experiencing the teaching unit described in part 1 of the chapter. The results show the potential of engaging young students' sophisticated statistical reasoning with some pedagogical support at an early stage and provide some design ideas for instructional sequences to lead young children to group comparisons.

In Chap. 14, Soldedad Estrella focuses on the challenging process of representing (modelling) for pupils in the first years of school. She makes a teaching proposal which involves the exploration of a set of raw data before young children can then go on to build their own representations to reveal and provide evidence of the behaviour of the data, its patterns, and relationships. Estrella first describes some concepts that support the teaching proposal and its aim to develop statistical thinking: meta-representational competence (MRC), some components of representation, transnumeration, statistical thinking, and data sense. She then goes on to detail the experiences of three 5-year-old preschool students (from a class of 27 students) and two 7-year-old primary pupils (from a class of 38 pupils) that participated in an open-ended data organization lesson. In both classes, the lesson was jointly designed by teachers in the school (a group of four preschool teachers and a group of four second Grade 4 teachers) that participated in a professional development course on statistics education which adopted the lesson study approach. Findings from the study indicate that strengthening teachers' reflections in lesson study groups promotes the connection between theory and teaching practice, enabling teachers to innovate in the statistics classroom and to get children involved in resolving exploratory data analysis situations. The richness of participating students' productions provided evidence of essential components of data representations and of increased understanding of data behaviour acquired by the children when freely developing their own representations. The chapter presents the diverse data representations produced by the children, details components (statistical, numerical, and geometric) of the different representations, and identifies transnumeration techniques they used, which helped them to gain deeper understanding of the characteristics of a data set and its relationships.

The intent of Chap. 15 authored by Lucía Zapata-Cardona was to explore young children's counting combinatorial strategies and to reflect on how these strategies could orient teachers' actions in the classroom when teaching combinatorics in the early years. To address this goal, a convenience sample of three young children (ages 6–8) were interviewed in a home setting while solving a combinatorial task centred on the process of combinatorial counting. The task was presented in verbal form and was accompanied by some manipulatives to help children visualize, explore, model, and solve the combinatorial task. Zapata-Cardona provides a thorough description of the combinatorial counting strategies the young children activated when solving the task, so as to illustrate the kind of questions and strategies that researchers and teachers could use to challenge young children's combinatorial reasoning and make them go beyond their initial strategies. One of the main ideas revealed through the investigation of the young children's strategies was the close relationship between their combinatorial reasoning and multiplicative reasoning, leading Zapata-Cardona to the conclusion that combinatorial reasoning could be stimulated from the moment children begin to work with multiplication rather than waiting for formal combinatorial instruction which usually occurs in secondary education. The author argues that teachers' strategies to support young children's combinatorial reasoning need to be grounded upon the parallel development of multiplicative reasoning; i.e. they should support young children's exploration of combinatorial counting processes through solving different formats of multiplicative situations. The chapter ends by presenting and discussing some strategies for teachers to support and challenge young children's combinatorial reasoning as drawn from the current study and the existing research literature on combinatorial development in the early years. These strategies include interesting tasks which to children to deal with combinatorial counting situations in a playful, attractive, and familiar way, manipulatives to support the modelling and exploration of combinatorial situations, and probing questions by the teacher to focus children's attention and to challenge their reasoning.

Modelling

In Chap. 16, Maria Meletiou-Mavrotheris, Efi Paparistodemou, and Loucas Tsouccas explore the educational potential of games for enhancing statistics instruction in the early years. Acknowledging the crucial role of teachers in any effort to bring about change and innovation, the authors conducted a study aimed at equipping a group of in-service primary teachers with the knowledge, skills, and practical experience required to effectively exploit digital games as a tool for fostering young children's motivation and learning of statistics. The study took place within a professional development programme focused on the integration of games within the early mathematics curriculum (Grades 1–3; ages 6–9), which was designed based on the Technological, Pedagogical and Content Knowledge (TPACK) framework (Mishra & Koehler, 2006) and was attended by six (n = 6) teachers. Following the TPACK model and action research procedures, the study was carried out in three phases: (i) familiarization with game-based learning; (ii) lesson planning; and (iii) lesson implementation and reflection. Each of the three phases supported teachers in strengthening the connections among their technological, pedagogical, and content knowledge. At the same time, various forms of data were collected and analysed in order to track changes in teachers' TPACK regarding game-enhanced statistics learning in the early years. Findings illustrate the usefulness of TPACK as a means of both studying and facilitating teachers' professional growth in the use of games in early statistics education. They indicate that the TPACK-guided professional development programme had a positive impact on all three perspectives of the participants' experiences examined: (i) attitudes and perceptions regarding game-enhanced learning; (ii) TPACK competency for using digital games; and (iii) level of transfer and adoption of acquired TPACK to actual teaching practice.

In Chap. 17, Lyn D. English describes two investigations which revealed 8-year-olds' statistical literacy in modelling with data and chance. These two investigations, one dealing with statistics and the other with probability, were implemented during the first year of a 4-year longitudinal study being conducted across grades 3 through 6 in two Australian cities. This was the participating students' first exposure to modelling with data. Children's responses to both investigations were explored in terms of how they identified variation, made informal inferences, created representations, and interpreted their resultant models. The responses indicate that these young students were developing important foundational components of statistical literacy. Using their understanding of variation as a foundation, they were able to make predictions based on their findings and to draw informal inferences, as well as generate and interpret a range of representational models to display data. This, English argues, points to the need for early statistics education to provide more opportunities for children to engage in modelling involving data and chance in order to capitalize on, and advance, their learning potential.

Concluding Remarks

Despite the importance of developing young learners' early statistical and probabilistic reasoning, the evidence base to support such development is scarce. An urgent need exists for scholarly publications, and a broader research agenda aimed at investigating the infiltration of probability and statistics into early mathematics teaching and learning practices and experiences. Thus, by focusing on this important emerging area of both research and practice, this publication fills a significant gap in the early mathematics education literature. To the best of our knowledge, this is the first international book to provide a comprehensive survey of state-of-the-art knowledge, and of opportunities and challenges associated with the early introduction of statistical and probabilistic concepts in educational settings, but also at home. While there are several manuscripts covering various aspects of early mathematics education, no other book focuses specifically on the disciplinary particularities of early statistics learning. With contributions from many leading international experts, this book provides the first detailed account of the theory and research underlying early statistics learning. It gives valuable insights into contemporary and future trends and issues related to early statistics education, informing best practices in mathematics education research and teaching practice.

Limerick, Ireland — Aisling Leavy
Nicosia, Cyprus — Maria Meletiou-Mavrotheris
Latsia, Nicosia, Cyprus — Efi Paparistodemou

References

Bakker, A. (2004). *Design research in statistics education: On symbolizing and computer tools*. Doctoral dissertation, Utrecht University.

Batanero, C., Arteaga, P., & Ruiz, B. (2010). Statistical graphs produced by prospective teachers in comparing two distributions. In V. Durand-Guerrier, S. Soury-Lavergne, & F. Arzarello (Eds.), *Proceedings of the Sixth Congress of the European Society for Research in Mathematics Education*. Lyon: ERME. Retrieved from www.inrp.fr/editions/editions-electroniques/cerme6/.

Ben-Zvi, D. (2006). Scaffolding students' informal inference and argumentation. In A.Rossman and B. Chance (Eds), *Proceedings of the Seventh International Conferenceon Teaching of Statistics* [On CD], Salvador, Bahia, Brazil, 2–7 July, 2006. Voorburg,The Netherlands: International Statistical Institute.

Clements, D., & Sarama, J. (2009). *Learning and teaching early math: The learning trajectories approach*. New York: Routledge.

Cobb, P., Confrey, J., diSessa, A., Lehrer, R., & Schauble, L. (2003). *Design Experiments in Educational Research. Educational Researcher, 32*(1), 9–13.

Common Core State Standards Initiative (2010). *Mathematics*. Washington, DC: Council of Chief State School Officers & National Governors Association Center for Best Practices.

Curcio, F. R. (1989). *Developing graph comprehension*. Reston, VA: NCTM.

English, L. (2012). Data modelling with first-grade students. *Educational Studies in Mathematics, 81*, 15–30.

Franklin, C. A., Kader, G., Mewborn, D., Moreno, J., Peck, R., Perry, M., & Scheaffer, R. (2007). *Guidelines for Assessment and Instruction in Statistics Education (GAISE) report: A pre-K–12 curriculum framework*. Alexandria, VA: American Statistical Association.

Garfield, J., & Ben-Zvi, D. (2008). *Developing students' statistical reasoning. Connecting Research and Teaching Practice*. The Netherlands: Springer.

Gil, E., & Ben-Zvi, D. (2011). Explanations and context in the emergence of students' informal inferential reasoning. *Mathematical Thinking and Learning, 13*, 87–108.

Gloy, K. (1995). *Die Geschichte des wissenschaftlichen Denkens: Verständnis der Natur*. München: Komet.

Hall, J. (2011). Engaging teachers and students with real data: Benefits and challenges. In C. Batanero, G. Burrill, & C. Reading (Eds.), *Teaching statistics in school mathematics:*

Challenges for teaching and teacher education (pp. 335–346). Dordrecht, The Netherlands: Springer.
Hourigan, M. and Leavy, A. M. (2016). Practical Problems: Introducing Statistics to Kindergarteners. *Teaching Children Mathematics, 22*(5), 283–291.
Ireland, S., & Watson, J. (2009). Building an understanding of the connection between experimental and theoretical aspects of probability. *International Electronic Journal of Mathematics Education, 4*, 339–370.
Konold, C., & Lehrer, R. (2008). Technology and mathematics education: An essay in honor of Jim Kaput. In L. D. English (Ed.), *Handbook of international research in mathematics education* (2nd ed., pp. 49–72). New York, NY: Routledge.
Leavy, A. M. (2015). Looking at practice: Revealing the knowledge demands of teaching data handling in the primary classroom. *Mathematics Education Research Journal, 27*(3), 283–309.
Leavy, A., & Hourigan, M. (2015). Motivating Inquiry in Statistics and Probability in the Primary Classroom. *Teaching Statistics, 27*(2), 41–47.
Leavy, A., & Hourigan, M. (2016). Crime Scenes and Mystery Players! Using interesting contexts and driving questions to support the development of statistical literacy. *Teaching Statistics, 38* (1), 29–35.
Leavy, A., & Hourigan, M. (2018). The role of perceptual similarity, data context and task context when selecting attributes: Examination of considerations made by 5–6 year olds in data modelling environments. *Educational Studies in Mathematics, 97*(2), 163–183.
Lesh, R., & Doerr, H.M. (2003). *Beyond constructivism: A models and modeling perspective on mathematics problem solving, learning and teaching*, Mahwah, NJ: Lawrence Erlbaum Associates, Inc.
Makar, K. (2014). Young children's explorations of average through informal inferential reasoning. *Educational Studies in Mathematics, 86*(1), 61–78.
Makar, K., & Ben-Zvi, D. (2011). The role of context in developing reasoning about informal statistical inference. *Mathematical Thinking and Learning, 13*13, 1–4.
Makar, K., Fielding-Wells, J., & Allmond, S. (2011, July). *Is this game 1 or game 2? Primary children's reasoning about samples in an inquiry classroom.* Paper presented at the Seventh International Forum for Research on Statistical Reasoning, Thinking, & Literacy. Texel, The Netherlands.
Meletiou-Mavrotheris, M., & Paparistodemou, E. (2015). Developing young learners' reasoning about samples and sampling in the context of informal inferences. *Educational Studies in Mathematics, 88*(3), 385–404.
Minois, G. (2002). *Die Geschichte der Prophezeiungen*, Düsseldorf: Albatros.
Mishra, P., & Koehler, M. J. (2006). Technological pedagogical content knowledge: A framework for teacher knowledge. *Teachers College Record, 108*(6), 1017–1054.
Moore, D. S. (1990). Uncertainty. In L. A. Steen (Ed.), *On the shoulders of giants: New approaches to numeracy* (pp. 95–137). Washington, DC: National Academy Press.
National Council of Teachers of Mathematics (2000). *Principles and standards for school mathematics*. Reston, VA: Author.
Paparistodemou, E., & Meletiou-Mavrotheris, M. (2008). Enhancing reasoning about statistical inference in 8 year-old students. *Statistics Education Research Journal, 7* (2), 83–106.
Paparistodemou, E. & Meletiou-Mavrotheris, M. (2010). Engaging Young Children in Informal Statistical Inference. In C. Reading (Ed.), *Data and Context in Statistics Education: Towards an Evidence-based Society. Proceedings of the Eighth International Conference on Teaching Statistics* (ICOTS8, July, 2010), Ljubljana, Slovenia. Voorburg, The Netherlands: International Statistical Institute.
Piaget, J., and Inhelder B. (1975). *The origin of the idea of chance in children.* Translated and edited L. Leake, Jr., P. Burrell, & H. Fischbein. NY: Norton.
Pratt, D. (2000). Making Sense of the Total of two Dice. *Journal of Research in Mathematics Education, 31*, 602–625.

Rubin, A. (2002). Interactive Visualizations of Statistical Relationships: What Do We Gain? *Proceedings of the Sixth International Conference on Teaching Statistics*. Durban, South Africa.

Rubin, A., Hammerman, J., & Konold, C. (2006). Exploring Informal Inference with Interactive Visualization Software. In A. Rossman, & B. Chance (Eds.), *Working Cooperatively in Statistics Education: Proceedings of the Seventh International Conference of Teaching Statistics (ICOTS-7)*, Salvador, Brazil.

Shaughnessy, J. M. (1992). Research in probability and statistics: Reflections and directions. In D. A. Grouws (Ed.), Handbook of research on mathematics teaching and learning (pp. 465–494). New York: Macmillan.

Shaughnessy, J. M. (2003). Research on students' understandings of probability. In J. Kilpatrick, W. G. Martin, & D. Schifter (Eds.), *A research companion to principles and standards for school mathematics* (pp. 216–226). Reston, VA: National Council of Teachers of Mathematics.

Shaughnessy J. M., Ciancetta M., Best K., & Canada D. (2004, April). *Students' attention to variability when comparing distributions*. Paper presented at the 82nd Annual Meeting of the National Council of Teachers of Mathematics, Philadelphia, PA.

Shaughnessy, J. M., Garfield, J., & Greer, B. (1996). Data handling. En A. J. Bishop, K. Clements, C. Keitel, J. Kilpatrick & C. Laborde (Eds.), *International handbook of mathematics education* (pp. 205–237). Dordrecht, The Netherlands: Kluwer Academic Publishers.

Watson, J. M. (2005). Variation and expectation as foundations for the chance and data curriculum. In P. Clarkson, A. Downton, D. Gronn, M. Horne, A. McDonough, R. Pierce & A. Roche (Eds.), *Building connections: Theory, research and practice (Proceedings of the 28th annual conference of the Mathematics Education Research Group of Australasia*, Melbourne, pp. 35–42). Sydney: MERGA. Retrieved from https://www.merga.net.au/documents/practical2005.pdf.

Westergaard, H. (1932), *Contributions to the history of Statistics*. P.S. King & Sons Ltd.: London.

Contents

Part I
Theory and Conceptualisation of Statistics and Probability in the Early Years

Chapter 1
Theorising Links Between Context and Structure to Introduce Powerful Statistical Ideas in the Early Years

Katie Makar

Abstract Recent literature in the early years has emphasised the benefits of introducing children to powerful disciplinary ideas. Powerful ideas in statistics such as variability, aggregate, population, the need for data, data representation and statistical inquiry are generally introduced in the later years of schooling or university and therefore may be considered too difficult for young children. However, at an informal level, these ideas arise in contexts that are accessible to young children. The aim of this chapter is to theorise important relations between children's contextual experiences and key structures in statistics. It introduces the notion of *statistical context–structures,* which characterise aspects of contexts that can expose children to important statistical ideas. A classroom case study involving statistical inquiry by children in their first year of schooling (ages 4–5) is included to illustrate characteristics of age-appropriate links between contexts and structures in statistics. Over time, engaging children in significant activities that rely on statistical context–structures can provide children with multiple opportunities to experience statistics as a coherent and purposeful discipline and develop rich networks of informal statistical concepts well before ideas are formalised. For teachers and curriculum writers, statistical context–structures provide a framework to design statistical inquiries that directly address learning intentions and curricular goals.

1.1 Introduction

Researchers have long argued that powerful mathematical ideas are accessible to young children (e.g. Alexander, White, & Daugherty, 1997; English & Mulligan, 2013; Greer, Verschaffel, & Mukhopadhyay, 2007; Perry & Dockett, 2008). Yet many approaches to teaching young children undervalue their capacity—and therefore limit their opportunities—to access powerful statistical ideas. Content is often disconnected from purposeful activity, and learning sequences tend to focus on small

---

K. Makar (✉)
The University of Queensland, Brisbane, Australia
e-mail: k.makar@uq.edu.au

A. Leavy et al. (eds.), *Statistics in Early Childhood and Primary Education*,
Early Mathematics Learning and Development,
https://doi.org/10.1007/978-981-13-1044-7_1

increments building from simple to complex. Incremental approaches tend to isolate and disconnect statistical ideas from their rich contextual and structural relations with other key ideas, making them less coherent from the students' perspective (Bakker & Derry, 2011).

Addressing the gap between the conviction that children can benefit from access to powerful statistical ideas and the operationalisation of this conviction is critical. How does one design age-appropriate learning experiences with complex content? In this paper, I theorise how the context of a problem can be a powerful scaffold for children to engage informally with powerful statistical ideas. The paper introduces the theoretical notion of *statistical context–structures,* which characterise aspects of problem contexts that can expose children to key statistical ideas and structures (concepts with their related characteristics, representations and processes). Using statistical context–structures to create repeated opportunities for children to experience informal statistical ideas has the potential to strengthen their understanding of core concepts when they are developed later. Exposure to informal concepts across a variety of problem contexts highlights their relationships to other core concepts, develops coherence of how statistical ideas work together, assists students to recognise contexts in which the ideas are appropriate and potentially useful, and improves the sense of relevance of statistical ideas.

The aim of this paper is to illustrate how a teacher in an early years classroom (children aged 4–5 years) used a personal problem context to informally introduce, scaffold and develop informal yet powerful statistical content. Over the course of two lessons, she used an inquiry approach and a context familiar to students to leverage initial conceptions of variability, aggregate, population, a need for data and the value of representation to record, analyse and communicate ideas about data.

1.2 Literature Review and Theoretical Framework

Statistical concepts that are isolated become atomistic and impoverished (Bakker & Derry, 2011). To develop rich statistical understandings, students must see how statistical concepts and structures are related to one another, to practices and conventions, to their prior knowledges and experiences, and their utility for solving problems. The focus of this literature review is on understanding links between students' reasoning in problem contexts and their reasoning about key structures in the discipline (mathematics or statistics).

Literature on informal learning environments has begun to establish how reasoning in context can strengthen students' valuing of mathematics and relationships between concepts. There has long been acknowledgement of a gap between students' formal and informal knowledge and reasoning (Confrey & Kazak, 2006; Raman, 2002; Sadler, 2004). Much of this is the result of teaching formal concepts before students have developed understanding of both their usefulness for solving problems and their connections to students' prior knowledge and belief structures. Because "mathematical ideas are fundamentally rooted in action and situated in activity"

Table 1.1 Mapping of statistical context–structures in Makar (2014)

Context entity	Statistical structures	Statistical context–structure and reasoning
Height	data	The measure of how tall a person is can be collected and recorded as height (cm) data
Height of a child	Single data point	A child is associated with their height data
Heights of students in the class	Aggregate	Collectively, the heights of the children in the class can be considered as an entity to investigate
Heights of children in the class differed	Variability	Because all heights in the class were not the same, the children had to grapple with how to manage the variability of the height data
Organised heights clumped in the middle	Distribution shape	When children invented ways to record and organise the data, they noticed that most heights were in the middle and fewer heights were high or low in value; this feature was stable across both classes
Typical height	Average	To find the typical height, children invented a point estimate to capture the most common height (mode) and an interval estimate to capture where "most" heights clumped. They used these estimates to predict (with uncertainty) the typical heights of children in other classrooms
Height of very tall child	Outlier	One child was substantially taller than the others and they considered this student to have atypical height. They reasoned that it was unlikely to see this height in other classes

(continued)

Table 1.1 (continued)

Context entity	Statistical structures	Statistical context–structure and reasoning
The heights of children in another class were collected and compared to their class	Sampling variability	Their surprise that the data in the class next door were similar to but different than their own class data prompted discussions about what aspects of their data were likely or unlikely to be encountered in other classes (e.g. similar values but different frequencies of each height; similar but possibly not exactly the same typical height)
The typical height of the children in one class was used to predict the typical height of children in another class and across Australia	sample-population inference	One Vietnamese child argued that her mother was considered short in Australia, but was of typical height in Vietnam. This prompted students to clarify that their classroom was not representative of other countries and that data would need to be collected from a country to find the typical heights there

(Confrey & Kazak, 2006, p. 322), learning concepts first informally as they are situated in problems allows students to build experiences over time with rich mathematical structures. These experiences with informal ideas also develop students' sense of the utility of mathematical ideas before their formalisation. "People extract information about the world more often than they are aware and that this knowledge exists in tacit form, influencing thought and behaviour while itself remaining mostly concealed from conscious awareness" (Litman & Reber, 2005, p. 440). For example, social practices (including mathematical conventions) can become adopted without the learner being conscious of what is being learned. Boekaerts and Minnaert (1999) argue that the active, non-threatening and explorative nature of informal learning can assist to develop and sustain students' learning in line with social goals and expectations elicited by the context, since "most informal learning contexts are more powerful for developing criteria for success, progress, and satisfaction, which are in accordance with the students' own need structure" (p. 542). Boekaerts and Minnaert further contend that informal learning can heighten students' valuing and learning goals because they perceive learning to be natural and spontaneous.

The theoretical framework in this chapter develops the idea of *statistical context–structures*. Statistical structures maintain consistent patterns (invariances), despite statistics being a field of variability. Statistical context–structures are conceptualised as a mapping between a connected web of statistical structures (concepts with their related characteristics, representations and processes) and contextual entities that stand in for the statistical structures, with relationships between the contextual entities corresponding to the relationships between the statistical structures. Reasoning about the contextual entities is analogous to reasoning about the statistical structures.

For example, the typical height of children in a classroom is a contextual entity that would allow students to reason about the concept of central tendency without explicitly learning about the statistical mean. Students' reasoning about the mean as a representative measure of Year 3 students' heights is still possible even though they have not formally learned what a mean is or how to calculate it. A key benefit is that their reasoning can include the relationship of the mean to other statistical concepts. A study by Makar (2014), for example, highlighted how Year 3 children (aged 7–8) reasoned about variability, distribution (shape, spread, centre, outliers) and sample-population inference as they wrestled with how to find the "typical height" of the children in their classroom. In the process, they invented and critiqued iterations of data displays of increasing sophistication resulting in a graph similar to a histogram. In this example, the children encountered multiple statistical context–structures (Table 1.1). None of the statistical structures they encountered were formalised, but by repeatedly reasoning about the context, the students gained important experiences with informal versions of advanced statistical structures on which they could later map onto the formal ideas (McGowen & Tall, 2010), while formally addressing the content for their own year level.[1] The role of the teacher was critical here to scaffold student learning through engineering learning experiences and using questioning to guide students' ideas. For example, the heights of the children in the class differed (see column 1, Table 1.1). Children were not formally taught the statistical structure "variability" (e.g. the concept of variability with its related terminology, characteristics, representations, measures and relationships with other statistical structures such as "distribution"), as this would not be appropriate content for 7–8-year-olds. Even without formally learning the statistical structure "variability" (see column 2), the children were able to work with variability in the context of managing the differing heights of the children in their class (see column 3). When children had to predict the typical height of Year 3 students in the class next door, they had to grapple with the variability of the height data in their class. Reasoning about differing heights in that context was analogous (and more age-appropriate) to reasoning about variability. The characteristics, representations and processes related to variability were, to the children, the characteristics, representations and processes needed for making sense of the differing heights.

In contrast, the mean is often taught as a calculation of a set of numbers to work out the "average" of that set. Multiple studies have highlighted how this approach has created an impoverished conceptualisation of central tendency as students neither

[1] In the Year 3 curriculum in Australia (Australian Curriculum: Mathematics, 2016), students would be expected to be able to identity an issue/question and relevant data to collect (ACMSP068), carry out a simple data investigation (ACMSP069) and interpret and compare data displays (ACMSP070).

see the mean as a representative value of a data set nor link it to related ideas of distribution, sampling or inference (Bakker & Derry, 2011; Konold et al., 2002; Mokros & Russell, 1995; Watson, 2006). Bakker and Derry (2011) have argued that an atomistic approach to learning in statistics, where ideas are taught in isolation, has resulted in a lack of coherence in students' statistical thinking. They contend that this has been one of the key challenges in statistics education. However, within rich well-engineered contexts, there are multiple and diverse ways and opportunities to work informally with foundational relationships among statistical structures.

1.3 Methodology

This article is based on a case study of a classroom of young children in the first year of schooling (called Foundation or Prep in Australia). Case study is beneficial to generate insights through "the complexities and contradictions" (Flyvbjerg, 2006, p. 237) of narrative as a problem is played out in practice. It creates opportunities for the researcher to wrestle with a theoretical problem through issues that arise, including serendipitously, in empirical details of the case.

> As an account of practice, explained analytically, case study is a valuable methodology for the research of educational practice, particularly given the scope for the representation of complex practice with multiple and bundled trajectories. Thus, while on the one hand the case attempts to represent complex practice; the case study is the analytical explanation, constructed and crafted to recount, analyse and generate … new ways of understanding complex practices. (Miles, 2015, pp. 315–316)

The case reported in this article used a retrospective analysis of data collected from a larger study that aimed to understand teachers' experiences over time in teaching mathematics through inquiry (e.g. Makar, 2012). At the time the lessons were conducted, the teacher and researcher were interested more generally in how young children respond to and are guided in inquiry. The retrospective analysis of the two lessons captured in this article allowed the researcher to study these lessons anew to seek insight into the way that the teacher and students utilised the problem context of the inquiry to scaffold the children's thinking about statistical concepts, representations and processes. In order words, the retrospective analysis was used by the author to identify the use of statistical context–structures and how the teacher used them to guide students' statistical reasoning.

1.3.1 Participants and Lessons

The participants in the case study were in a prep class (about 20–25 children, aged 4–5 years old) in a suburb of a major city in Australia (prep is equivalent to kindergarten in most countries). The teacher was highly experienced in teaching with inquiry but this was her first time teaching this age of class (previously she taught

Year 3, ages 7–8 years). The data in this paper relied on classroom videos from two 40 min lessons taught on consecutive days at the end of the second month of the school year (in Australia, the school year runs from late January to mid-December). In the first lesson, the teacher introduced the question, "Do most students in Prep L have blue eyes?" and as a class the students sought a method to find out. Iterations of investigation and discussion were used to build on children's experiences and resulting ideas, scaffolded by the teacher. Children individually followed methods that made sense to them, observed their peers' work and discussed their ongoing progress with the teacher and/or as a class. In the second lesson, children continued their progress towards answering the inquiry question using iterative cycles of investigation work and whole class discussion. The lesson wrapped up by counting children with each eye colour.

1.3.2 Data Collection and Analysis

Video data are not objective, nor do they capture all of what is happening in a class (Roschelle, 2000). The choice of placement is deliberate and depends on the research aims. In this study, there were two key placements of the camera—stationary or roving. In either case, the choices that were made were based on seeking insights into students' ideas and the teacher's interaction with them. The camera was used in a stationary mode (on the tripod) if the focus was on the whole class, for example, during sessions when students were seated altogether on the carpet (e.g. when lessons were introduced or during sharing sessions). This allowed for the researcher to gain both general context for the timeline of events and also captured individual contributions by the teacher and students. In particular, this was a critical aspect of data collection to focus on the teacher's questions and how she guided the learning, as well as students' articulation of their thinking at a particular stage of the lesson. Together, this focus on the teacher and students' sharing allowed for the evolution of ideas to be traced to when they were first introduced. The camera was in roving mode (on or off the tripod) when students were working at their tables. During working sessions, the camera either followed the teacher as she interacted with students or it captured students working at one of the tables.

The data were analysed retrospectively using a video analysis process adapted from Powell, Francisco and Maher (2003). The process included seven stages: (1) intent viewing, (2) describing the video data, (3) identifying critical events, (4) transcribing, (5) annotating, (6) constructing a storyline and (7) composing narrative (p. 413). In the initial three stages, the videoed lessons were observed and a video log was created with timestamps, screen-captured images and short-running descriptions of what was happening. Critical events were marked in the video log as rich segments for potential analysis to help focus the observation. These first three stages provided an overall picture of the lesson to ensure that the data were fit for purpose to move to the fourth stage (transcription). The transcript was used to select and annotate excerpts and construct a preliminary (but disjointed) storyline. The author met with the teacher of the lesson to discuss the storyline, clarify the researcher's observations and focus the direction of the narrative. The resulting narrative was developed by iteratively reviewing, editing and elaborating the initial storyline including a second consultation with the teacher.

1.4 Results

The results section will use data from a prep class (ages 4–5) as they investigated the question, *Do Most Children in Prep L have Blue Eyes?* This question came from a comment made in the class by one of the children during an activity about their own eye colour. In setting up this question, the teacher used this problem context to informally introduce five key statistical ideas and structures: (1) acknowledging variability as an issue to resolve; (2) recognising that the individual and the aggregate are related, but not the same; (3) distinguishing what the population is for the investigation; (4) being aware of the need for data and evidence; and (5) valuing representations as ways to record, analyse and communicate results from data in solving problems. The data across the two lessons are presented chronologically in order to illustrate the development of students thinking over the lessons, although the entirety of the lessons is not presented. The critical role that the teacher played is highlighted to scaffold and progress reasoning using the statistical context–structures.

1.4.1 Informally Introducing Variability, Aggregate and Population

In introducing the inquiry question, the teacher Ms Louarn asked students to express their initial thoughts about whether most students in the class had blue eyes. Because this question is about a characteristic of the class as a whole, it is a question about the aggregate. Ms Louarn encouraged students to share their ideas and emphasised when students observed that there were different eye colours in the class (variability). At the same time, she nudged their anecdotal comments towards thinking about the aggregate.

Oliver:	Some people have green eyes too.
Ms Louarn:	They certainly do. So, do you think that more people in prep would have green eyes or blue eyes?
Oliver:	Green eyes.
Ms Louarn:	You think lots of people would have green eyes. What do you think, Kai? …
Kai:	The lessest have green eyes
Ms Louarn:	Less. Is that what you're saying? So you think fewer people in prep have green eyes than blue eyes. [Lesson 1; starting at video timestamp 1:04]

Oliver's response could either have been an observation, or perhaps a counter-example to the question. That is, his point that "Some people have green eyes too" may have been an answer to the investigation question (Do most students in Prep L have blue eyes) using anecdotal evidence. To encourage Oliver to think about the aggregate question, Ms Louarn incorporated his response into the investigation question to ask him again. His response, again green eyes, was acknowledged before she moved on to another response. The teacher emphasised two key points: first, that there was variability in the class in relation to eye colour (linking difference between individuals with the variability of the aggregate), and second, that there was a lack of consensus about which eye colour in the class was the most common (an aggregate question). This second point suggested a need for evidence (data), a point Ms Louarn would return to. The problem under investigation allowed for students to reason about variability because not all eye colours were the same. It also allowed them to reason about characteristics of the aggregate (whether the majority of the class had blue eyes) as opposed to individuals, giving them experience reasoning about the aggregate.

As students continued to share, the opportunity arose to clarify the population under investigation when students mentioned their parents' eye colours.

Ava:	I think most of the people in this class, they have brown eyes.
Ms Louarn:	Do you know anybody with brown eyes?
Ava:	… Um, my mum does, my dad doesn't.
Ms Louarn:	Are your mum and dad in prep?
Ava:	No.
Ms Louarn:	It's great to know mum and dad's eyes. Let's just think about children in prep at the moment. … Kai?
Kai:	My dad has green eyes.
Ms Louarn:	Yes, so sometimes our parents have different eyes from us, and obviously you have got brown eyes and you're saying your dad has got green eyes. We are just going to talk about people in prep at the moment. [1:56]

Ava introduced a third eye colour, brown, as a possible answer to the inquiry question. She also went further to bring in others she knew, like her mother, who had brown eyes. This allowed Ms Louarn to press further to informally clarify the population that was the target of their inquiry. The response from Kai suggested that this point was not yet acknowledged by the children. Note, however, that the variability of eye colour was a tacit assumption within the problem space.

By this stage, early in the lesson, the children had begun to experience several statistical context–structures through discussing the question, *Do most students in Prep L have blue eyes?* Four statistical structures that they encountered at an informal level (recognised by adults as data, variability, aggregate, population) were not experienced in isolation, and they were experienced by the children within the problem context (their personal context), as context–structures. That is, when children reasoned about "eyes", they were reasoning about "data". As context–structures, the statistical structures were considered in relation to one another (e.g. different colours of eyes created a challenge to consider a question about "eyes" as an aggregate; the aggregate in question did not include their parents, who were outside the population). Variability, aggregate and population were also considered in relation to the statistical idea that data are evidence, which is the focus in the next section.

1.4.2 Suggesting a Need for Data

Throughout the sharing session, the teacher guided the discussion within the familiarity of the context, while concurrently and informally emphasising statistical relationships. It would have been possible for her not to emphasise these aspects by exploring, for example, children's eye colours in relation to their parents or encouraging general sharing about people who children knew had various eye colours. Ms Louarn also could have curtailed the discussion above by asking the children to sort their eye colour drawings into categories or stacking them like a bar graph. However, the teacher instead used the investigation to begin to informally develop statistical ideas, the need for evidence and the important role that data play in answering a statistical question.

Following the discussion above, Ms Louarn moved to elicit from the children an approach to address the inquiry question. Some of the seeds of this investigation had already been sown: the lack of consensus about which eye colour was most common, discussions of evidence (individual anecdote and aggregate) and suggestion of the population of focus. Students shared their ideas as Ms Louarn recorded them. Most children focused on initially just looking at their peers' eyes. For example, Will said, "We, um, we could go and look at eyes. We should go and look in the eyes". After this idea was repeated by other children, the teacher confirmed with a show of hands that most in the class agreed that they would go around and look at everyone's eyes in the class.

At this point, the children had (with assistance) suggested that in order to find out whether most children in the class had blue eyes, they would need to look at the eyes of the children in their class. Although this may seem obvious to an adult, this was an initial and tentative link between the question and a suggestion that evidence was needed to check if this claim was true. At this age, they were not yet thinking about how just looking at everyone's eyes could help them to answer the inquiry question. They were yet to recognise a need for data: to record their observations as they looked at eyes or to analyse their recordings to determine an answer.

Ms Louarn:	Who's got a different idea?
Mila:	I will look at, um, um, everyone's colours eyes, and I will, um, um, make a picture.
Ms Louarn:	Ah! Mila has got an interesting thing, she says she is going to look at everybody's eyes and then she is going make a picture. What sort of picture you would make Mila?
Mila:	(unintelligible) then I'm gonna to paint all of the eyes and then, I am gonna, um, um, and then I'm gonna put them in my, and then I'm going to make my own shop, and then I am gonna make lots of different colours of friends!
Ms Louarn:	So, I think this what you said. That are going to find out what colour eyes everybody's got and you're going to draw a picture of their eyes. Is that what you said? (Mila smiles and nods) That's a really an interesting idea. I'd like to think about that. (To the class) Do you think that might help us remember, whose eyes that we've got?
Students:	yes
Ms Louarn:	That's a great idea. We go and look at everybody's eyes and then we draw a picture, so that we can remember the colour of everybody eyes. Thank you Mila, I like that idea. [7:57]

Mila's mention of a drawing gave Ms Louarn an opportunity to reframe her suggestion as a way of recording their observations, emphasising the benefit of recording as a way to remember and keep track of whose eyes were observed. Sienna built on Mila's idea and suggested using the drawings to find out what everyone's eye colours were (and they'd be done).

Ms Louarn:	Ah! So you are suggesting that if we look at the pictures of ourselves that we could find out from them what colour eyes people have got. That's a good idea too. And what you would do after that? So you would look at ourselves over there, and then what would you do?
Sienna:	Then you look if you're right and if they're right. And you can see that they are right. [10:33]

Using Sienna's mention of their drawings, Ms Louarn privileged Sienna's idea to emphasise the benefit of using representations (rather than just "looking" at eyes); she further elaborated to suggest to students that these recorded drawings would still require another step. Jack further built on Sienna's idea, suggesting how having the drawings would allow them to go further to count.

Ms Louarn:	Yes, Jack?
Jack:	Look at everybody's eyes, look at my eyes and see if umm, count how many eyes is blue or not.
Ms Louarn:	Well, Jack just said something very interesting. So he is going to look at eyes as well, but then, then we can count the eyes when we make a picture, that is good idea! [11:31]

Three tentative statistical ideas were initiated in the discussion, ideas to build on over the course of the lessons: (1) a need for data (e.g. Will: "look at eyes") to answer the inquiry question; (2) the benefit of recording (e.g. Mila: "make a picture") to remember; and (3) recording was not enough, there was a need to analyse the data (e.g. Jack: "count how many eyes are blue or not"). These three ideas, in context, maintained a coherence of experiencing data as a statistical structure, with its characteristics (as an observation), representation (recording for memory) and processes (data collection was not enough; analysis was needed to answer the question).

1.4.3 Recording and Analysing the Data

The teacher decided to let them begin even though their plan was only partially constructed. Several children walked around and observed their peers' eyes and reported to Ms Louarn. Her response was to emphasise a need to record.

Thanh:	I found 8 blue eyes.
Ms Louarn:	You found 8 blue eyes! How are you going to remember that next time?
Thanh:	Try and remember?
Ms Louarn:	You're going to try and remember. And so do you think if you found 8 blue eyes, do you think more people in prep have blue eyes? (Student shakes head no and then shrugs shoulders.) [15:26]

After a few minutes, most children were at least looking at eyes. For some, they saw this as collecting evidence, and for others they were likely mimicking their peers.

A few children drew pictures of children's eyes, their own and/or others', with eyes coloured (Fig. 1.1). For students who were colouring only eyes (and not other facial features), they appeared to have moved towards an image of the eyes as the relevant aspect of the context to record (as opposed to other facial features). This abstraction of the eyes suggested a move towards seeing the recording as data. Even if only one child did so purposefully, others often followed. The discussions then became important to connect these practices to their utility in solving the problem.

Fig. 1.1 Children's recordings of eye colours in the class

Ms Louarn temporarily stopped the class as they were working and reiterated the problem, "We're going to find if it is true that most people in prep have blue eyes". She suggested a number of options that students were undertaking to find out. After another 10 min, Ms Louarn asked students to bring their ideas back to the circle on the carpet, including drawings if they had them.

Ms Louarn:	[Children] did what they said they were going to do: Look at the eyes, some people said make a picture of the eyes, and some people said counting the eyes. So some people have done that. Would someone like to put their hand up and tell us what they found out about our question? What did you do Aisha?
Aisha:	Um I didn't get to do the Bec's hair. (She shows her drawing with two people's faces including hair, nose, eyes and mouth).
Ms Louarn:	… So you've got two people there. Are you going to draw a picture of everybody in the class and a picture of their eyes?
Aisha:	I don't know if I will be able to fit them on here.

Ms Louarn:	But is that your idea? (Aisha nods) I think that is a really clever idea. Aisha would draw a picture of everybody in the class and she would draw the colour of their eyes and that's a good way of making a picture isn't it? Tomorrow when we come back she will be able to remember it all. Thank you Aisha I think that is a really clever idea. You're right it might take a little while … but it's a great idea. [30:48]

Other students shared who had drawn the full face, hair and eyes of one or more people. Sienna had drawn eyes and numbers next to them (Fig. 1.2).

Fig. 1.2 Sienna's representation of eye colours

Ms Louarn:	What have you got there Sienna? Show everybody what you've done. And can you tell us all about that.
Sienna:	It's a list about people who have brown eyes and blue eyes and green eyes. Um most people do have the same colour eyes. I couldn't draw everyone's eyes.
Ms Louarn:	Why was that? Did you run out of time?
Sienna:	Yes.
Ms Louarn:	Is that what happened you ran out of time. (Sienna nods) So how many have you done so far? How many people have blue eyes?
Sienna:	(Sienna counts each individual blue eye and the teacher asks clarification if it is 12 eyes or 12 people. She counts again) … 1, 2, 3, 4, 5, 6 …
Ms Louarn:	So you got 6 people with blue eyes. … Whose eyes have you got there Sienna? (Sienna recalls the names.) Right, Sienna tomorrow that's going to be my first question so I want you to have a think between now and tomorrow, what can you do on your drawing—which is sensational by the way—to remember whose eyes they are? [35:01]

An emphasis throughout the lesson was on enculturating students into an expectation of representing and providing evidence of their investigation towards addressing the inquiry question, *Do most students in Prep L have blue eyes?* This consistent focus allowed students to enrich the connection between the problem context (responding to the inquiry question using their everyday knowledge) and relevant statistical structures (evidence which relied on data, representation, aggregate and analysis). For example, slowly through the lesson, more students adopted the practice of using eyes (rather than entire drawings) labelled with names to represent the students in the class. This strengthened the relationship between children's eyes (context) and structures (eyes as data, moving towards aggregate).

Sienna's acknowledgement showed emerging awareness that the drawings of eyes were contextual representations of data. This context–structure link allowed her to discuss "eyes" as "data". Ms Louarn recapped the ideas that had been presented and encouraged the other students to think about some of these ideas as they continued working towards addressing the inquiry question. The pattern continued the following day, periods of working interspersed with sharing; through iteration, most children adopted practices of drawing people or eyes recorded as data, as the teacher continually emphasised the benefits of observing, recording and counting to focus on the aggregate question.

1.5 Discussion

The focus of this paper was to examine the use of problem context as a proxy for working with statistical structures in a class of young children. It was not to provide evidence of individual success in understanding the links between the context and the statistics, but at an informal level provide children with a low-stakes opportunity to be exposed to and engage in reasoning with powerful statistical structures.

In the lessons presented, Australian children in Foundation Year (also called "Prep", which is similar to kindergarten, ages 4–5) sought to evaluate a peer's claim that most children in the class had blue eyes. The key structural elements of statistics that were informally introduced—variability, aggregate, population, data and representation—are critical as foundations for understanding any statistical concepts and practices. As these ideas were informally introduced, they became part of the problem space in subsequent discussions. The familiarity of the shared context of eye colour gave the children a way to reason about concepts—concretely and informally—through the context of the problem. The focus on data as evidence throughout the lessons allowed for discussion of informal versions of several statistical structures by allowing the context to stand in for those structures. This reasoning was similar to what would be done in later years using the more abstract statistical structures as part of that discussion. One mapping is given in Table 1.2 of the contextual elements that students experienced through the familiar context and the related statistical structure.

Although the statistical structures themselves and the links between the context and the related statistical structures were unknown to the children, their reasoning about the context (or emerging reasoning, or mimicking) paralleled more formal statistical reasoning that would be developed over time. For example, focusing on only considering the eye colours of the children in the classroom (rather than their parents) was explained in relation to the inquiry question about eye colour in their class. Their classmates were the population relevant to the inquiry, and the children's reasoning about their classmates' eye colours was analogous to reasoning more abstractly about a population. The statistical structures encountered by the children were not limited to this specific context. That is, although statistics is a field based on variability, the patterns and invariances within variability expose important statistical structures within the field. For example, relationships between data and population hold regardless of the context.

The context of the problem was content-rich and complex, allowing for multiple statistical concepts, relationships, tools and structures that had analogies in the context to be used not in isolation, but in relation to each other (holistically) and purposefully to solve a problem (cf. Bakker & Derry, 2011). Furthermore, it provided opportunity to enculturate the children in statistical practices. Ms Louarn, the teacher, played a key role in her questioning and privileging of focus ideas. She used the children's ideas to generate, build on and challenge their emerging strategies. Asking questions and critiquing ideas were also seen as valued practice (not the emphasis of this paper). The norms that were developed in the classroom allowed for productive interactions as children became accustomed to what was valued and normalised as part of the classroom culture. For example, in publicly sharing their ideas, children influenced peers to shift inefficient practices (e.g. drawing an entire person), provided ideas when others were stuck and generated opportunities for feedback (e.g. comparing one's own drawing with those shared in the circle).

Being aware of statistical context–structures is a valuable framework for teachers. By identifying elements of the problem context that stand in for statistical structures,

Table 1.2 Mapping statistical context–structures in the lessons

Context element	Related statistical structure	Links between contexts and context–structures
Eye colour	Data	It was necessary to observe children's eye colours to answer the inquiry question
Multiple eye colours	Variability	The variability of eye colour was the problem to be managed (otherwise no investigation would be needed)
Children in the class	Population	The eye colours of people outside of the class, like parents, were not relevant
Drawing of self	Single data point	Children drew themselves or a friend; these drawings represented a single data point
Drawings of eyes	Data representation	The need to record eye colour (and not hair colour) as evidence focused children on salient aspects to represent or ignore
Counting eyes	Data analysis	Counting provided a way to compare groups (blue- vs. brown-eyed children) to answer the inquiry question
Drawings of collections of eyes	Collection of data points	When students drew collections of eyes, their drawing represented collection of data points
Questions about the class	Focus on aggregate	The inquiry question required students to look beyond single or multiple individuals to consider collective qualities of the aggregate

teachers can become sensitised to problems that would likely engage in content aligned with the teacher's goals. This is often a challenge in inquiry when it can appear as though the content cannot be determined in advance. This will also assist the teacher in developing questions that will emphasise (through privileging and revoicing) or develop (through questioning) desired content out of children's ideas.

Acknowledgements The data collection and writing of this chapter were funded by the Australian Research Council (LP0990184 and DP170101993, respectively). The author wishes to thank Debra for designing and teaching the lessons and for her insights during the retrospective analysis.

References

Alexander, P. A., White, C. S., & Daugherty, M. (1997). Analogical reasoning and early mathematics learning. In L. English (Ed.), *Mathematical reasoning: Analogies, metaphors and images* (pp. 117–147). Mahwah, NJ: Lawrence Erlbaum Associates.

Bakker, A., & Derry, J. (2011). Lessons from inferentialism for statistics education. *Mathematical Thinking and Learning, 13*(1–2), 5–26.

Boekaerts, M., & Minnaert, A. (1999). Self-regulation with respect to informal learning. *International Journal of Educational Research, 31*(6), 533–544.

Confrey, J., & Kazak, S. (2006). A thirty-year reflection on constructivism in mathematics education in PME. In A. Gutiérrez & P. Boero (Eds.), *Handbook of research on the psychology of mathematics education: Past, present and future* (pp. 305–345). Rotterdam: Sense Publishers.

English, L. D., & Mulligan, J. T. (2013). Perspectives on reconceptualizing early mathematics learning. In L. English & J. Mulligan (Eds.), *Reconceptualizing early mathematics learning* (pp. 1–4). Dordrecht, The Netherlands: Springer.

Flyvbjerg, B. (2006). Five misunderstandings about case-study research. *Qualitative Inquiry, 12*(2), 219–245.

Greer, B., Verschaffel, L., & Mukhopadhyay, S. (2007). Modelling for life: Mathematics and children's experience. In W. Blum, P. Galbraith, H. W. Henn, & M. A. Niss (Eds.), *Modelling and applications in mathematics education: The 14th ICMI study* (pp. 89–98). Dordrecht, The Netherlands: Kluwer Academic Publishers.

Konold, C., Robinson, A., Khalil, K., Pollatsek, A., Well, A., Wing, R., & Mayr, S. (2002, July). *Students' use of modal clumps to summarize data*. Paper presented at the Sixth International Conference on Teaching Statistics (ICOTS-6), Cape Town, South Africa.

Litman, L., & Reber, A. S. (2005). Implicit cognition and thought. In K. J. Holyoak & R. G. Morrison (Eds.), *The Cambridge handbook of thinking and reasoning* (pp. 431–453). Cambridge: Cambridge University Press.

Makar, K. (2012). The pedagogy of mathematical inquiry. In R. Gillies (Ed.), *Pedagogy: New developments in the learning sciences* (pp. 371–397). Hauppauge NY USA: Nova Publications.

Makar, K. (2014). Young children's explorations of average through informal inferential reasoning. *Educational Studies in Mathematics, 86*(1), 61–78.

McGowen, M. A., & Tall, D. O. (2010). Metaphor or met-before? The effects of previous experience on practice and theory of learning mathematics. *Journal of Mathematical Behavior, 29,* 169–179.

Miles, R. (2015). Complexity, representation and practice: Case study as method and methodology. *Issues in Educational Research, 25*(3), 309–318.

Mokros, J., & Russell, S. J. (1995). Children's concepts of average and representativeness. *Journal for Research in Mathematics Education, 26*(1), 20–39.

Perry, B., & Dockett, S. (2008). Young children's access to powerful mathematical ideas. In L. English (Ed.), *Handbook of international research in mathematics education* (2nd ed., pp. 75–108). New York: Routledge.

Powell, A. B., Francisco, J. M., & Maher, C. A. (2003). An analytical model for studying the development of learners' mathematical ideas and reasoning using videotape data. *Journal of Mathematical Behavior, 22*(4), 405–435. https://doi.org/10.1016/j.jmathb.2003.09.002.

Raman, M. (2002). Coordinating informal and formal aspects of mathematics: Student behaviour and textbook messages. *Journal of Mathematical Behavior, 21,* 135–150.

Roschelle, J. (2000). Choosing and using video equipment for data collection. In A. E. Kelly & R. A. Lesh (Eds.), *Handbook of research design in mathematics and science education* (pp. 709–729). Amsterdam: Kluwer Academic Publishers.

Sadler, T. (2004). Informal reasoning regarding socioscientific issues: A critical review of research. *Journal of Research in Science Teaching, 41*(5), 513–536.

Watson, J. M. (2006). *Statistical literacy at school: Growth and goals*. Mahwah, NJ: Erlbaum Associates.

Chapter 2
Probabilistic Thinking and Young Children: Theory and Pedagogy

Zoi Nikiforidou

Abstract Over the last decades, there has been a lot of interest in exploring young children's early probabilistic thinking, considering educational, cognitive and mathematical dimensions in children's learning and development. Today, probability is incorporated in many mathematical and statistical curricula and the ongoing research on children's probabilistic competencies has produced remarkable and educationally valuable conclusions. The aim of this chapter is to critically review key theoretical models of probabilistic thinking that cover the period of early childhood and to highlight a number of pedagogical implications while introducing probabilistic concepts in early childhood educational contexts. The traditional Piagetian claim that children during the preoperational period find it difficult to differentiate certainty and uncertainty seems to be replaced by findings that support children's capacity to engage with notions of probability. Recent research underlines how intuitions and experience, informal mathematical knowledge, probability literacy as well as curriculum development and task design play a significant role in shaping and enhancing preschoolers' probabilistic thinking, not only while they are young but with a lifelong perspective.

2.1 Setting the Scene: Probability, Literacy and Children

From early in life, children experience and interact with the world around them while making sense of the possible, random and impossible. They develop their understanding of the world through causal and statistical reasoning (Gopnik & Schulz, 2007), by making connections and using information and cues from around them, in order to predict and expect outcomes, when possible. Learning about the world requires learning about probabilistic relationships (Yurovsky, Boyer, Smith, & Yu, 2013) in framing what is likely and what is not. On many occasions, children through expe-

Z. Nikiforidou (✉)
Liverpool Hope University, Liverpool, UK
e-mail: nikifoz@hope.ac.uk

A. Leavy et al. (eds.), *Statistics in Early Childhood and Primary Education*,
Early Mathematics Learning and Development,
https://doi.org/10.1007/978-981-13-1044-7_2

rience and cognitive processing develop an understanding of probabilities as part of the development of their scientific, mathematical and social knowledge.

The nature of probability has three main approaches. The classical interpretation of probability is simply the fraction of the total number of possibilities in which the event occurs. Laplace (1814, 1951: 6–7) noted: '*The theory of chance consists in reducing all the events of the same kind to a certain number of cases equally possible, that is to say, to such as we may be equally undecided about in regard to their existence, and in determining the number of cases favorable to the event whose probability is sou*ght'. The second interpretation is the frequentist, where the possibilities of events may be assigned unequal weights and probabilities can be computed a posteriori. In this case, probability is based on the long-run behaviour of random outcomes (Konold, 1991). The third approach is subjective probability where probability tends to be 'a degree of belief', where biases, heuristics and intuitions interplay. *'Probability does not consist of mere technical information and procedures leading to solutions. Rather, it requires a way of thinking that is genuinely different from that required by most school mathematics. In learning probability, students must create new intuitions'* (Fischbein & Schnarch, 1997, p. 104). Thus, probability can become very complex and sophisticated and embraces a way of thinking in enabling us to cope with randomness, uncertainty and unpredictability through computational and subjective ways.

Gal (2005) highlights that many school curricula focus on the classical and/or frequentist views of probability instead of considering the big picture. He puts forward the notion of probability literacy, as '*the ability to access, use, interpret, and communicate probability-related information and ideas, in order to engage and effectively manage the demands of real-world roles and tasks involving uncertainty and risk*' (Gal, 2012, 4). His model suggests that five knowledge bases as well as supportive dispositions form probability literacy. Gal (2005) lists the five knowledge elements of probability literacy as the exploration of big ideas, like, variation, randomness, predictability/uncertainty; the estimation of probabilities; the use of language to communicate chance; the understanding of the context where probabilities are applied and the consideration of critical questions when dealing with probabilities. Nevertheless, according to Gal (2005) the dispositional elements are equally important building blocks of probabilistic literacy. These are: critical stance; beliefs and attitudes and personal sentiments regarding uncertainty and risk.

Similarly, Borovcnik (2016) defines probability literacy as '*the ability to use relevant concepts and methods in everyday context and problems*' (p. 1500). Probability literacy is essential in modern times, and recently, there has been growing interest in identifying ways and approaches in incorporating probabilities in education. Batanero et al. (2016) confirm the recognition of probability literacy by educational authorities globally, but also encourage more attention towards practical and pedagogical issues in implementing probability in curricula. The main aim of including probability in schools relates to its usefulness for daily life, its instrumental role in other disciplines and its important role in decision-making as a skill for competent and knowledgeable future citizens (Gal, 2005; Fisk, Bury, & Holden, 2006). Most studies focus on school-aged children and adolescents, but there is limited investigation capturing early experiences during preschool years. In this direction, the aim of this chapter is twofold: to present an overview of key theories and models on the development of probabilistic thinking, with an emphasis on early childhood, and to address pedagogical factors while implementing probabilities in early years. The first part identifies key characteristics and concepts that young children engage with at ages 3–7, based on theory and research, and the second part reviews important curriculum-related aspects in embedding probabilities in early years practice.

2.2 Children's Probabilistic Thinking: From Piaget and Fischbein to Contemporary Studies

The development of probabilistic thinking was traditionally and systematically explored by Piaget and Inhelder (1975). Within the Piagetian theory of cognitive development, it is recognised that the evolution of the idea of chance and probabilities is '*a kind of synthesis between operations and the fortuitous*' (p. 216). In the book 'The origin of the idea of chance in children' (1951), Piaget recognised that all occurrences in our daily lives are complex and proposed three developmental stages of the idea of chance. The first stage, which relates to early childhood, the preoperational period before the age of 7 or 8, is characterised by prelogical reasoning and limited cognitive capacity to understand irreversibility, deduction, random mixing and random distribution. Through a number of studies, Piaget concluded that children during this period do not distinguish the possible from the necessary and mark the development of the idea of chance during the second stage, when more advanced logical and arithmetical operations appear.

During this first stage, children base their judgements regarding random draws on phenomenism, passive induction and egocentrism. In a heads and tails game (with crosses and circles), where 10–20 counters were thrown at once, children were asked to predict the outcomes. Piaget would substitute the counters with fixed counters showing a cross on both sides aiming at recording children's reactions. Children in stage I would accept what they saw (phenomenism) as possible, whereas children in stage II would refuse to accept the coincidence of all counters showing a cross. Children also at stage I would try to justify this occurrence as part of their subjective and personal beliefs in that, for example, one can get crosses only if they toss the counters in a certain way (egocentrism). Also, young children would judge the outcome based on the immediately preceding facts (passive induction); if crosses came out previously, then crosses are likely to come out again.

For Piaget, children at the preoperational stage find it difficult to combine multiple interactions of phenomena as well as their irreversibility or independence. He tested this through another task, with the use of a rectangular box based on a sloping pivot and equal numbers of two sets of coloured balls. The two sets of balls are separated by a divider at one end of the box, and then children would be asked to predict after a number of tippings of the box the position and arrangement of the balls. Children in stage I would initially expect the balls after the first tipping to land in their original positions, and even if there was an element of mixture (one white ball ending up in the other coloured set of balls), they would seek for uniformities. This indicates that children at these ages have difficulty in understanding the idea of irreversibility and randomness.

Fischbein (1975) provided a more educational approach to probabilistic thinking in children and emphasised the role of intuitions in developing understandings of probabilistic notions. He focused on the importance of the intuitive endowment of the child and defined intuitions as '*forms of immediate cognition in which the justifying elements, if any, are implicit*' (p. 5). He underlined that when facing probabilistic events our behaviour requires specific intuitions, in the sense of predictions and responses 'guessed at' that are characterised by immediacy, globality, extrapolative capacity, structurality and self-evidentness. These intuitions are long-verified mechanisms, stabilised by social learning and personal experience. Fischbein (1975) provided different categories of intuitions and proposed that there should be separation between the concept of probability as an explicit, correct computation of odds and the intuition of probability, as a subjective, global evaluation of odds. He agreed that there is a developmental pattern in the emergence of probabilistic thinking in that through age and experience children develop more profound understandings. Therefore, he recommended the necessity to '*train, from early childhood, the complex intuitive base relevant to probabilistic thinking*' (p. 131).

Primary intuitions are cognitive acquisitions formed by experience before systematic instruction and are found in the preschool child. Secondary intuitions are formed by scientific instruction, mainly through school and formal education and transfer social experience to scientific truth. Another dichotomy in intuitions, proposed by Fischbein (1975), is between affirmatory and anticipatory intuitions. Affirmatory intuitions are based on the feeling of certitude of events; thus, anticipatory intuitions are global views on the solution to a situation which precede the problem-solving process. After several studies, Fischbein (1975) reached the conclusion that, in contradiction to Piaget, the intuition of chance is present before the age of 6–7. He argued that young children can indicate the capacity to evaluate chance and estimate the odds in a probabilistic manner, as they develop their primary and preoperational intuitions through daily experiences and subjective or perceptual considerations. This is confirmed in cases where the number of possibilities is small, where the nature of the problem is clear, where rewards for correct answers are present and where prior instruction in the concepts of chance and probability is enhanced. Thus, intuitions are key in the learning process and construction of probabilistic knowledge.

On a number of occasions, there has been a misinterpretation in that Piaget and Fischbein were contradictory. This is not the case; instead, they attempted to study and explore the development of probabilistic thinking in young children through different lenses. Piaget had a more developmental approach focusing on the role of intellectual ability, while Fischbein emphasised the role of intuitive thinking and pedagogy. More recently, and particularly over the last 20 years, there has been an ongoing interest in exploring the characteristics of young children's probabilistic thinking in formal and informal contexts of learning.

Jones et al. (1997) were pioneers in proposing a framework for assessing and nurturing children's thinking in probability from early ages. The aim of their framework was to provide aspects of children's probabilistic thinking in a comprehensive way that would support curriculum designers and pedagogues to implement probabilities in mathematical curricula and instruction. They examined four core constructs: sample space, probability of an event, probability comparisons and conditional probability.

Sample space: The sample space, Ω, is a key construct in understanding randomness (Nunes et al. 2014) and is about the set of all possible outcomes in a given situation. For example, the sample space (Ω) for rolling an ordinary dice would be Ω = (1, 2, 3, 4, 5, 6).
Probability of an event: The probability of an event A is the likelihood that event will occur P (A). With young children, this construct could be explored through guessing in a set task what is more or less likely to occur and reasoning why. For example, what is more likely to come up, if you turn over a card from a set of 6 hidden cards, where 5 are red and 1 is yellow?
Probability comparisons: Probability comparisons reflect the capacity to determine and justify: a. which probability situation is most likely to produce the desired outcome and b. whether two or more probability situations offer the same or a fair chance for the desired outcome. For example, if green wins, which of the following two spinners should I choose to increase the likelihood of winning: the one with ¾ in blue and ¼ in green or the one with ¼ blue and ¾ green?
Conditional probability: Conditional probability provides a way to reason about the outcome of an experiment, based on partial information or on additional information; the probability for an event A given B is denoted by P(A | B). For example, in a box, there are 4 black and 2 white bears. If we shake the box and a bear is drawn, what colour is it likely to be? If this bear is not repositioned in the box and there is a second draw, what colour is this second bear likely to be?

Through observations and interviews with children at different ages, Jones et al. (1997; 1999) proposed 4 levels of probabilistic thinking. They defined as probabilistic thinking '*children's thinking in response to any probability situation*' (Jones et al., 1999, p. 488). By probability situation, or a situation involving uncertainty, they consider an activity or random experiment where more than one outcome is possible; thus, the actual outcome cannot be predetermined but only inferred. In detail, Level 1 is associated with subjective thinking; here, children make intuitive judgements based on their imagination or personal preferences, consistent with reasoning that is subjective or influenced by irrelevant aspects. Level 2 is transitional between subjective and naive quantitative thinking, where students often make inflexible attempts to quantify probabilities. Level 3 involves the use of informal quantitative thinking in that students use more generative strategies in listing the outcomes of two-stage experiments and in coordinating and quantifying thinking about sample space and probabilities. Finally, Level 4 incorporates numerical thinking and students demonstrate the use of valid numerical measures to describe the probabilities.

Table 2.1 Jones et al. (1997) framework for assessing probabilistic thinking—*Level 1: Subjective*

Concept	Type of response
Sample space	• lists an incomplete set of outcomes for one-stage experiments
Probability of an event	• predicts most/least likely event based on subjective judgements • distinguishes 'certain', 'impossible', possible events in a limited way
Probability comparisons	• compares the probability of an event in two different sample spaces, usually based on various subjective or numeric judgements
Conditional probability	• following a particular outcome, predicts consistently that it will occur next time, or alternatively that it will not occur again (overgeneralises)

Level 1 is age-related to early childhood (Table 2.1). In the same direction, Way (2003) recorded that around 5 years and 8 months, during the non-probabilistic thinking stage that she proposed, children show minimal understanding of randomness and are strongly reliant on visual comparisons. Children under 6 years may possess intuitive notions of probability, but these are unstable. Likewise, Nikiforidou and Pange (2010) found that 4-year-olds rely on visual comparisons and distinguish 'impossible' and 'possible' events in a limited way. Also, Sobel et al. (2009) found that children's probabilistic inferences develop into early elementary school, but preschoolers might also have some understanding of probability when reasoning about causal generalisation.

Examples from recent studies confirm that preschoolers have a sophisticated understanding of probability concepts. For example, Kushnir and Gopnik (2005) found in their study that children aged 4–6 apply the probabilistic element of the frequency of co-occurrence when developing causal relationships. Boyer (2007) used a computerised decision-making task to find that 5–6-year-olds select the more probable outcome by demonstrating intuitive sensitivity to probability. Girotto and Gonzalez (2008) found through three different studies that when preschoolers, around 5 years old, are faced with uncertain events, they are able to integrate a new piece of information in making inferences and as such indicate adaptation to posterior probability. Fisk et al. (2006) found that children aged 4–5 would commit the conjunction fallacy while participating in tasks involving choice between the more likely of two events, a single event and a joint event (conjunctive or disjunctive). Moreover, Girotto et al. (2016) found that in probabilistic choice tasks, 5-year-olds made optimal choices, whereas 3–4-year-olds based their responses on randomness and/or superficial heuristics. Such studies, as well as others, provide insights on preschoolers' probabilistic reasoning in diverse probabilistic situations. These, in turn, can inform practice and ways of fostering children's probabilistic literacy in educational contexts.

2.3 Probability in Early Childhood Educational Practice

2.3.1 Curriculum Design: Constructivism and Proposed Instructional Models

Early years education provides the foundation for fundamental conceptual understanding, knowledge and dispositions needed for further learning. In his idea of the *spiral curriculum*, Bruner (1960) addresses the role of probability and underlines: '*If the understanding of number, measure, and probability is judged crucial in the pursuit of science, then instruction in these subjects should begin as intellectually honestly and as early as possible in a manner consistent with the child's forms of thought. Let the topics be developed and redeveloped in later grades*' (pp. 53–54). He emphasises the necessity of introducing the equally important concepts of number, measure and probability as early as possible, in a way that relates to the child's cognition. These can then be revisited and reconstructed through time and progression.

Cobb (2007) agrees that mathematical learning is an interactive as well as a constructive process. It is a process where prior knowledge and experiences are used as the foundation for constructing and reorganising conceptual and theoretical ideas. In a similar direction, Sharma (2014) believes that probabilistic thinking can be developed slowly and systematically through carefully designed activities in appropriate learning environments. She favours the learning context that challenges students to explore and reflect on any discrepancies they observe and the one that facilitates evaluations and justifications in both verbal and representational modes.

Jones et al. (1999) took a socio-constructivist position in their study, supporting that probability knowledge can arise from students' attempts to solve problems, to build on and reorganise their informal knowledge, and to resolve conflicting points of view. Under this position, social processes are important mechanisms through which participants negotiate meaning and co-construct knowledge in collaborative learning environments (Cobb, 2007). The instructional sequence argued begins with the presentation of a meaningful task or problem and continues with an invitation to solve that problem in multiple ways, which leads to the sharing, justifying and discussing of those problem-solving strategies in small or large group discourses (Garfield & Ben-Zvi, 2009). This links to the predict–observe–explain (White & Gunstone, 1992) strategy that probes understanding, especially in science education. First, the students must predict the outcome of some event and reason about their prediction; then they must describe what they see happening; and finally, they must reconcile any conflict between prediction and observation.

Likewise, Sharma (2014) proposes a possible teaching sequence to explore probability, based on the example of a die rolling game:

1. Posing a task: introduce the task in a meaningful context
2. Making predictions: individually and next in pairs encourage students to discuss and record their predictions

3. Playing the game: encourage students in pairs/small groups to interact with the game and record the actual outcomes
4. Planning explorations: the whole class shares and discusses their ideas
5. Data collection and analysis: in groups, students collect and record data
6. Further exploration: representation of dice outcomes in various ways (i.e. tree diagrams, tables) (pp. 81–82).

Constructivist approaches to teaching and learning consider intuition and prior knowledge as a starting point for further learning (Gelman & Brenneman, 2004). Nikiforidou et al. (2013) found that the linkage between informal and formal probability learning in the preschool classroom can be enabled if the subject content and the cognitive capacity of children match. Young children not only know some mathematics before reaching formal schooling, but they are ready and eager to learn more of it (Greenes, 1999). Children encounter probabilistic judgements and relationships in their daily routines and develop an informal understanding of what is likely, possible, uncertain or random. It is these initial understandings and personal experiences that can be the stepping stone in instruction. As a matter of fact, children learn through physical and social interactions, before school, and formulate informal knowledge and understanding (Ginsburg, Lee, & Boyd, 2008). HodnikČadež and Škrbec (2011) propose that the probability contents in the preschool and early school period should be related to using everyday probability language, answering probability or likelihood questions about specific data, answering probability or likelihood questions about specific situations and collecting and reflecting on empirical data.

2.3.2 *Aspects of Instruction and Task Design in Probability Learning*

Manipulatives play a key role in children's mathematical understanding as they offer ways of connecting mathematical ideas to real-world experiences (McNeil & Jarvin, 2007). Manipulatives, both concrete and virtual, enable children to experience consciously and unconsciously mathematical thinking through their senses (Swan & Marshall, 2010); through exploration, manipulation, interaction and observation. Their design and how they are introduced in practice are key in children's meaning-making and reasoning. However, their presence only is not adequate for meaningful learning to occur. Instead, their effectiveness depends on how they are embedded in comprehensive, well-planned activities (Sarama & Clements, 2009), which gradually build on more advanced knowledge through play, exploration, repetition and stimulation (DeVries, Zan, Hildebrandt, Edmiaston, & Sales 2002). Falk et al. (2012) support that children's implicit probabilistic knowledge can be strengthened by devising hands-on educational measures and interactions through a playful way. Furthermore, HodnikČadež and Škrbec (2011) agree that concrete experiences and experimentation are key in teaching probabilistic concepts in preschool children.

An example of such experiences can originate from picture story books. Picture books have been found to can act as means for the construction of knowledge and higher understanding in mathematics instruction (Elia, van den Heuvel-Panhuizen, & Georgiou, 2010). They provide opportunities for meaningful connections between young children's prior knowledge and the content presented. Kinnear (2013) in her study on children's statistical reasoning used picture books as a task and data context, with children aged 5. She found that children responded to the uncertainty created by an unresolved problem in the story and by making predictions if the book generated personal interest either through the illustrations, the characters or the mystery presented.

In probability learning, manipulatives, concrete and technological, could be dice, spinners, cards, board games, tinker cubes, urns, boxes or bags composed by variant ratios and proportions of items, stories and scenarios, visual stimuli, props and tools, toys. Batanero et al. (2016) observe that as these physical devices can be acted upon, they are increasingly used in probability education aiming to induce chance events (e.g. by rolling, spinning, choosing) and the development of key probabilistic concepts. Some examples are presented in Table 2.2. However, as Pratt (2011) recommends, more research is needed in exploring the role of these artefacts in the development of new curricula and the linkage between probability and real-world phenomena. He mentions: '*...it is debatable whether there is much advantage in maintaining the current emphasis on coins, spinners, dice and balls drawn from a bag... now that games take place in real time on screens, probability has much more relevance as a tool for modelling computer-based action and for simulating real-world events and phenomena*' (Pratt, 2011, p. 892).

Falk et al. (2012) emphasise that the structure of the problem is a key learning factor when using probabilistic tasks. They found that young children from the age of 4 can be introduced to probability through playful ways. Furthermore, Schlottmann and Wilkening (2012) underline that task complexity, in relation to linguistic, memory and meta-cognitive demands, can define children's probabilistic thinking. Skoumpourdi et al. (2009) supported that the important factors in the nature and structure of the particular probabilistic task or problem situation for preschoolers are: a meaningful context, the manipulation of concrete materials, the facilitation of rich discussions, the reflective process and children's informal knowledge of probability. Thus, if there is a play element, materials, discourse and simplicity in the task, children can interact with probabilistic notions.

Another important pedagogical factor when introducing probability in early childhood relates to the significance of questions initiated by the teacher. According to Sharma (2014), the teacher plays an important role in posing questions that prompt students' thinking and reasoning. Through open-ended questions, students get the opportunity to deepen their perceptions and share them with others. Similarly, according to Friel et al. (2001, p. 130), the questions have to '*provoke students' understanding of the deep structure of the data presented*'. Sharma (2014) also recommends the use of some sentence beginners to help students write/express their responses.

Table 2.2 Examples of probabilistic tasks for preschoolers

Materials	Task	Key concept
 – a bag/box – different sums of items	After introducing children to the materials, we ask them to place the items in the bag and mix them up. Without seeing, we ask them '*If one item is selected* (either by themselves, or a puppet) *what do you think will come out*?' Children can record their answers in 2 stages: prior to the draw (their predictions) and after the draw (actual outcome). Discussion can be facilitated in comparing and analysing the data. Variations to the distribution of the sample space are encouraged	– sample space – likelihood of events
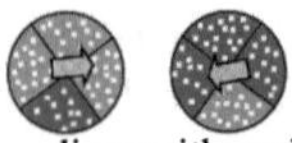 – discs with variations in the sample space	After introducing children to the materials, we ask them the following questions: '*If I want to bring orange, which spinner should I choose?*', '*If I want to bring blue for the next 5 times, which spinner should I choose?*' Again, children can record their predictions and actual outcomes for further discussion	– probability comparisons – sample space – likelihood of events

These could be, for example, 'From the table it can be seen that… because…'. Way (2003) also noticed that teachers may build awareness of the relationship between the sample space and the likelihood of events through the repetition of games and the use of guiding questions.

Technology and its role in statistics and probability education is a growing field of interest (Batanero et al., 2016; Tishkovskaya & Lancaster, 2012). Chance et al. (2007) discuss how recent and ongoing developments in using technology in teaching statistics correlate with changes in course content, pedagogical methods and instructional formats. Batanero et al. (2016) agree that technology provides a big opportunity for probability education, and Borovcnik and Kapadia (2009) underline that probabilistic software offers more efficient, graphically oriented possibilities to supply experience with randomness. For example, Paparistodemou et al. (2008) examined the strategies through which children aged 5.5 and 8 years engaged in constructing a fair spatial lottery game. They found that the microworld enhanced children's deterministic and stochastic thinking when exploring fairness and randomness. Haworth et al. (2010) found that in designing digital games, additional visual representations like decision trees that represent probabilistic reasoning support children's thinking processes. In another study, Paparistodemou et al. (2002) built a probability game to study young children's understanding of random mixture. Children as young as six could make sense of random mixtures in this game-like environment. Nowadays, the discussion has moved beyond technology itself, towards ways in which technological programs and tools can support the teaching and learning process of probability.

2.4 Conclusion

Probability can be approached not only through mathematical calculations but also through subjective intuitions. Children, from a really young age, through experience and experimentation construct knowledge and dispositions towards probabilistic concepts. They encounter situations where uncertainty and randomness apply. It was initially Piaget and Fischbein who explored the origins of probabilistic thinking in young children, and subsequent research has revealed ways through which children think and act within probabilistic contexts. Developmentally, children as young as four show engagement with notions of probabilistic thinking, and it is argued that the early childhood classroom can set foundations for probability literacy.

In this direction, it is proposed that there are some pedagogical implications to be considered when implementing probability in the preschool setting. These implications derive from both more generic approaches characterising early childhood pedagogy and more specific features applying to probabilistic thinking and reasoning in young children. Prior knowledge, intuitions, meaningful tasks in connection to children's personal worlds are important. Simple concepts of probability can be explored through discussion, group work, collaborative learning, concrete experiences and co-construction of knowledge. The concepts presented in this chapter (sample space, probability of an event, probability comparisons and conditional probability) are based on the framework proposed by Jones et al. (1997).

These concepts can be approached through instructional sequences, like predict–observe–explain. In these, children can engage with problem-solving situations underpinned by possibilities and probabilities. The learning experience can be enriched through a number of ways; the use of manipulatives (like dice, spinners, boxes), scenarios and story picture books (that encourage inferencing about specific data and situations), technological tools and software (like microworlds, digital games), discourse (like the use of open-ended questions), repetition and collaboration. However, further research is needed to investigate in more detail approaches through which probabilistic thinking can be fostered in a child-centred way. In particular, the role of manipulatives, physical and virtual, as transitional objects that enable doing and thinking—action and perception—needs more exploration. Other possible directions for future research could involve the role of the practitioner, the role of technology, the connection of probability to statistics and other fields.

To sum up, children's probabilistic competence is more profound than previously thought. The early childhood classroom can be the starting point of a spiral curriculum that introduces probability, aiming at probability literate future citizens.

References

Batanero, C., Chernoff, E. J., Engel, J., Lee, H. S., & Sánchez, E. (2016) Research on teaching and learning probability. In: *Research on teaching and learning probability. ICME-13 topical surveys* (pp. 1–33). Cham: Springer.

Borovcnik, M. (2016). Probabilistic thinking and probability literacy in the context of risk. *Educação Matemática Pesquisa, 18*(3), 1491–1516.

Borovcnik, M., & Kapadia, R. (2009). Research and developments in probability education. *IEJME-Mathematics Education, 4*(3), 111–130.

Boyer, T. W. (2007). Decision-making processes: Sensitivity to sequentially experienced outcome probabilities. *Journal of Experimental Child Psychology, 97,* 28–43.

Bruner, J. (1960). *The process of education*. Cambridge, MA: Harvard University Press.

Chance, B., Ben-Zvi, D., Garfield, J., & Medina, E. (2007). The role of technology in improving student learning of statistics. *Technology Innovations in Statistics Education, 1*(1). Available at: http://escholarship.org/uc/item/8sd2t4rr. Accessed 5 September 2017.

Cobb, P. (2007). Putting philosophy to work: Coping with multiple theoretical perspectives. In F. K. Lester (Ed.), *Second handbook of research on mathematics teaching and learning* (pp. 3–38). Charlotte, NC: Information Age.

DeVries, R., Zan, B., Hildebrandt, C., Edmiaston, R., & Sales, C. (2002). *Developing constructivist early childhood curriculum: Practical principles and activities*. New York: Teachers College Press.

Elia, I., van den Heuvel-Panhuizen, M., & Georgiou, A. (2010). The role of pictures in picture books on children's cognitive engagement with mathematics. *European Early Childhood Research Journal, 18*(3), 275–297. https://doi.org/10.1080/1350293X.2010.500054.

Falk, R., Yudilevich-Assouline, P., & Elstein, A. (2012). Children's concept of probability as inferred from their binary choices—Revisited. *Educational Studies in Mathematics, 81,* 207–233.

Fischbein, E. (1975). *The intuitive sources of probabilistic thinking in children*. Dordrecht: Reidel.

Fischbein, E., & Schnarch, D. (1997). The evolution with age of probabilistic, intuitively based misconceptions. *Journal for Research in Mathematics Education, 28*(1), 96–105.

Fisk, J. E., Bury, A. S., & Holden, R. (2006). Reasoning about complex probabilistic concepts in childhood. *Scandinavian Journal of Psychology, 47,* 497–504.

Friel, S. N., Curcio, F. R., & Bright, G. W. (2001). Making sense of graphs: Critical factors influencing comprehension and instructional applications. *Journal for Research in Mathematics Education, 32*(2), 124–158.

Gal, I. (2005). Towards "probability literacy" for all citizens: Building blocks and instructional dilemmas. In G. A. Jones (Ed.), *Exploring probability in school. Challenges for teaching and learning* (pp. 39–63). Dordrecht, The Netherlands: Kluwer.

Gal, I. (2012). Developing probability literacy: Needs and pressures steeming from frameworks of adult competencies and mathematics curricula. In *Proceedings of the 12th International Congress on Mathematical Education* (pp. 1–7), Seoul, Korea.

Garfield, J. B., & Ben-Zvi, D. (2009). Helping students develop statistical reasoning: Implementing a statistical reasoning learning environment. *Teaching Statistics, 31*(3), 72–77.

Gelman, R., & Brenneman, K. (2004). Science learning pathways for young children. *Early Childhood Research Quarterly (Special Issue on Early Learning in Math and Science), 19*(1), 150–158.

Ginsburg, H. P., Lee, J. S., & Boyd, J. S. (2008). Mathematics education for young children: What it is and how to promote it. *Social Policy Report of the Society for Research in Child Development, 22,* 3–23.
Girotto, V., & Gonzalez, M. (2008). Children's understanding of posterior probability. *Cognition, 106,* 325–344. https://doi.org/10.1016/j.cognition.2007.02.005.
Girotto, V., Fontanari, L., Gonzalez, M., Vallortigara, G., & Blaye, A. (2016). Young children do not succeed in choice tasks that imply evaluating chances. *Cognition, 152,* 32–39.
Gopnik, A., & Schulz, L. (2007). *Causal learning: Psychology, philosophy, computation.* New York: Oxford University Press.
Greenes, C. (1999). Ready to learn: Developing young children's mathematical powers. In J. Copley (Ed.), *Mathematics in the early years* (pp. 39–47). Reston: National Council of Teachers of Mathematics.
HodnikČadež, T., & Škrbec, M. (2011). Understanding the concepts in probability of pre-school and early school children. *Eurasia Journal of Mathematics, Science & Technology Education, 7*(4), 263–279.
Haworth, R., Bostani, S., & Sedig, K. (2010). Visualizing decision trees in games to support children's analytic reasoning: Any negative effects on gameplay? *International Journal of Computer Games Technology, 1,* 1–12. https://doi.org/10.1155/2010/578784.
Jones, G. A., Langrall, C. W., Thornton, C. A., & Mogill, A. T. (1999). Students' probabilistic thinking in instruction. *Journal for Research in Mathematics Education, 30*(5), 487–519.
Jones, G., Langrall, C., Thornton, C., & Mogill, T. (1997). A framework for assessing and nurturing young children's thinking in probability. *Educational Studies in Mathematics, 32,* 101–125.
Kinnear, V. A. (2013). *Young children's statistical reasoning: a tale of two contexts.* Ph.D. thesis, Queensland University of Technology.
Konold, C. (1991). Understanding students' beliefs about probability. In E. von Glassersfeld (Ed.), *Radical constructivism in mathematics education* (pp. 139–156). Holland: Kluwer.
Kushnir, T., & Gopnik, A. (2005). Young children infer causal strength from probabilities and interventions. *Psychological Science, 16*(9), 678–683.
Laplace, P. S. (1951). *A philosophical essay on probabilities* (F. W. Truscott & F. L. Emory, Trans.). New York, NY: Dover (Original work published 1814).
McNeil, N. M., & Jarvin, L. (2007). When theories don't add up: Disentangling the manipulatives debate. *Theory into Practice, 46*(4), 309–316.
Nikiforidou, Z., & Pange, J. (2010). The notions of chance and probabilities in pre-schoolers. *Early Childhood Education Journal, 38*(4), 305–311. https://doi.org/10.1007/s10643-010-0417-x.
Nikiforidou, Z., Pange, J., & Chadjipadelis, T. (2013). Intuitive and informal knowledge in preschoolers' development of probabilistic thinking. *International Journal of Early Childhood, 45*(3), 347–357. https://doi.org/10.1007/s13158-013-0081-6.
Nunes, T, Bryant, P, Evans, D, Gottardis, L, & Terlektsi, M. (2014). The cognitive demands of understanding the sample space. *ZDM—International Journal on Mathematics Education, 46*(3), 437–448. https://doi.org/10.1007/s11858-014-0581-3.
Paparistodemou, E., Noss, R., & Pratt, D. (2002). Exploring in sample space: Developing young children's knowledge of randomness. In B. Phillips (Ed.), *Proceedings of the 6th international conference on teaching statistics*, CapeTown. Voorburg, The Netherlands: International Statistics Institute.
Paparistodemou, E., Noss, R., & Pratt, D. (2008). The interplay between fairness and randomness in a spatial computer game. *International Journal of Computers for Mathematical Learning, 13*(2), 89–110. https://doi.org/10.1007/s10758-008-9132-8.
Piaget, J., & Inhelder B. (1975). *The origin of the idea of chance in children* (L. Leake, Jr., P. Burrell, & H. Fischbein, Trans., & Ed.). New York: Norton.
Pratt, D. (2011). Re-connecting probability and reasoning about data in secondary school teaching, In *Proceedings of the 58th World Statistics Conference* (pp. 890–899), Dublin.
Sarama, J., & Clements, D. H. (2009). "Concrete" computer manipulatives in mathematics education. *Child Development Perspectives, 3*(3), 145–150.

Schlottmann, A., & Wilkening, F. (2012). Judgment and decision making in young children. In M. K. Dhami, A. Schlottmann, & M. R. Waldmann (Eds.), *Judgment and decision making as a skill: Learning development and evolution* (pp. 55–83). New York: Cambridge University Press.

Sharma, S. (2014). Teaching probability: A socio-constructivist perspective. *Teaching Statistics*, 78–84.

Skoumpourdi, C., Kafoussi, S., & Tatsis, K. (2009). Designing probabilistic tasks for kindergartners. *Journal of Early Childhood Research, 7*(2), 153–172. https://doi.org/10.1177/1476718X09102649.

Sobel, D. M., Sommerville, J. A., Travers, L. V., Blumenthal, E. J., & Stoddard, E. (2009). The role of probability and intentionality in preschoolers' causal generalizations. *Journal of Cognition and Development, 10*(4), 262–284.

Swan, P., & Marshall, L. (2010). Revisiting mathematics manipulative materials. *Australian Primary Mathematics Classroom, 15*(2), 13–19.

Tishkovskaya, S., & Lancaster, G. (2012) Statistical education in the 21st century: A review of challenges, teaching innovations and strategies for reform. *Journal of Statistics Education, 20*(2). Available at: http://www.amstat.org/publications/jse/v20n2/tishkovskaya.pdf. Accessed 25 September 2017.

Way, J. (2003). *The development of children's notions of probability*. Ph.D. thesis, University of Western Sydney.

White, R. T., & Gunstone, R. F. (1992). *Probing understanding*. Great Britain: Falmer Press.

Yurovsky, D., Boyer, T., Smith, L. B., & Yu, C. (2013). Probabilistic cue combination: Less is more. *Developmental Science, 16*(2), 149–158.

Part II
Learning Statistics and Probability

Chapter 3
Emergent Reasoning About Uncertainty in Primary School Children with a Focus on Subjective Probability

Sibel Kazak and Aisling M. Leavy

Abstract The classical, frequentist and subjective interpretations of probability are the three main perspectives on the quantification of uncertainty. While the first two are emphasised in most school curriculum materials, the subjective notion of probability either is neglected or has minimal mention. Yet, it is closely related to what people commonly use for everyday reasoning. In this chapter, we combine a critical literature review of children's reasoning about uncertainty from both qualitative and quantitative perspectives with an analysis of empirical data. We explore the types of language 7–8-year-old children use to predict and describe outcomes and how they reason about the likelihood of outcomes of chance events using subjective probability evaluations before and after experiment and simulation. Data show that children used chance language relatively accurately to describe the likelihood of chance events and most of them had a quantitative understanding of equal likelihoods. Modifying predictions based on experiment and simulation results seemed to be intuitive for young children.

3.1 Theoretical Perspective

Articulation of uncertainty is an essential component of the statistical thinking process in which decisions and predictions are made on the basis of data in everyday context. Probability is the measurement we use to quantify the uncertainty about an outcome. The classical, frequentist and subjective interpretations of probability are the three main perspectives on the quantification of uncertainty. Classical probability refers to the ratio of the number of favourable cases in an event to the total number of

S. Kazak (✉)
Faculty of Education, Pamukkale University, Denizli, Turkey
e-mail: skazak@pau.edu.tr

A. M. Leavy
Department of STEM Education, Mary Immaculate College-University of Limerick, Limerick, Ireland
e-mail: Aisling.Leavy@mic.ul.ie

A. Leavy et al. (eds.), *Statistics in Early Childhood and Primary Education*, Early Mathematics Learning and Development,
https://doi.org/10.1007/978-981-13-1044-7_3

equally likely cases. This ratio is called theoretical probability in mathematics textbooks and computed based on an analysis of sample space. The frequentist approach defines probability of an event as the limiting value of its relative frequency in a large number of trials. In school curricula, this is often called experimental probability, which is estimated based on observed results from an experiment or simulation. Subjective probability is considered as a personal degree of belief (or strength of judgment) and changes based on personal judgment and information available about a given outcome. While the first two conceptions of probability are emphasised in school curriculum materials, the subjective notion of probability either is neglected or has minimal mention despite it being closely related to what people commonly use for everyday reasoning.

Our understanding of children's ideas of probability has been informed greatly by the work of Piaget and Inhelder (1975) who argued that children's conception of probability depended on their understandings of part–whole relationships. They also indicated that children's quantitative reasoning about probabilities developed with their understanding of the sample space and the combinatorial operations according to their developmental stages. Prior to the formal operational period, children tend to evaluate the likelihood of events on the basis of subjective judgments based on prior experiences or personal beliefs (Jones, Thornton, Langrall & Tarr, 1999). On the other hand, Fischbein (1975) considered children's intuitions based on individual experiences important as their development of formal conceptions of probability could be mediated through specific instruction and social interactions. While his approach offers insights into the development of children's probability estimates with the appropriate instructional support, we are particularly interested in subjective probability estimates.

The importance of chance language is reflected in its inclusion in international curriculum documents. Where probability is introduced in the primary curriculum, the emphasis is on describing events and discussing likelihoods using chance language. The Australian Primary curriculum for Year 1 (ages 6–7) recommends students 'identify outcomes of familiar events involving chance and describe them using everyday language such as 'will happen', 'won't happen' or 'might happen" (Australian Curriculum, Assessment and Reporting Authority [ACARA], 2014). Similarly in grades 3–4 (ages 9–10) in Ireland, the curriculum emphasises using the vocabulary of uncertainty and chance (possible, impossible, might, certain, not sure, chance, likely, unlikely, never, definitely), ordering events in terms of likelihood of occurrence, and identifying and recording outcomes of simple random processes (NCCA, 1999). In contrast, countries that delay the introduction of probability until secondary education commence with a focus on numerical probabilities and may or may not focus on language. The UK curriculum for Key Stage 3 (ages 11–14) suggests that students 'record, describe and analyse the frequency of outcomes of simple probability experiments involving randomness, fairness, equally and unequally likely outcomes, using appropriate language' (DoE, 2013). However, the Common Core Standards in the USA delay the introduction of probability until grade 7 (ages 12–13) and make no reference to the language of chance (Common Core Standards Initiative, 2010).

Within this chapter, we combine critical literature review of these three different perspectives on children's reasoning about uncertainty with qualitative analysis of empirical data in order to understand their ways of reasoning about uncertainty from a subjective probability point of view. The research questions are:

(1) What types of language do 7–8 year olds use to predict and describe outcomes?
(2) How do children reason about the likelihood of outcomes of chance events using subjective probability evaluations before and after experiment and simulation?

3.2 Review of the Literature

3.2.1 Children's Use of Chance Language

Comprehension of chance language is an important aspect of statistical literacy. While not exposed to formal quantitative measures of chance until the middle grades, children interact with thc language of chance in their everyday lives. For example, they frequently refer to fairness and chance in schoolyard games and in computer games.

Watson and Moritz (2003) identify the lack of research on children's abilities to evaluate and determine chance phrases. The few studies carried out with school-age children reveal that older children show greater facility in interpreting chance language (Green, 1982; Mullet & Rivet, 1991). The latter study of 9–15 year olds demonstrated their ability to discriminate different probability terms such as "certain", "likely" and "doubtful" with older children producing more accurate numerical estimates. This supports the finding of Hoffner, Cantor, and Badzinski (1990) that older children (grade 4) were able to differentiate between similar probability terms (e.g. "probably" and "definitely") better than younger children (pre-K and grade 1). A common finding among several studies is the difficulty that children of all ages have differentiating between the terms 'unlikely' and 'cannot happen' (Green, 1982) and delineating between 'very likely', 'moderately likely' and 'certain' (Green, 1982). The work of Fischbein, Nello and Marino (1991) also reveals difficulties with the terms "impossible", "possible" and "certain". This research challenges assumptions around the complexities of the term 'possible' and suggests that understandings of possible occur earlier than understandings of certainty. Furthermore, the tendency to equate 'rare' with 'impossible' and 'highly frequent' with 'certain' is highlighted alongside the caution that children 'tend to substitute mathematical meanings with subjective expectations' (Fischbein et al., 1991, p. 528).

Children demonstrate difficulty interpreting equal likelihoods. Some equate equal likelihood with the presence of uncertainty (Watson, 2005), whereas others interpret it as a lack of preciseness as indicated by their selection of statements such as 'It may happen or may not' (Green, 1982). Children also use the term equal likelihoods to refer to the chance of any outcome considered probable (Amir & Williams, 1999).

Moreover, individuals using causal reasoning with the aim of predicting outcome of a single trial—called as outcome approach (Konold, 1989), tend to interpret equal likelihoods (50% chance) as meaning 'I don't know'.

3.2.2 Children's Reasoning About Uncertainty

With regard to articulating uncertainty quantitatively, Piaget and Inhelder (1975) argued that the ability to relate the part (favourable outcomes) with the whole (all possible outcomes) was essential to children's development of the probability concept. Accordingly, research with young children considered understanding probability as ratio crucial for the conception of probability. In probability comparison tasks involving alternative choices, Falk (1982) conjectured that children (from ages 5–6) begin to make a choice based on 'more target elements', move to focus on 'less non-target elements' and then attempt to combine both quantities by considering 'greater difference in favour of target outcome'. Referring to the limitations of tasks using the simple choice paradigm, Acredolo, O'Connor, Banks and Horobin (1989) investigated children's (ages of 7, 9 and 11) probability estimations for different sets of tasks as they rated the likelihood of a target event on a non-numerical scale. They argued that even young children could use both numerator and denominator information to estimate probabilities in given tasks.

As earlier research on probability learning focussed primarily on the classical notion of probability, we have still little or minimal understandings of young children's conceptions of probability from the frequentist and subjective approaches. In a study using a model-based approach to reasoning about uncertainty (Horvath & Lehrer, 1998), young children (ages 7–8) engaged in aspects of experimental probability, but the focus was on relating the sample space with experimental outcomes, which was apparently difficult for most of these children without appropriate instructional support. When predicting the most and least likely outcomes of rolling two dice, children tended to rely on previously obtained empirical outcomes rather than use a model of sample space. In another study, it was suggested that experimental data could promote the development of children's probabilistic thinking (Kafoussi, 2004). For example, when making a prediction in a probability comparison problem, a group of children (aged 5) conducted an experiment and used the empirical outcomes to resolve different views, in particular concerning a non-probabilistic idea, i.e., "The blue color [is easiest to come out], because this is higher up in the bag." (p. 36). Kafoussi also found that in equally likely situations young children could spontaneously make their judgments without the need for an experiment. Moreover, Jones et al. (1999) pointed out that children tend to place "too much faith in small samples of experimental data when determining the most or least likely event" (p. 150) in early levels of probabilistic reasoning and tend to recognise the need for larger sample as they begin to develop informal quantitative reasoning.

While Jones et al. (1999) described both experimental and theoretical probabilities in their framework, subjective probability was not addressed as a construct. They used the term 'subjective judgement' as an early notion before quantitative reasoning used in predictions or probability comparisons. In relation to the development of ideas related to subjective probability, Huber and Huber (1987) demonstrated the emergence of ability in using personal knowledge or beliefs in comparisons of subjective probabilities starting from age 5. Moreover, Acredolo et al. (1989) considered the relations between children's subjective probability estimates and theoretical probabilities. However, there was no opportunity for posterior evaluation of these probability estimates with new information, such as experiment results. The research described below is intended to address this gap.

Next, we explain the method of the study we conducted with children; present and discuss the findings in relation to previous studies; and state our conclusions.

3.3 Methodology

3.3.1 Participants

Participants were 16 children (8 boys, 8 girls) from an Irish primary school. They were third-grade students aged 7–8 and had not studied probability in school mathematics. Children came from a variety of socio-economic and ethnic backgrounds and six of them were second language English speakers.

3.3.2 The Jellybean Task

The task was adapted by the researchers from a study by Acredolo et al. (1989). Each pair of children was read a scenario:

> A teacher celebrated her class's good behaviour by offering every child a jellybean. To keep children from fighting for their favourite colours, she decided that each would have to take a jellybean from a well-shaken bag without looking. Erin's favourite colour jellybean is green. We are going to try and see what chance Erin has of taking a green jellybean from the bag of jellybeans. But, Erin isn't allowed to look!

Before being presented with jellybeans, the 'happy face scale' was introduced (see Fig. 3.1) as a tool to determine what children believe the likelihood of drawing a green jellybean from a specific bag is without requiring any calculation. Instead of a probability scale with fractions, decimals or percentages, a non-numerical one was preferred for subjective probability estimates due to the age of the children. A protocol was developed to explain the use of the scale:

> This is a happy face scale. If we think something *will definitely happen* we mark the happy face [point to happy face]. If something will *definitely not happen* we mark the sad face [point to sad face]. Sometimes things can be in the middle – that is when we are not sure. It *might happen or it might not* [point to neutral face].

Practice was provided using the scale. A bag of white jellybeans was presented and children asked: What is the chance of Erin taking a white bean from the bag of white jellybeans? Can you mark it on the happy face scale? What is the chance of Erin drawing a red bean from the bag of white jellybeans? Can you mark it on the happy face scale?

Working in pairs, children were then asked to determine the likelihood of a green jellybean being drawn from different transparent sandwich bags of green and pink jellybeans. The total number of jellybeans in each bag remained constant for each pair; however, it differed between pairs so that each pair got a selection of bags of 6, 8 or 10 jellybeans (see Table 3.1). For each bag, the children were asked to mark their predicted likelihood on the 'happy face scale' (see Fig. 3.1).

3.3.3 *Setting and Procedure*

The setting was a university teaching room. Children worked in pairs and were seated at a table with two pre-service teachers. One pre-service teacher took the role of interviewer and was responsible for asking questions using a pre-prepared interview protocol. The other recorded children's responses on the task. The researcher was present during all interviews and observed the application of correct protocols. All tasks were completed in approximately 30 min.

3.3.3.1 The Jellybean Task: Providing Estimates

Bags of jellybeans were presented in random order. For example, pair 7 who were assigned bags of 10 beans received them in the order: Bag 1 (2*G*8*P*), Bag 2 (3*G*7*P*), Bag 3 (1*G*9*P*), Bag 4 (4*G*6*R*), Bag 5 (5*G*5*R*) and Bag 6 (9*G*1*R*). For each bag presented, the following questions were posed:

Protocol questions for estimations:

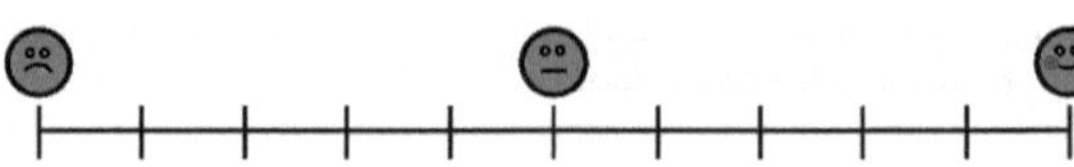

Fig. 3.1 'Happy face scale' to record the predicted likelihood of a green jellybean occurring

Table 3.1 Sample combinations of green (*G*) and pink (*P*) jellybeans in an assortment of bags

	Bag of 6	Bag of 8	Bag of 10
Bag 1	2*G* 4*P*	2*G* 6*P*	2*G* 8*P*
Bag 2	3*G* 3*P*	3*G* 5*P*	3*G* 7*P*
Bag 3	1*G* 5*P*	1*G* 7*P*	1*G* 9*P*
Bag 4	4*G* 2*P*	4*G* 4*P*	4*G* 6*P*

- How many jellybeans are there altogether?
- How many green jellybeans are there? How many pink jellybeans are there?
- What is the chance of Erin getting a green jellybean?
- How would you feel if you were Erin? Why would you feel like this?
- Where would you mark the happy face scale? Explain why you marked it here [pointing to the mark made by child]. Why did you not mark it here [pointing lower] or here [pointing higher]?
- What words would you use to describe the chances of picking a green jellybean?
- Is it impossible that Erin gets a green jellybean? Is it certain that Erin gets a green jellybean? Why? Why not?

3.3.3.2 The Jellybean Task: Carrying Out Experiments

The following was read:

> Let's pretend we are Erin and see what happens if we actually try getting a bean from the bag of jellybeans. Let's see if we get a green jellybean. But, we cannot look! We are going to take a jellybean out (without looking). Then we'll record the colour on this page. We'll put the jellybean back in the bag and shake the bag. We are going to do this 24 times and see how many green jellybeans Erin gets! Will you help me?

Two bags of jellybeans were identified for use in experiments. For each bag, the jellybeans were removed from the transparent sandwich bag and placed in a brown paper bag (to prevent children seeing the jellybeans). One child picked a jellybean and the other child marked the colour on a graph; the bean was replaced. The experiment was carried out 24 times and the outcome recorded for each draw. Children were asked what they noticed about the graph they constructed. Referring to the outcomes of the experiment (as illustrated on the graph), the same series of questions as those posed in the 'protocol questions for estimations' (see Sect. 3.3.3.1) were presented. The second bag of jellybeans was selected and the experiment repeated. Thus for both experiments, children were given the opportunity to update their probability estimates (on the happy face scale) based on the evidence collected.

3.3.3.3 The Jellybean Task: Carrying Out Simulations

The following was read:

> We have this really cool computer programme that can pretend to pull jellybeans from the bag of jellybeans. Look – here is the bag with 1 green bean and the rest are pink. [point to image on screen]. The computer has a button called 'run' [point to button] and it makes the computer pull the beans out 1000 times really fast. Just like running really fast. It puts its hand in and out of the bag 1000 times. It shows us what it gets each time [point to the table]. Then it makes this really pretty graph over here that show us how many green and how many pink jellybeans Erin would get if we did it 1000 times.

Two simulations were run on laptops using *TinkerPlots* (Konold & Miller, 2011), for the same two bags of jellybeans as each pair had in the previous experiments, and the ensuing graphs discussed. Children were directed to examine the graph constructed by *TinkerPlots*. Reminded that the graph represented the 1000 times the computer pretended to pull out jellybeans, the same series of questions as those posed in the 'protocol questions for estimations' (see Sect. 3.3.3.1) were presented.

3.4 Results and Discussion

We first present children's use of chance language and then describe their reasoning about uncertainty as they articulated their subjective probability estimations on the happy face scale. In describing the results, we will focus on three pairs of students: Fred and Dillon (bags of 10), Ken and Deniz (bags of 6), Katia and Narin (bags of 8).

3.4.1 Children's Use of Chance Language

3.4.1.1 Children's Use of Specific Chance Language

In their estimations, Fred and Dillon used terms such as 'teency bit', 'sort of a chance' and 'bad/terrible chance' to refer to low probability situations (2*G*8*P*, 1*G*9*P*). Their language shifted to 'might' when the likelihood of a green increased (3*G*7*P*, 4*G*6*P*). The use of comparative language revealed their tendency to compare likelihoods between different bags of jellybeans. For example, when they were first presented with 2*G*8*P* and then 3*G*7*P*, Fred stated that there was a 'bit more of a chance' of getting a green in the second bag. Again, when 4*G*6*P* followed 1*G*9*P*, Dillon stated there was 'more of a chance' on this occasion. Ken and Deniz also used chance language in their estimations. They used 'slight chance' (2*G*4*P*), 'good chance' (3*G*3*P*, 4*G*2*P*) and 'really small chance' (1*G*5*P*) to refer to the chance of getting green. When engaging in experiments, Fred and Dillon's use of chance language to compare outcomes was evident. Dillon stated that Erin would be 'happier' with the 3*G*7*P* situation (resulting in 13 pink and 11 green) than the 2*G*8*P* situation (resulting in 15 pink and 9 green) as 'the last time she got 9 and this time she got 11'. This was qualified by Fred who stated Erin would be 'just really happy not super happy' based on his observation that there were 'still more pinks than greens'. We can see the positioning of 'super happy' as an indication of greater likelihood than 'really happy' signalling the ability to use qualifiers to discern likelihoods within one end of the probability continuum (in this case >50%). A similar effort was made by Ken to differentiate between likelihoods within the lower end of the probability continuum. When describing how Erin would feel as a result of the 1*G*5*P* experiment (resulting in 5 green and 19 pink), he said 'she is sad but not the saddest because there is a small chance of getting a green'.

Katia and Narin, both non-native English speakers, used the term 'chance' less frequently than other children. Katia used 'maybe' in each of her four estimations; however, its use ceased in the experiments and simulations. When presented with the 2*G*6*P* situation, she stated that 'maybe' green could occur but there was a 'small chance'. When presented with 3*G*5*P*, she thought that 'maybe pink' because there were more pink. Her language when presented with 1G7P revealed her consideration of the small possibility of a green occurring when she said 'no chance.… Maybe but maybe not, I think not'. It was not until asked to estimate the equal likelihood scenario (4*G*4*P*) that she used 'maybe' to refer to both colours occurring when she stated 'equal chance—maybe green and maybe pink'. Hence for Katia, 'maybe' may have also referred to the *possibility* of one event occurring over the other. Interestingly, once discussing the outcomes from experiments and simulations, Katia no longer used 'maybe' and increased her use of the term 'chance' to refer to the possibility of a green jellybean occurring. In the experiments, she referred to the 'tiny chance' ($\#P = 19$, $\#G = 11$) and 'really good chance' ($\#P = 10$, $\#G = 15$) of a green occurring; similarly, she stated 'chances are small' (2*G*6*P* and 1*G*7*P*) and 'equal chance' (4*G*4*P*) of a green occurring in the simulations.

Narin is a contrast as she did not make estimates of the likelihood of a green jellybean being drawn. On each occasion, she stated that she was 'not sure' and 'we don't know' (2*G*6*P*), 'maybe she could [get a green] maybe not, I don't know' (3*G*5*P*), 'not impossible' (1*G*7*P*) and 'I don't know' (4*G*4*P*). While she did not know the chance of getting a green, she was very focused on the happiness levels of Erin in relation to the constitution of the bags of jellybeans. Interestingly, similar to her partner Katia, her language changed during the experiment and simulation situations. In these situations, Narin immediately used chance language to refer to 'very small' and 'little chance' arising from the experiment ($P = 19$, $G = 11$). In the simulations, she described the possibility of drawing a green as 'not impossible' (3*G*5*P*) and 'equal chance cause it's the best bag' (4*G*4*P*).

3.4.1.2 Understandings of Terms "Impossible", "Possible" and "Certain"

Analysis of the language provides additional insights into the findings of Fischbein et al. (1991) regarding difficulties of children with the terms "impossible", "possible" and "certain".

The situations presented always involved some combination of green and pink jellybeans, and hence, there was not a situation where a jellybean was 'certain' or 'impossible' to occur. Analysis of the transcripts from Narin and Katia reveals that they did not spontaneously use 'certain' or 'impossible' to describe the possibility of getting a pink or green jellybean. However, Narin's use of the term 'not impossible' to refer to the chance of getting a green jellybean when providing an estimation (1*G*7*P*) and describing a simulation (3*G*5*P*) indicates she understood the meaning of 'impossible' and possibly used it as an anchor to situate the rest of her language. Responses when asked if certain colours were certain or impossible to occur revealed understandings of these chance terms. When asked if the occurrence of a green bean was impossible, Katia stated 'no there's 2 in the bag' (2*G*6*P*) and 'no there's 3 in the bag but more pink' (3*G*5*P*). Similarly, when asked if it was certain that a green would be drawn (1*G*7*P*), Katia replied 'No, not at all!' The following transcript arising from the simulation for 1*G*7*P* provides insights into their relatively rich understandings of the terms certain and impossible:

Teacher:	From the 1000 times around how many times did Erin get a green jellybean?
Katia:	Tiny, this small [uses fingers to show a tiny space]
Teacher:	So where might you mark on the happy face scale? [Both Katia and Narin mark the sad face]
Teacher:	Katia why did you mark the sad face?
Katia:	Because there are so many pink! Erin would be sad
Teacher:	What about you Narin why did you mark the sad face?
Narin:	Because she would be so so sad, it's awful
Teacher:	But is it impossible to get a green?
Narin:	No
Katia:	No. You might but there's only one in the bag
Teacher:	So are you really really sad or are you somewhere in between?
Narin:	Really really sad! No chance of getting that one green!
Teacher:	So are you certain it will be a pink bean?
Narin:	Em …
Teacher:	What does 'certain' mean?
Narin:	I will [get a green]!
Teacher:	Excellent, so is it certain you will get a pink?
Narin:	Maybe one chance of green, maybe!

The responses of Deniz, however, may provide some support for Fischbein et al.'s (1991) observation of the tendency to equate 'rare' with 'impossible' and 'highly frequent' with 'certain'. Deniz spontaneously used the terms 'no chance of green' (1*G*5*P*), 'definitely get a green' (4*G*2*P*), 'impossible to get a green' (1*G*5*P* experiment resulting in 5*G* and 19*P*) and 'definitely get a green' (4*G*2*P* experiment resulting in 19*G* and 7*P*). While his judgments were always in the correct direction of the probability continuum, he appeared to discount the presence of the lesser appearing colour, even when engaged in experiments. One possible explanation for his use of such definitive language may be that he was drawing on his use of these terms from real-world contexts. For example, 'impossible' in child parlance may imply extreme difficulty or 'rareness' of an event occurring, rather than the probabilistic meaning (children say it is 'impossible' to get to a higher level of a computer game as it took days to advance; or, that they have no chance of winning a game). This finding may provide support for the contention that children 'tend to substitute mathematical meanings with subjective expectations' (Fischbein et al., 1991, p. 528).

3.4.1.3 Understandings of Equal Likelihoods

Katia did not demonstrate difficulty describing equal likelihood situations and referred to 'equal chance … because there are equal amounts' (4*G*4*P*) and her partner Narin also used the term 'equal chance'. Katia justified her selection of the middle value on the happy face 'because there are the same amounts of each, she can pick maybe green and maybe pink'. Thus, for Katia, we can see that her use of 'maybe' does not indicate uncertainty or a lack of preciseness as found in studies by Watson (2005) and Green (1982). The discussion between Katia and Narin explaining why they marked the neutral face arising from the 4*G*4*P* simulation (Fig. 3.2) indicates their ability to coordinate the possibility of getting both a pink and a green and utilise chance language:

Teacher:	Katia why did you mark the neutral face?
Katia:	There are four and four, so Erin might be happy with what she gets and she might be sad
Teacher:	Narin why did you mark the neutral face?
Narin:	Because we don't know what she will get, it's somewhere in between
Teacher:	So is it impossible to get a green?
Narin:	No
Katia:	No. You might but there's equal chance
Teacher:	Do you think it's certain?
Narin:	No it's not, and not impossible either

While Ken and Deniz marked their estimates for the 3*G*3*P* bag of beans at the neutral position, their description 'very small chance' (Deniz) and 'an okay chance' (Ken) does not indicate a rich understanding of equal likelihood. Following the 3*G*3*P* simulation, Ken stated 'there is a good chance of her getting a green, it is almost the same chance as her getting a pink'. Again, they both marked the neutral position on the happy face scale. This may indicate while both may quantitatively understand the notion of equal likelihood (as expressed by their marks on the scale), they may be still developing the associated chance language to describe equal likelihoods. There was no equal likelihoods situation presented for Fred and Dillon.

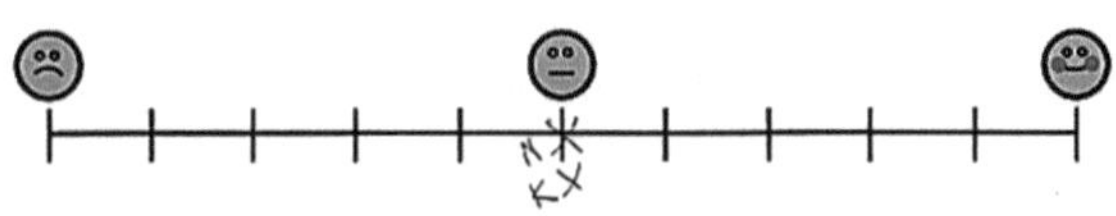

Fig. 3.2 Katia and Narin's marks to show their probability estimations based on the simulation results ($n = 1000$) for 4*G*4*P*

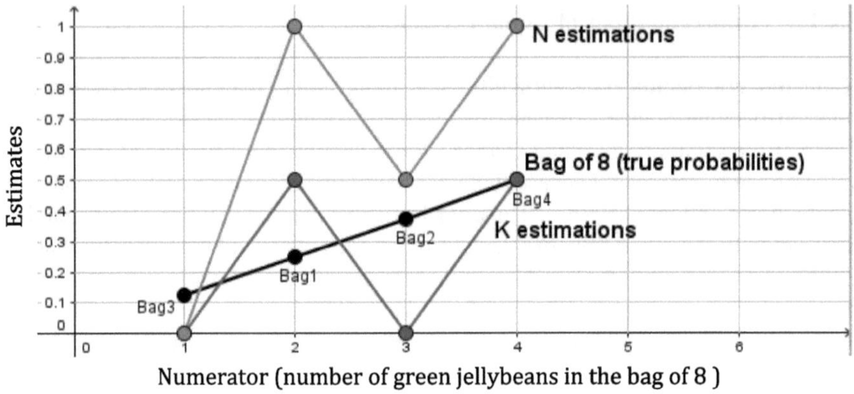

Fig. 3.3 Probabilities estimated by Narin (in blue) and Katia (in red) following conversion to a 0–1 scale and true probabilities (in black)

3.4.2 *Children's Reasoning About Uncertainty*

3.4.2.1 Children's Initial Probability Estimations

In the initial estimations of probability of getting a green jellybean in four different bags (2*G*6*P*, 3*G*5*P*, 1*G*7*P*, 4*G*4*P* presented in this order), the markings of Katia and Narin on the happy face scale were not necessarily consistent with the true probabilities (Fig. 3.3). When presented with the 2*G*6*P* situation (bag1), Katia and Narin overestimated the probability of getting a green jellybean.

In the following dialogue, Katia noted that there were more pink jellybeans than green ones and associated the term "maybe" with the neutral face (probability 0.5). Narin, on the other hand, focused on the possibility of getting a green jellybean from the bag and marked the happy face (probability 1).

Teacher:	Why did you choose there Katia?
Katia:	Because two green and more pink, maybe he can pick the pink
Teacher:	Very good and what about you Narin?
Narin:	She can get green so she will be happy

Next in the 3*G*5*P* situation (bag2), Katia underestimated the probability (Fig. 3.3) as she focused on the number of pink jellybeans ("Because there's too many pink, Erin will be sad when she picks out a pink."). Narin's estimation was close to the true probability but she marked the neutral face (probability 0.5) because "I felt like picking it. I don't know why". She did not seem to have a reasonable explanation for her prediction. In general, Narin's estimations tended to be higher than Katia's estimations with an exception of the 1*G*7*P* situation (bag3) in which both children marked the sad face (probability 0), very close to the true probability ($P(G) = 0.13$). While Katia mentioned only the number of green jellybeans ("Because there's only one of the green ones. Erin will feel like there's no chance of getting it"), Narin used the expression "sad" because there was "one green and more pink". In the situation of equal probability (4*G*4*P*), Katia marked the neutral face (probability 0.5) because "there are equal amounts", whereas Narin interpreted the equal amounts differently when she said "She can get green so she will be happy because there are equal amounts". Narin seemed to focus on the possibility of getting a green jellybean and thought "there's a big chance of picking the green".

Deniz and Ken's subjective probability evaluations were overall consistent with the true probabilities (Fig. 3.4). With the exception of Deniz's equal estimations for bags 1 and 2 (probability 0.5), they assigned higher estimates as thc number of green jellybeans in the bag increased. In the case of small probability (1*G*5*P*, bag3), both children's estimations were very close to the true probability ($P(G) = 0.17$) because Deniz believed that there was a "very small chance" and Ken explained that "there is one green so definitely one chance". When there was one more green in the bag (2*G*4*P*, bag1), they predicted a higher probability and marked the neutral face (probability 0.5) because they thought "there is a chance she might get a pink". Ken also said "Maybe she could get a green". When there was an equal probability (3*G*3*P*, bag2), they did not pay attention to the equal amounts of green and pink jellybeans in the bag. Even though Deniz marked the middle in the scale (probability 0.5), his reason was related to having a "slight chance of getting a green" rather than equal chances. Ken on the other hand believed that there was "a good chance, better than the last" when he marked next to the neutral face (probability 0.6). When there was more green jellybeans than pink ones (4*G*2*P*), Ken considered that there were still two purple ones as he marked halfway between the neutral and happy faces (probability 0.75). However, Deniz was more confident that Erin would get a green because "it's easy, hardly any pinks" when he marked the happy face (probability 1).

Fred and Dillon's subjective probability evaluations were both quite consistent with and very close to the true probabilities as seen in Fig. 3.5. As the number of green jellybeans in the bag increased, they assigned higher estimates, except for Dillon's equal estimations for bags 1 and 2 (probability 0.5). Their qualitative descriptions of the chances for each bag were also consistent with the assigned probabilities: "bad/terrible chance" for 1*G*9*P*, "a teency bit/sort of a chance" for 2*G*8*P*, "a bit more of chance" for 3*G*7*P* and "more of a chance" for 4*G*6*P*. Both children's estimations were the same as the true probability ($P(G) = 0.1$) in the 1*G*9*P* situation. This could be due to the number of green jellybeans (1) in a bag of ten and the happy face scale divided into ten equal parts because Dillon said "it's bad of a chance for her because there's only one green" and chose the first marking after the sad face (probability 0).

Although some children's subjective probability estimates tended to be close to the true probabilities, they were mostly based on intuitions rather than on the part–whole relationship as expected in relation to previous research findings (Piaget & Inhelder, 1975; Falk, 1982).

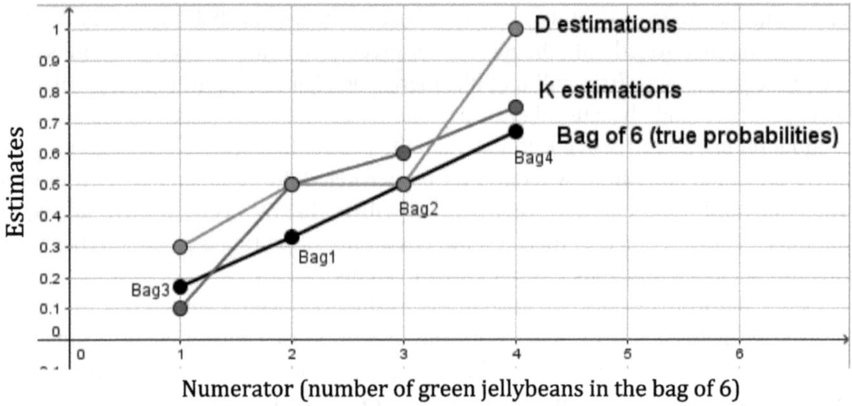

Fig. 3.4 Probabilities estimated by Deniz (in blue) and Ken (in red) after converted to a 0–1 scale and true probabilities (in black)

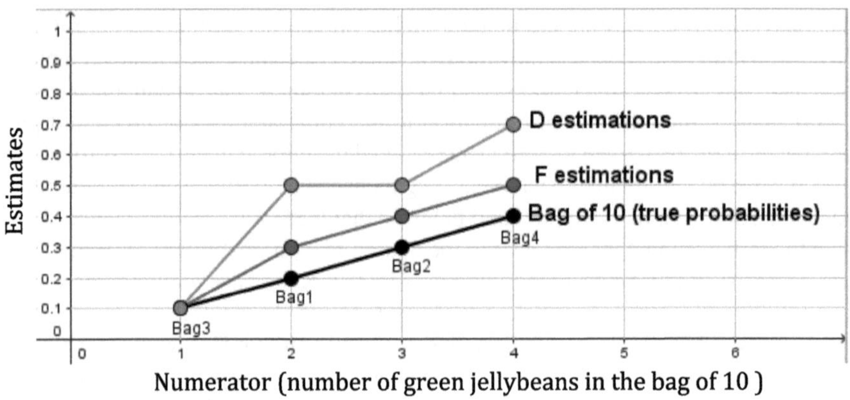

Fig. 3.5 Probabilities estimated by Dillon (in blue) and Fred (in red) after converted to a 0–1 scale and true probabilities (in black)

3.4.2.2 Children's Probability Estimations After the Experiment and the Simulation

When engaging in subjective probability evaluations, children tended to use data, collected from physical experiments ($n = 24$) and computer simulations ($n = 500$ or 1000) of randomly drawing a jellybean out of a specific bag, to update their previous probability estimations. For instance, after drawing 24 jellybeans from bag2 (*3G7P*) Fred and Dillon changed their initial probability estimates considerably based on the results shown in the student generated dot plot in Fig. 3.6 ($\#P = 13$, $\#G = 11$) due to the high number of green jellybeans from a small sample. This is consistent with Jones et al.'s (1999) observation of children's belief in representativeness of small samples. When the sample size increase to 500 using *TinkerPlots* (TP), both children updated

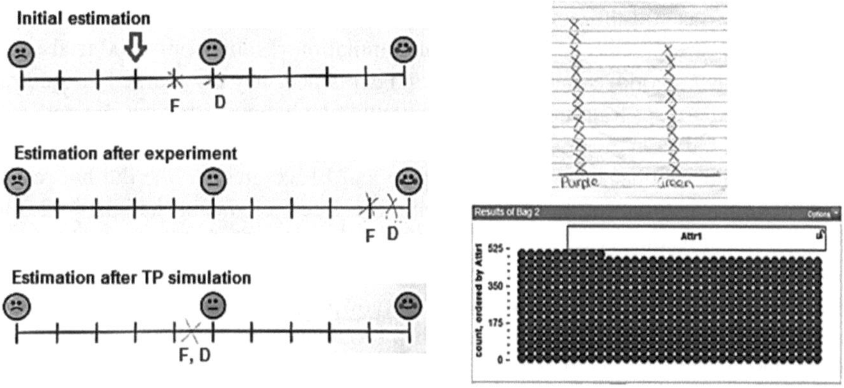

Fig. 3.6 On the left, Fred and Dillon's probability estimations marked on the scale (red arrow shows the true probability) for $3G7P$; on the right dot plot of experiment results at the top and TP plot of simulation results at the bottom

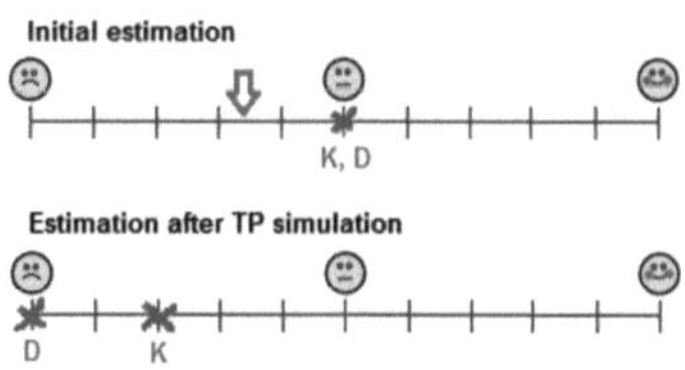

Deniz: Not a great chance that she will get a green

Ken: Not a great chance but not a very bad one either

Fig. 3.7 Ken and Deniz's estimations on the scale before (red arrow shows the true probability) and after TP simulation results for $2G4P$ and their description of simulation results

their probability estimates towards to the true probability value (slightly below the neutral face, probability 0.5) based on the results shown on the TP plot in Fig. 3.6. Children seemed able to use the visual proportion of green and pink jellybeans on the plot without any computation to update their probability estimations.

Analysis of the transcripts from Ken and Deniz reveals that the two children responded quite differently based on the data. While Ken tended to adjust his probability estimates relatively reasonably after new information, Deniz had a tendency to make extreme estimations, except for the equal probability situation ($3G6P$). For example, while both children marked the neutral face (probability 0.5) initially for $2G4P$, with the simulation results Deniz chose to mark the sad face (probability 0) and Ken's marking was also at the lower end (probability 0.2) but closer to the true probability (Fig. 3.7). Their descriptions of simulation results also indicate the contrasting probability evaluation.

In the $4G4P$ situation (bag 4), Katia had already noticed the 'equal amounts' of green and pink jellybeans in the bag in her initial estimation (marking neutral face, probability 0.5). Similarly, children in Kafoussi's (2004) study could spontaneously make judgements about equally likely cases. After the experiment results ($\#P = 10$, $\#G = 14$), Katia again used the bag content to explain her marking (neutral face):

"because there is still the same amount of each colour in the bag so Erin could get either a pink or a green". After seeing the simulation results from 1000 trials, she remarked "That's good, equal amounts!" and explained why she marked the neutral face again: "there are four and four, so Erin might be happy with what she gets and she might be sad". On the other hand, Narin's initial estimation was marked at happy face (probability 1) because "there's 4 green and I love green". She did not revise her probability estimation since there were more green jellybeans than pink ones in the experiment results. After seeing TP simulation results, Narin eventually changed her estimation to the neutral face (probability 0.5) by saying "there's equal chance". However, her reasoning, "Because we don't know what she will get, it's somewhere in between", seems to be consistent with the outcome approach (Konold, 1989), rather than based on equal proportions of two colours in the data.

3.5 Conclusions

Our analyses of both children's use of chance language and probability estimates on a non-numerical scale in the presented tasks provided new insights into children's reasoning about uncertainty from a subjectivist probability point of view. While children used chance language relatively accurately to describe the likelihood of chance events, there was also evidence of the influence of real-world use of chance language for some children (e.g. Dillon and Deniz). We observed a similar finding of the tendency to equate 'rare' with 'impossible' and 'highly frequent' with 'certain' (Fischbein et al., 1991) when some children (Narin, Katia and Deniz) were inclined to quantify their subjective estimate of probabilities as 0 for the least/less likely outcomes with bags 1*G*7*P* and 3*G*5*P* and as 1 for the high likely outcome with bag 4*G*2*P*. A quantitative understanding of equal likelihoods was evident (for most) as in Kafoussi's (2004) study in which young children could evaluate the likelihood of equally likely events without experimenting. However, for some the associated probabilistic language to describe equal likelihoods was in the early stages of development (Ken and Deniz).

In children's initial probability estimations, the quantification of very small and equal likelihoods tended to be more consistent with the true probabilities in general even though they were not able to relate the part with the whole, which is considered essential for understanding of probability by Piaget and Inhelder (1975). This shows the importance of building on children's intuitions (Fischbein, 1975) when developing the formal conception of probability. However, some children were inclined to mark the neutral position (probability 0.5) when they were not sure about the outcome (e.g. Ken, Katia, Narin). This seems to be consistent with associating '50% of chance' with 'I don't know' as seen in the outcome approach (Konold, 1989).

Modifying predictions based on previous results appeared to be intuitive for young children. In our analysis, we examined the change in children's estimations when presented with additional data arising from the experiment and simulation results. We observed stable estimations when probabilities are very small or equal. We would argue that in other cases, small sample data could mislead children as they were likely to believe in small samples of experimental data (Jones et al., 1999). Our focus on both children's probability estimates and chance language also suggested a close link between the updated estimates and the use of chance language.

One instructional recommendation is the use of comparison situations (i.e. would you prefer Bag A or Bag B of jellybean) as Fred and Dillon demonstrated rich use of probabilistic language when they compared bags when constructing estimates. In probability teaching at early years, we should start encouraging children to link their qualitative reasoning about uncertainty to the quantitative reasoning with appropriate tools for subjective estimates of probability. There is already the potential to develop children's conception of probability in that area but it is often neglected in school curricula. The happy face scale used in this study is a useful tool that can bridge these two ways of reasoning about uncertainty. For its effective use in both teaching and research, we suggest asking children to articulate why they marked that specific location on the scale initially and to explain why they updated their estimates on the scale after collecting some data. Children's subjective evaluations of likelihood of random events should not be ignored or seen as simply an obstacle. Rather these intuitions should be developed into more powerful ideas with appropriate learning tasks. More research is needed to examine whether trials and simulations support development of children's reasoning about uncertainty and the revision of probability estimates.

In the task design, we considered events for which other types of probability estimates (classical and frequentist) were possible to be used. Since these children had not fully developed the concept of ratio yet, they primarily relied on their intuitive understanding of the relationship between the favourable and unfavourable outcomes, or their personal beliefs to make their subjective estimates. In view of Huber and Huber's (1987) and our findings, this task design based on the urn model (i.e. drawing from a jellybean from a bag) has potential to provide a concrete situation for studying young children's subjective probability estimates with some stability. Further research is needed to examine the stability of the probability estimates of young children in problems where it is impossible or impractical to calculate a true probability.

References

Acredolo, C., O'Connor, J., Banks, L., & Horobin, K. (1989). Children's ability to make probability estimates: Skills revealed through application of Anderson's functional measurement methodology. *Child Development, 60*(4), 933–945.

Amir, G. S., & Williams, J. S. (1999). Cultural influences on children's probabilistic thinking. *Journal of Mathematical Behavior, 18*(1), 85–107.

Australian Curriculum, Assessment and Reporting Authority. (2014). *Foundation to year 10 curriculum: Statistics and Probability* (ACMSPO24). Retrieved June 10, 2017 from http://www.australiancurriculum.edu.au/mathematics/curriculum/f-10?layout=1.

Department for Education, DoE. (2013). *The national curriculum in England: Key stages 3 and 4 framework document.* Available at: https://www.gov.uk/government/publications/national-curriculum-in-england-secondary-curriculum. Accessed June 10, 2017.

Falk, R. (1982). Children's choice behaviour in probabilistic situations. In D. R. Grey, P. Holmes, V. Barnett, & G. M. Constable (Eds.), *Proceedings of the first international conference on teaching statistics* (pp. 714–726). University of Sheffield: Teaching Statistics Trust.

Fischbein, E. (1975). *The intuitive sources of probabilistic thinking in children.* Dordrecht, The Netherlands: Reidel.

Fischbein, E., Nello, M. S., & Marino, M. S. (1991). Factors affecting probabilistic judgements in children and adolescents. *Educational Studies in Mathematics, 22*(6), 523–549.

Green, D. R. (1982). *Probability concepts in school pupils aged 11–16 years.* Unpublished Ph.D. Loughborough University. Available from https://dspace.lboro.ac.uk/dspace-jspui/bitstream/2134/7409/2/D177387.pdf.

Hoffner, C., Cantor, J., & Badzinski, D. M. (1990). Children's understanding of adverbs denoting degree of likelihood. *Journal of Child Language, 17*(1), 217–231.

Horvath, J., & Lehrer, R. (1998). A model-based perspective on the development of children's understanding of chance and uncertainty. In S. P. LaJoie (Ed.), *Reflections on statistics: Agendas for learning, teaching, and assessment in K-12* (pp. 121–148). Mahwah, NJ: Lawrence Erlbaum.

Huber, B. L., & Huber, O. (1987). Development of the concept of comparative subjective probability. *Journal of Experimental Child Psychology, 44*(3), 304–316.

Jones, G., Thornton, C., Langrall, C., & Tarr, J. (1999). Understanding students' probabilistic reasoning. In L. V. Stiff & F. R. Curcio (Eds.), *developing mathematical reasoning in grades K-12: 1999 yearbook* (pp. 146–155). Reston, VA: National Council of Teachers of Mathematics.

Kafoussi, S. (2004). Can kindergarten children be successfully involved in probabilistic tasks? *Statistics Education Research Journal, 3*(1), 29–39.

Konold, C. (1989). Informal conceptions of probability. *Cognition and Instruction, 6,* 59–98.

Konold, C., & Miller, C. D. (2011). *TinkerPlots2.0: Dynamic data exploration.* Emeryville, CA: Key Curriculum.

Mullet, E., & Rivet, I. (1991). Comprehension of verbal probability expressions in children and adolescents. *Lang Communication, 11,* 217–225.

National Council for Curriculum and Assessment, NCCA. (1999). *Primary school curriculum: Mathematics.* Dublin, Ireland: Stationary Office.

National Governors Association Center for Best Practices, & Council of Chief State School Officers. (2010). *Common core state standards for mathematics: Grade 7 statistics and probability.* Retrieved from http://www.corestandards.org/Math/Content/7/SP.

Piaget, J., & Inhelder, B. (1975). *The origin of the idea of chance in children.* New York: W. W. Norton & Company Inc.

Watson, J. M. (2005). The probabilistic reasoning of middle school students. In G. A. Jones (Ed.), *Exploring probability in school: Challenges for teaching and learning* (pp. 145–168). New York: Springer.

Watson, J. M., & Moritz, J. B. (2003). The development of comprehension of chance language: Evaluation and interpretation. *School Science and Mathematics, 103*(2), 65–80.

Chapter 4
Variation and Expectation for Six-Year-Olds

Jane Watson

Abstract Watson (2005) made the claim that contrary to the traditional order of introduction in the school curriculum, where measures associated with expectation (e.g. mean) were introduced years before measures associated with variation (e.g. standard deviation), children began to develop the concept of variation before that of expectation. This study explores the primitive understanding of the two ideas by seven 6-year-olds as they worked through four interview protocols devised for older students. The protocols included drawing ten lollies from a container of 100, 50 of which were red, creating a pictograph from concrete materials to show how many books some children had read; interpreting from a movable bar chart with information on how children travel to school; and explaining maximum daily temperatures for their city. These contexts were then used to ask the students to make predictions, for example related to the number of red lollies out of 10, who would most want a book for Christmas, how a new child would come to school, and the highest maximum temperature for a year. Across the contexts, students were asked to create or manipulate representations of information (data). At no time were the words "variation", "expectation", or "data" used with the children. Videos, transcripts, and written artefacts were analysed to document demonstration of understanding of the two concepts in relation to data. Evidence of appreciation of variation occurred much more frequently than evidence of appreciation of expectation.

4.1 Introduction and Background

The history of statistics as part of the school mathematics curriculum is relatively short compared with topics related to geometry and algebra. Approximately 30 years ago, the National Council of Teachers of Mathematics (NCTM) published its *Curriculum and Evaluation Standards for School Mathematics* (1989), including statistics and probability at all grade levels from kindergarten. In introducing what

J. Watson (✉)
University of Tasmania, Hobart, Australia
e-mail: Jane.Watson@utas.edu.au

A. Leavy et al. (eds.), *Statistics in Early Childhood and Primary Education*,
Early Mathematics Learning and Development,
https://doi.org/10.1007/978-981-13-1044-7_4

might be termed the "practice of statistics" for grades K-4, the NCTM suggested the inclusion of experiences for students to

- collect, organize, and describe data;
- construct, read, and interpret displays of data;
- formulate and solve problems that involve collecting and analyzing data;

as well as to

- explore concepts of chance. (p. 54)

At about the same time, Moore (1990) wrote his seminal manuscript on "uncertainty". Among the many significant points advanced were three that provide background for this chapter. In recognising the importance of the new curriculum (NCTM, 1989), he said, "However, because of the emphasis that these recommendations place on data analysis, it is easy to view statistics in particular as a collection of specific skills (or even as a bag of tricks)" (p. 95). Second, although not the first to see the importance of context to the field of statistics, as Rao (1975) had made the point earlier, Moore amplified the word "data", which was key in the NCTM's *Standards* (1989). Moore emphasised that "… data are not merely numbers, but *numbers with a context* … Teachers who understand that data are numbers in a context will always provide an appropriate context when posing problems for students" (p. 96). Third, in summing up his message about uncertainty, Moore (1990) focussed on the fundamental concept underlying data and data analysis: variation. "The core elements of statistical thinking" were

1. The omnipresence of *variation* in processes …
2. The need for *data* about processes …
3. The design of *data production* with variation in mind …
4. The *quantification* of variation …
5. The *explanation* of variation. (p. 135)

Shortly after this, Green (1993) started asking questions related to Moore's claims for variation: "What do students understand of variability and how does this originate?" and "What are the essential experiences needed to develop a full appreciation of variability?" (pp. 227–228).

Furthering the issues raised by the questions of Green (1993), Shaughnessy (1997), echoing the opinion of Moore (1990) above, suggested that one of the problems associated with the lack of focus on variability in the classroom was the procedural nature of teaching combined with the complex computations needed to calculate the standard deviation, which was the measure of variation used by statisticians. Because the belief was common that a "measure" was needed for every concept in mathematics, variation had to wait for the standard deviation, which required, for example, the square root. In the meantime, the arithmetic mean was obtained by a simple procedure using addition and division, and hence, *expectation* became the focus of the curriculum.

Expectation arises out of the variation in data when data are summarised, perhaps with a measure of centre or a measure of association. Konold and Pollatsek (2002) used the metaphor of "signals within noisy processes" to characterise expectation and variation. Both sets of terms also apply when considering trials of random processes related to probability models. Often in authentic settings, expectations are materialised as predictions of outcomes from data in a context involving variation. In other settings, however, stated expectations or predictions may be the catalyst leading to consideration of the variation creating them.

Shaughnessy (1997) followed his supposition about the cause of the delay in focussing on variation in the curriculum with a call to "investigate students' conceptions of variability and try some research approaches that uncover what our students can do in problem solving in chance and data, rather than merely documenting what they are unable to do" (p. 18).

More recently, the *Guidelines for Assessment and Instruction in Statistics Education (GAISE) Report* (Franklin et al., 2007) brought variation to the forefront in its description of the four steps of statistical problem-solving.

I. Formulate Questions, Anticipating Variability—Making the Statistics Question Distinction
II. Collect Data, Acknowledging Variability—Designing for Difference
III. Analyse Data, Accounting of Variability—Using Distributions
IV. Interpret Results, Allowing for Variability—Looking beyond the Data. (pp. 11–12)

The nature of variability that *GAISE* sees as relevant at Level A (the lowest of three levels across the school years) includes measurement variability, natural variability, and induced variability but it only sees these considered within a data set at Level A (p. 15). *GAISE* also speaks of the helping young children distinguish variability from error and how these notions are used to explain outliers, gaps, and clusters (p. 33). Although the *Common Core State Standards for Mathematics* (Common Core State Standards Initiative, 2010) recognises variability as an essential starting point for the study of statistics, statistics and probability are not included in the curriculum until Grade 6.

The most recent *Australian Curriculum: Mathematics* (Australian Curriculum, Assessment and Reporting Authority [ACARA], 2016) recognises both variation and expectation in its early years, Foundation to Year 2. In these years, "[c]hildren have the opportunity to access mathematical ideas … by developing an awareness of the collection, presentation and variation of data and a capacity to make predictions about chance events" (p. 8). This chapter expands on this extract by considering both variation and prediction (i.e. expectation) for both "data" and "chance". Further, reflecting Moore (1990), it is the uncertainty created by data, whether collected from surveys, experiments, or random devices, which means that a prediction made from the results of the data collected must be expressed with a corresponding degree of uncertainty. Statisticians may express this uncertainty in confidence intervals or p values, but the goal for young children is to use the evidence they have collected or the information available to them to express informally how confident they are in their concluding expectation.

Although research on school students' understanding of the practice of statistics has grown rapidly since it was introduced in curriculum documents, the focus has been on older students (e.g. Konold & Pollatsek, 2002; Lehrer, Kim, & Schauble, 2007; Watson, Callingham, & Kelly, 2007). This focus is likely to reflect the assumptions associated with the requirements to draw conventional graphs and calculate statistical measures. Research on young children's statistical understanding has been relatively sparse. Recently, the work of English (2010, 2012), Hourigan and Leavy (2015), and Kinnear (2013) has focussed on representations created to model situations based on picture books. In particular, English (2012) and Kinnear and Clark (2014) engaged children in activities based on picture books in the context of recycling and rubbish collection that brought attention to variation in contexts leading to making predictions. In this very concrete, yet imaginary context, children were able to consider variation in the data presented and make predictions, that is, state expectations, that were reasonable. In considering the relationship of variation and expectation across the grades Watson (2005) claimed that appreciation of variation developed before expectation, opposite to the focus of curricula suggested by Shaughnessy (1997). In continuing to consider the relationship of variation and expectation in detail, this study gathers more evidence to investigate this claim for quite young children.

In the light of the evolving appreciation of the importance of introducing statistical and probabilistic notions in early childhood, of using meaningful contexts in which to do so, and of developing a foundation based on understanding variation, this chapter presents the outcomes of interviews with seven 6-year-old children in their beginning year of formal schooling. The interviews sought to present the children with meaningful contexts that would allow them to display their naïve understanding of data, variation, and expectation. How would the students respond to the data presented to them or imagined? Would they recognise or create variation and how would they deal with it? Would their predictions reflect meaningful expectations from the data and variation experienced?

4.2 Methodology

4.2.1 *Interview Protocols*

As a part of previous projects focussing on the statistical understanding of students in Grades 3–9, four interview protocols had been developed, three related to the data section of the curriculum and one to the chance section (Watson & Kelly, 2005; Watson & Moritz, 1999, 2001). Three of the protocols included the use of concrete manipulative materials. The first was based on a container containing 100 lollies of different colours, of which 50 were red. The lollies questions used with students are shown in Fig. 4.1. The second involved creating a pictograph to show how many books some children had read using cut-out images of books and of the named

Lollies Interview Protocol

1. Suppose you have a container with 100 lollies in it. 50 are red, 20 are yellow, and 30 are green. The lollies are all mixed up in the container.You pull out 10 lollies.

a) How many reds do you expect to get? __________ Why?

b) Suppose you did this several times. Do you think this many would come out every time?
Why do you think this?

c) How many reds would surprise you? ___________Why do you think this?

2. Suppose six of you do this experiment.

a) What do you think is likely to occur for the numbers of red lollies that are written down? ______, ______, ______, ______, ______, ______
Why do you think this?

3. Look at these possibilities that some students have written down for the numbers they thought likely.

6, 9, 8, 6, 8, 7
3, 7, 5, 8, 5, 4
5, 5, 5, 5, 5, 5
2, 3, 4, 3, 4, 2
10, 10, 10, 10, 10, 10

a) Which one of these lists do you think best describes what might happen? _________ Why do you think this?

b) Which of these do you think describes least well what might happen? __________ Why do you think this?

4. Suppose that 6 students did the experiment. What do you think the numbers will most likely go from and to?

a) From __________ (lowest) to __________ (highest) number of reds.
Why do you think this?

b) Now try it for yourself: ______, ______, ______, ______, ______, ______

5. Suppose you did this experiment lots and lots of times, wrote down the number of reds, put them back, mixed them up.

a) Can you show a way to record the result you might get?

b) Now use the graph [grid provided] below to show what the number of reds might look like for the 40 students.

Fig. 4.1 Lollies interview protocol

children. The books questions are shown in Fig. 4.2. The third protocol employed a large moveable bar chart, which could be manipulated to show the number of children who arrived at school by four different means of transport. The transport questions are shown in Fig. 4.3. The fourth protocol involved speculating about the maximum daily temperature in their city given the information that the average daily maximum temperature for the year was 17 °C. The weather questions are shown in Fig. 4.4. The materials used in the first three protocols are shown in Fig. 4.5: the container with 100 lollies of which 50 were red, 30 green, and 20 yellow; the images of books and named children; and the moveable bar chart with the bars set for the initial question asked.

The results of analysing the responses to these protocols have been published for older children based on other research questions, at times including some of the

Books Interview Protocol

We have some cards here, to represent some children, and some cards for the books they have read. *(Show information sheet)*

- **Now suppose that Anne read 4 books, Danny read 1, Lisa read 6, and Terry read 3.**

a) How can you make a picture of the information given on the books they've read? You can use the cards anyway you like.
Why did you do it that way?

b) If someone came into the room, what could they tell from your picture?

c) **OK, now suppose Andrew read 5 books.**
Can you use some more cards to show that Andrew read 5 books?

- **Now, suppose Jane read 4 books.**

d) Can you use some more cards to show that Jane read 4 books?

e) **Now, suppose Ian hasn't read any books.**
Can you show that Ian hasn't read any books?

f) **Now, suppose everyone went to the library and read one more book each.**
Can you change your picture to show that everyone reads one more book each?

g) If someone came into the room, what could they tell from your picture now?

h) From your picture, can you tell who likes reading the most? How?

i) How would you tell how many books they've read all together?

j) Who do you think is most likely to want a book for Christmas?
Why do you think that?

- **Suppose Paul came along.**

k) Can you tell how many books he might have read?

- **Now suppose Helen came along.**

l) Can you tell how many books she might have read?

Fig. 4.2 Books interview protocol

responses of the 6-year-old children (e.g. Kelly & Watson, 2002; Watson & Kelly, 2005; Watson & Moritz, 1999, 2001). The data reported in this chapter are a comprehensive summary of all exchanges with the 6-year-old children in the four contexts to gauge the starting points for their appreciation of variation and expectation.

Table 4.1 summarises the focus of the protocols on the relationship of variation and estimation in the four contexts for the data in the context. Some contexts were expected to be more difficult, and the protocols were ordered as presented in Table 4.1 because of the increasingly complex contexts associated with decreasing concrete hands-on contact with materials. Two of the protocols (books and weather) were shortened from use with older children by eliminating more complex explanations of representations at the end of the protocols. The lollies protocol was exactly as developed (e.g. Shaughnessy, Watson, Moritz, & Reading, 1999) and the transport protocol had one extra question, c), added near the beginning.

Transport Interview Protocol

Mr. Smith is a school teacher in Hobart. This is a graph of how people in Mr. Smith's class travel to school.

a) Have you seen a graph like this before?
b) What can you tell from the graph?
c) Which way do most children come to school?
d) How many children come to school by car?
e) How many more children walk than ride bicycles to school?
f) How many children are in Mr. Smith's class?

- **Suppose a new student moves into Mr. Smith's class.**

g) How do you think the new student is most likely to come to school? Why do you think that?

- **Suppose today, everyone was told the bus wouldn't be coming tomorrow.**

h) Can you make the graph show how it might look after everyone arrives at school tomorrow if the bus didn't come? Why have you made the graph like that?

- **Now we'll put the graph back the way it was before.**

i) Suppose it rains tomorrow morning.
j) Can you make the graph show how it might look after everyone arrives at school tomorrow if it was raining? Why have you made the graph like that?

Fig. 4.3 Transport interview protocol

Table 4.1 Data, variation, and expectation in the protocols

Protocol	Data	Initial encounter	Subsequent link
Lollies	Number of reds in a draw of ten [actual lollies]	Expectation (prediction)	Variation (in prediction, in data collected)
Books	How many books students had read [physical representations]	Variation (in data)	Expectation (predictions from data representation)
Transport	How many students travel by four modes to school [graphical representation]	Variation (in data)	Expectation (predictions when data conditions change)
Weather	Single value of average temperature [no representation]	Expectation (implied in context and average)	Variation [predictions of variation in data to fit expectation (average)]

4.2.2 Participants

The seven children (five boys and two girls) were in a preparatory class (before Grade 1) in a government school with a teacher who had implemented an innovative

Weather Interview Protocol

Some students watched the news every night for a year, and recorded the daily maximum temperature in Hobart. They found that over the whole year the average highest temperature in Hobart was 17 °C.

a) What does this tell us about the temperature in Hobart?
b) Do you think all the days had a highest temperature of 17 °C? Why or why not?
c) What do you think the highest temperature in Hobart might be for 6 different days in the year?

_______, _______, _______, _______, _______, _______

Why did you make these choices?
d) For the whole year, what do you think the highest and lowest daily temperature in Hobart would be?

highest __________ lowest _____________
e) For the month of January, what do you think the highest and lowest daily temperature in Hobart would be?

highest __________ lowest _____________
f) For the month of July, what do you think the highest and lowest daily temperature in Hobart would be?

highest __________ lowest _____________
g) How would you describe the temperature for Hobart over a year in a graph or picture?

Fig. 4.4 Weather interview protocol

Fig. 4.5 Materials for the protocol for lollies (left), books (centre), and transport (right)

mathematics programme but who had not yet introduced material related to chance and data that year. The children were chosen by the teacher, from a class of 25, as articulate and willing to talk with "visitors from the university". Each interview took place individually in a quiet room for approximately 45 min, including all protocols. Students showed interest in all questions and did not appear to experience fatigue. Parental permission was obtained and the interviews were video-recorded, from which transcripts were produced.

4.2.3 *Analysis*

For the purpose of this chapter, the data from the interviews were reanalysed specifically with respect to three aspects of the students' developing understanding:

- DATA—the children's interaction with the data related to the contexts presented in the protocols;
- VARIATION—the children's capacities to (a) recognise and/or describe variation in data presented or created and (b) include acknowledgement of variation within predictions made; and
- EXPECTATION—to use the variation implicit or explicit in the context to make predictions that reflect meaningful expectations.

At this age and lack of experience, it was not the aim to classify the responses to the protocol questions hierarchically but to document the interaction of the basic concepts of variation and expectation in the contexts exposing the students to data. Following the example of Russell (1990) in exploring how "children construct their ideas about data" (p. 158), the analysis goes beyond the data to explore how children use data to construct ideas about variation and expectation. A descriptive account is presented to illustrate how children are capable at quite young ages to engage with these big ideas intuitively, although often without the ability to provide statistical justifications. The terminology of "data", "variation", and "expectation" was not used during the interviews, and the language suggested in Figs. 4.1, 4.2, 4.3, 4.4 was closely followed.

4.3 Results

The results are presented for each protocol with a summary at the end for the three aspects of student understanding explored across the protocols: data, variation, and expectation.

4.3.1 *Lollies Protocol*

Expectation was the main contextual motivation in the lollies protocol (Figs. 4.1 and 4.5) with interest in the contribution variation made in the predictions of students or in their explanations of the outcomes they obtained from their trials. All of the students understood the setting and the drawing out of the lollies that created the data with which they worked, although the word "data" was not used. Most of the questions were based on predictions of outcomes of drawing lollies from the container, although the questions were posed in a manner to allow recognition of variation in the outcomes.

Given the contents of the container, the initial expectations for the number of reds in ten draws (no replacement) were reasonable: 4, 5, or 6, with qualifications of "or more", "about", or "maybe" for four responses, recognising potential variation. The reasons, however, were not based on proportional reasoning:

- There are some reds on top and bottom—in the corner.
- 5+5=10, 5 of one colour and 5 of another.
- 5+5=10, one more makes 6, and 4 is 10.

One student appeared to have an intuition about the proportion but did not have the language to express it: "Because there's 50, and 5 … like 10". All students said either "No" or "Don't know" when asked if repeated draws would produce the same result. Responses reflected appreciation of variation in sampling, for example, "Might get a different number every time" or "If it's mixed up I might get 4 yellow, 3 red, and 3 green". When asked how many reds would be a surprise, five said a higher number such as "maybe 10", with two saying "6" or "6 or 5". Justification for these answers generally reflected other possibilities or "don't know" with the response for "6" being "it's my favourite number".

When asked to predict the outcomes for the number of reds in six separate trials, six of the seven responses contained no repeated numbers of reds, whereas one had "4" listed twice. Four predictions were consistent with a mean of 5 reds, with two sets considered high and one set low. In terms of variation, four were judged as wide and three as reasonable.[1] Only one set of values was both centred on 5 and with reasonable variation.

Asked which of five outcomes of six draws (see Fig. 4.1) would best describe the most likely outcome, four chose the best response, "3, 7, 5, 8, 5, 4", whereas one each chose "all 10s", "all 5s", and "2, 3, 4, 3, 4, 2". The four reasons for the best response were similar to "Mixed up, different amounts" with only one specifically mentioning "5". The reasoning for "all 10s" was "you could get the same number", whereas for "all 5s" it was "there are more red". Asked which set described the likely outcomes least well, students either replied "all 5s" or "all 10s" with intuitive reasoning reflecting the list ("it's got heaps of 10s") or the contents of the container ("not enough 10s").

When asked the range of outcomes for six trials, responses varied from "0–10" to "2–8" and "3–9" with five responses including "10". Reasons were generally idiosyncratic, for example, "I can fit 10 in my hand" or "2+8=10", or reflecting single outcomes, for example, "I might grab them all from the red part".

Asked to show a way to record the results from many trials, six displays are shown in Fig. 4.6. An oral response of the other student was "ask everyone—get a clipboard". Drawings (ii), (iii), and (vi) represented the setting of the trials, whereas (i) and (v) recorded numbers for the outcomes. Drawing (iv) was accompanied by the explanation, "write each number up to 8 and then write the number of people next to each of the numbers that got that many".

[1] Criteria for categorising the predictions and type of variation are given in Shaughnessy et al. (1999).

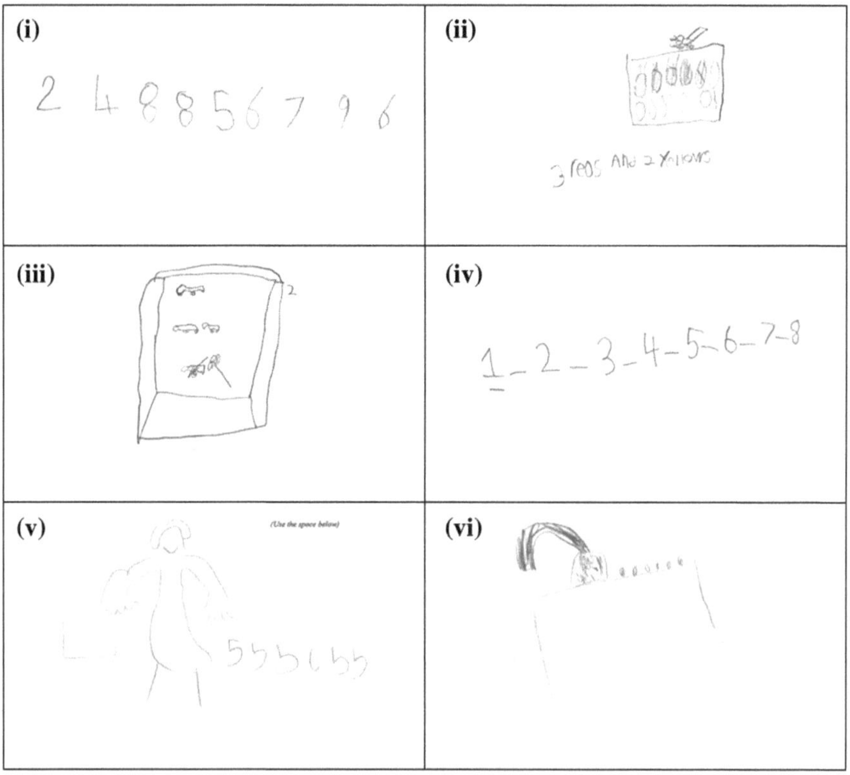

Fig. 4.6 Representations for data collection for lollies protocol

When provided with grid paper and asked to show how many reds 40 students might draw from the container, five appeared to understand the task as they were colouring in the squares, although they did not necessarily fill the squares from the bottom of the grid. None were urged to complete the task for all 40 students. Two students explicitly said they had not seen a graph like the grid before. Two graphs are shown in Fig. 4.7. The choices of squares to fill reflected either "possible" outcomes or "numbers I like", with no further explanation.

4.3.2 Books Protocol

The books protocol (Figs. 4.2 and 4.5) gave students the opportunity to use concrete materials to represent a data set. Of interest was how they used their displays to show the variation in the data and ultimately how they would make predictions about implications from within or outside the data displayed.

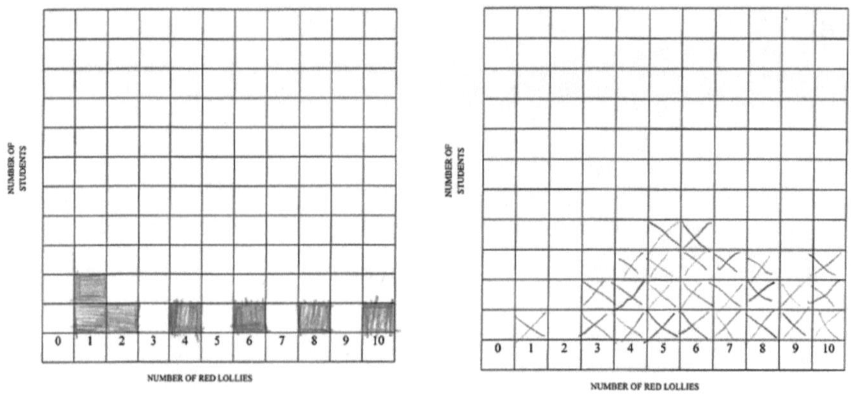

Fig. 4.7 Grids filled in for number of reds in repeated draws of ten lollies for "other" students

When presented with the cards for the books and the characters, six of the seven students could distribute the cards next to the characters in such a way that the numbers of books read could be compared for the characters. Two students distributed the books to one side of the character, two students distributed the books vertically below the character, and two students distributed the books around the character (non-overlapping). The remaining student ignored the data information and piled cards on top of each other next to the character. When asked what someone new to the room could tell from their displays, some responses repeated the information displayed or said "she could tell by counting". One student made up a story about the girl with the most books "winning" and if "someone came along and stole one of her books she'll only have 3 left".

When Andrew, having read five books, and Jane, having read four books, were introduced, all students added the characters and their books as before, including the student who put the books in piles. Given a card for Ian with no books, one student puts him to the side, whereas the others put him with the other characters (with no books). When asked to show an additional library book for each character, all students added a book to each character but those with books scattered around the character had trouble keeping track of which had received an extra book and the student with piles of books got confused about which character owned which book and missed out one character.

Asked what a new visitor could now tell from the display, responses varied widely from "people are reading books", to "did you put one more on each person?", to again reading a count of how many books each character had read. Asked who likes reading the most, all said Lisa because she had the most books; none expressed any uncertainty in the suggestion. Asked how they could tell how many books the characters had read altogether, some just said "count them", whereas others tried to do so, with mixed success and only one reaching the correct total of 33. Asked who was most likely to want a book for Christmas, again five students said Lisa because she "reads the most" or "likes reading". The remaining two students gave different imaginary accounts, "Ian, because he's only got one book" and "Terry, he's got 5 books—dinosaur one, a skeleton one, and a giraffe one and he wants one on plants so he can see how they grow".

Finally, students were asked to predict how many books two new students, Paul and Helen, might have read. Responses were quite varied. One boy would not predict for either new character saying, "Don't know, my sister always makes me guess, I have to put up with it!" The rest made predictions, including "0, because it was his first time in the library and he doesn't know how to choose books", "10, I just think he would", and "3, because one of my sisters is 3". None of the responses used the information in the display (data) to make a prediction about the new students.

4.3.3 Transport Protocol

At the beginning of the transport protocol (Fig. 4.3), four of the students said they had seen a bar graph like the moveable bar graph in Fig. 4.5 before; three had not. Initial questions checked if students could read the graph and distinguish the variation presented. Two students required initial help in reading the graph, but then all said that "most" children came by bus. When asked how many came by car, six replied "6" and one who had trouble initially said "5". The two questions requiring basic mathematical calculations caused difficulty. Five of the students worked out that 4 more of Mr. Smith's class came by bus than car with two counting the lines between the bars and two justifying their answer with "3 + 4 = 7". Only one student, however, obtained the total of 18 children in the class, by counting on his fingers. Others required help and one responded, "Just go down there [to the class] properly and count and the ones I count can stand up and when I go back down they can sit down". One student suggested 10 because that was the largest number on the vertical axis of the bar graph.

Asked to predict how a new child would come to school, students provided a wide range of responses and explanations, as shown in Table 4.2. Only one response was based on the bar graph, whereas the others were based on the students' own experiences or imagined scenarios in the context.

Students were then asked to predict how Mr. Smith's class would arrive if there were no bus and to move the bars on the graph to show their predictions. One student pushed bus down to zero and moved car to 7, bike to 7, and walking to 5 for the total of 18. Others required help to make the adjustments necessary to adjust for the 7 in the bus.

After returning the bar graph to its original position (Fig. 4.5), students were asked to adjust the graph again to show how Mr. Smith's class would get to school if it were raining tomorrow. Again this was not an easy task for the students, with none making all of the adjustments for the correct total without prompts. Two students adjusted car and bus upward appropriately but did not initially move bike and walk to zero. Two others moved bike and walk to zero but had difficulty adjusting car and bus upward by the correct numbers, one agreeing that some of Mr. Smith's class did

Table 4.2 How a new child would come to school

ID	Transport (g)	Why?
Prep1	Bus	Most people come on the bus
Prep2	He would just do one of these (points to modes) and if I was coming to school I might go in the car or bus	If I walk it might take too long and if I ride my bicycle I could have an accident and go on the road and a car is coming and the bicycle might get wrecked to pieces
Prep3	Car	I think by car because they had never been to high school before … [I: Supposing they had been to school for several months now, how do you think they will come to school now?] By bus. [I: Why?] Because they have learnt where the bus stop is at the school and they have learnt where the bus stop is at their home and then they just go in the bus down to school from their home
Prep4	Bus	I reckon about bus … Because its mum probably goes to work and stuff … And her dad too probably
Prep5	(Pushes the car up one)	I come to school by car
Prep6	Car	Because if you went on a bike you might get hit and walking you might get hit, and those two the safest (car/bus)—[I: So, how would the new kid get to school?] I catch the bus and I think the new kid catches the bus
Prep7	Walk	Because they might not want to go with their mum

not come to school that day. Adjusting the variation in the data to fit the prediction in this context was very difficult, even with the concrete representation of the bar graph to help.

4.3.4 Weather Protocol

The weather protocol was the most difficult for the 6-year-old students, and many parts used with older students were not included for these students. The initial question (see Fig. 4.4), worded for the possible suggestion of variation by the students, only elicited one response about the temperature that appeared to acknowledge variation, "it's a little bit cold, lower than today". Two responses in the context reflected an interpretation of the expectation: "quite hot at 17 °C" and "going to be hot for the whole year". Other responses commented on the TV news and it being wrong with the weather. One student was "not sure" what the average temperature meant.

When asked the more explicit question about all days of the year having 17 °C as the highest temperature, five said "no", one said "sometimes", and one said "yes, maybe". Explanations included the following, acknowledging variation.

- No. The temperature always changes.
- No. You get summer, spring, winter, autumn, and summer again. You get hot, mild/cool, cold, mild/cool, and hot again.
- No. One day it might be cold, the next day it might be colder.
- No. Every single thing is different, so they do different things every single day.
- Sometimes. Sometimes it's raining.

Predicting the temperatures for six different days of the year revealed an acknowledgement of variation but not necessarily appropriate values for temperatures in the city. Three of the responses were within ranges reasonable for the city, for example "11–30" (°C), whereas three others had maxima of "70" °C or higher and one had a range of "5–10" °C (too low for the city). Similar responses were given for January and July.

Only four students were asked to draw a representation of the temperature over a year. Their representations are shown in Fig. 4.8. Figure 4.8a represents the variation from a sunny day and to a rainy day, in Fig. 4.8b the circle represents "the land with how hot it would be written on it" (perhaps from seeing a weather map on television), Fig. 4.8c shows a beach and the student described how "hot it is when we go to the beach", and Fig. 4.8d shows a picture of the student in the sun and she explained what she wore "when it was hot or cold". Figure 4.8a, d, and perhaps Fig. 4.8b, and the accompanying explanations, recognise the variation present in the weather.

4.3.5 Summary

Students' familiarity with the context within which the data were presented or created influenced their ability to comprehend the questions asked. Lollies, books, and transport contexts were all quite familiar, with weather related to temperature less so. For lollies, the "data" were based on the actual sweets in a container; for books, the data were cards representing books and children in a one-to-one match with the context; for transport, the data were represented in a moveable bar chart, although never individually; and for weather, no data were presented and hence they needed to be created by the students.

For the lollies protocol, students appreciated the variation in outcomes but did not have the language to explain random behaviour. Only one student used the same number twice in reporting the number of red lollies in six different trials. For the others, this may reflect a naïve view that the numbers were being "used up" as they were chosen, or because the students were given a choice, they would be "fair" in choosing as many different numbers as possible. A similar tendency appeared for Grade 3 in a related study with 56% of students not including repeated values, whereas this decreased to 27% for older students (Kelly & Watson, 2002). In a

Fig. 4.8 Representations for the temperature over a year

different context of predicting outcomes following the presentation of the data in a table, Kinnear and Clark (2014) found 5-year-olds more likely to reuse numbers from the table.

For the books protocol, when asked to make predictions about the characters based on the data visible in their displays, six students could do so with reasons based on the data (e.g. "most" and "least"). When asked about characters outside those visible, however, they did not use the visible data to inform a prediction. In the transport protocol, predicting how a new child outside of the data displayed would come to school, posed a similar difficulty as for predicting how many books a new character would have read. In the weather context, students clearly understood about changes in weather conditions but struggled with actual numerical temperatures, which is not surprising at their age. Working backward from a specific expectation expressed as an "average", however, was very difficult in the context.

4.4 Discussion

The four protocols, initially devised for older students, helped distinguish the limits of understanding of variation and expectation for these 6-year-olds. The results support the view that recognising and discussing variation in data in their experience are very natural to 6-year-olds, even though they may not be able to explain its origin.

Dealing with variation also generally develops before the ability to express meaningful expectation related to that variation. In the two protocols that began with variation (books and transport), students' expectations, expressed as predictions, were often not based on reasoning associated with the data but instead with imaginary situations, within or outside the context. In the two protocols initiated with expectation, the lollies task was easier because the concrete materials were in front of the student and the prediction was based on "visible" data. Being presented with the fixed expectation in the weather protocol was more difficult because it was a single value associated with a less familiar context (temperature). For lollies, there was variation in the predictions made, based on variation in the lollies seen in the container. The predictions had well-understood boundaries (0–10 red), whereas for weather, the variation was in the data without boundaries as such, and with which students had much less familiarity. This made the task more difficult but the students understood enough about the context to suggest numbers for temperatures.

In Kinnear's (2013) study, the responses where students gave predictions or explanations based on the context of the protocol but not based on the data presented were called abductive reasoning. In her study, the context was a picture book including a plot, which some of her 5-year-old students used to make predictions, rather than examining the actual data provided in the context. Similar examples from the current protocols include discussing where red lollies may be in the container or how many would fit in the hand, suggesting that a character is not familiar with the library for selecting a book, discussing the amount of time it takes to reach school or familiarity with bus routes, and providing general characteristics of weather and seasons. For these protocols, however, there were also other responses that were based completely on imagination, not context, such as, "my sister is 3" or choosing "numbers I like". Studies such as these with young children suggest there is a progression in thinking from what might be called imaginary reasoning outside of the context presented, to abductive reasoning using only the context presented, to the beginning of statistical (or inferential) reasoning using the data within the context in decision-making (Ben-Zvi, Aridor, Makar, & Bakker, 2012; Makar & Rubin, 2009). Asking for predictions for books read or transport to school for children outside of the visible data set could be considered precursors to introducing samples and populations, elements of inferential reasoning. Ben-Zvi et al. and Makar and Rubin, however, were working with students in Grades 4–6 and also focussing on acknowledging uncertainty in decision-making. Students in this study were not questioned about the certainty of the responses given, although the impression gained from some answers was that they were guesses, indicating that certainty was not an issue. More research with young children should shed light on this suggested pathway and propose ways of scaffolding children into the practice of statistics.

The predominance of variation throughout the interviews, which students had virtually no trouble recognising or creating, supports the views of Moore (1990) and Shaughnessy (2003) that variation is in fact the foundation of all statistical enquiry. In terms of expectation, it is variation that either creates the prediction or provides supporting evidence that the expectation is reasonable, supporting Watson's (2005) claim that appreciation of variation is the starting point for children's engagement with the practice of statistics.

Acknowledgements These students were interviewed as part of Australian Research Council Grant A00000716, "The development of school students' understanding of variation". Ben Kelly assisted with the interviews. Thank you to the teacher, Denise Neal, for creating a classroom environment where the children were very happy to talk to "visitors from the university".

References

Australian Curriculum, Assessment and Reporting Authority. (2016). *Australian curriculum, version 8.2*. Sydney, NSW: ACARA.

Ben-Zvi, D., Aridor, K., Makar, K., & Bakker, A. (2012). Students' emergent articulations of uncertainty while making informal statistical inferences. *ZDM Mathematics Education, 44,* 913–925.

Common Core State Standards Initiative. (2010). *Common core state standards for mathematics*. Washington, DC: National Governors Association for Best Practices and the Council of Chief State School Officers. Retrieved from http://www.corestandards.org/assets/CCSSI_Math%20Standards.pdf.

English, L. D. (2010). Young children's early modelling with data. *Mathematics Education Research Journal, 22*(2), 24–47.

English, L. D. (2012). Data modelling with first-grade students. *Educational Studies in Mathematics, 81,* 15–30. https://doi.org/10.1007/s10649-011-9377-3.

Franklin, C., Kader, G., Mewborn, D., Moreno, J., Peck, R., Perry, M., & Scheaffer, R. (2007). *Guidelines for assessment and instruction in statistics education (GAISE) report: A preK-12 curriculum framework.* Alexandria, VA: American Statistical Association. Retrieved from http://www.amstat.org/education/gaise/.

Green, D. (1993). Data analysis: What research do we need? In L. Pereira-Mendoza (Ed.), *Introducing data analysis in the schools: Who should teach it?* (pp. 219–239). Voorburg, The Netherlands: International Statistical Institute.

Hourigan, M., & Leavy, A. (2015). A meaningful *driving question* motivates kindergartners to engage in all five stages of the PPDAC data cycle. *Teaching Children Mathematics, 22*(5), 283–291.

Kelly, B. A., & Watson, J. M. (2002). Variation in a chance sampling setting: The lollies task. In B. Barton, K. C. Irwin, M. Pfannkuch, & M. O. J. Thomas (Eds.), *Mathematics education in the South Pacific* (Proceedings of the 26th Annual Conference of the Mathematics Education Research Group of Australasia, Vol. 2, pp. 366–373). Sydney, NSW: MERGA.

Kinnear, V. (2013). *Young children's statistical reasoning: A tale of two contexts.* Doctoral dissertation, Queensland University of Technology. Retrieved from http://eprints.qut.edu.au/63496/.

Kinnear, V. A., & Clark, J. A. (2014). Probabilistic reasoning and prediction with young children. In J. Anderson, M. Cavanagh, & A. Prescott (Eds.), *Curriculum in focus: Research guided practice* (Proceedings of the 37th Annual Conference of the Mathematics Education Research Group of Australia, Sydney, July, 2014, pp. 335–342). Adelaide, SA: MERGA.

Konold, C., & Pollatsek, A. (2002). Data analysis as the search for signals in noisy processes. *Journal for Research in Mathematics Education, 33,* 259–289.

Lehrer, R., Kim, M., & Schauble, L. (2007). Supporting the development of conceptions of statistics by engaging students in measuring and modeling variability. *International Journal of Computers for Mathematical Learning, 12,* 195–216.

Makar, K., & Rubin, A. (2009). A framework for thinking about informal statistical inference. *Statistics Education Research Journal, 8*(1), 82–105. Retrieved from http://iase-web.org/documents/SERJ/SERJ8(1)_Makar_Rubin.pdf.

Moore, D. S. (1990). Uncertainty. In L. A. Steen (Ed.), *On the shoulders of giants: New approaches to numeracy* (pp. 95–137). Washington, DC: National Academy Press.

National Council of Teachers of Mathematics. (1989). *Curriculum and evaluation standards for school mathematics*. Reston, VA: Author.

Rao, C. R. (1975). Teaching of statistics at the secondary level: An interdisciplinary approach. *International Journal of Mathematical Education in Science and Technology, 6,* 151–162.

Russell, S. J. (1990). Counting noses and scary things: Children construct their ideas about data. In D. Vere-Jones (Ed.), *Proceedings of the 3rd International Conference on the Teaching of Statistics*. Voorburg, The Netherlands: International Statistical Institute. Retrieved from http://iase-web.org/documents/papers/icots3/BOOK1/A2-6.pdf.

Shaughnessy, J. M. (1997). Missed opportunities in research on the teaching and learning of data and chance. In F. Biddulph & K. Carr (Eds.), *People in mathematics education* (Proceedings of the 20th annual conference of the Mathematics Education Research Group of Australasia, Vol. 1, pp. 6–22), Waikato, New Zealand: MERGA.

Shaughnessy, J. M. (2003). Research on students' understandings of probability. In J. Kilpatrick, W. G. Martin, & D. Schifter (Eds.), *A research companion to Principles and Standards for School Mathematics* (pp. 216–226). Reston, VA: National Council of Teachers of Mathematics.

Shaughnessy, J. M., Watson, J., Moritz, J., & Reading, C. (1999, April). School mathematics students' acknowledgment of statistical variation. In C. Maher (Chair), *There's more to life than centers*. Presession Research Symposium, 77th Annual National Council of Teachers of Mathematics Conference, San Francisco, CA.

Watson, J. M. (2005). Variation and expectation as foundations for the chance and data curriculum. In P. Clarkson, A. Downton, D. Gronn, M. Horne, A. McDonough, R. Pierce, & A. Roche (Eds.), *Building connections: Theory, research and practice* (Proceedings of the 28th annual conference of the Mathematics Education Research Group of Australasia, Melbourne, pp. 35–42). Sydney: MERGA. Retrieved from https://www.merga.net.au/documents/practical2005.pdf.

Watson, J. M., Callingham, R. A., & Kelly, B. A. (2007). Students' appreciation of expectation and variation as a foundation for statistical understanding. *Mathematical Thinking and Learning, 9,* 83–130.

Watson, J. M., & Kelly, B. A. (2005). The winds are variable: Student intuitions about variation. *School Science and Mathematics, 105,* 252–269.

Watson, J. M., & Moritz, J. B. (1999). Interpreting and predicting from bar graphs. *Australian Journal of Early Childhood, 24*(2), 22–27.

Watson, J. M., & Moritz, J. B. (2001). Development of reasoning associated with pictographs: Representing, interpreting, and predicting. *Educational Studies in Mathematics, 48,* 47–81.

Chapter 5
The Impact of Culturally Responsive Teaching on Statistical and Probabilistic Learning of Elementary Children

Celi Espasandin Lopes and Dana Cox

Abstract The objective of this chapter is to explore the impact of culturally responsive teaching on statistical and probabilistic learning of elementary children. In this study, the learning of mathematics and statistics is centred on solving problems with themes derived from the children's culture and the context in which they live. We also aim to understand and describe indicators of the development of different forms of combinatorial, probabilistic and statistical reasoning that young children acquire throughout their second and third year of schooling. The data presented in this chapter emerged during a longitudinal research exploring the temporal dimension of experience. The methodology chosen generates rich detail and allows for the segmentation of data, while focusing on the act of listening to those who are willing to express their reasoning. This enables the researcher to discern human action and take into account the social practices, the subjective experiences, identity, beliefs, emotions, values, contexts and complexity of the participants. The results of the research include the description of the thought processes that emerge as children engage in using mathematics and statistics during their second and third year of elementary school showing that children need to experience problematizing activities involving diverse situations for development of probabilistic and statistical reasoning.

5.1 Introduction

Research on what and how probability and statistics should be taught and learned in childhood is still limited in Brazil, the USA and elsewhere in the international arena. In particular, there is a need to better describe and understand children's ways of knowing related to probability and statistics and how children reason about data.

C. E. Lopes (✉)
Av. Gessy Lever, 915—casa 383, Valinhos, São Paulo 13272-000, Brazil
e-mail: celi.espasandin.lopes@gmail.com

D. Cox
Depto of Mathematics, 301 S. Patterson Ave, Oxford, OH 45056, USA
e-mail: dana.cox@miamioh.edu

A. Leavy et al. (eds.), *Statistics in Early Childhood and Primary Education*,
Early Mathematics Learning and Development,
https://doi.org/10.1007/978-981-13-1044-7_5

The objective of this chapter is to explore the impact of culturally responsive teaching (Gay, 2010) on statistical and probabilistic learning of elementary children. Culturally responsive teaching is defined as using the cultural characteristics, experiences, and perspectives of diverse students as conduits for teaching them more effectively. In this chapter, we consider data collected in the context of a larger longitudinal study. In the course of that study, teachers began making modifications to the standard curriculum to respond to children's gesture, reasoning and lived mathematics related to probability and statistics. We had an opportunity to study the impact of these modifications and identify structural elements and triggers for mathematical and statistical learning.

Specifically, the present study aims to understand and describe indicators of the development of different forms of reasoning that children acquire throughout their second and third years of schooling in relation to the generation of possibilities, ideas of chance, and the process of data collection, tabulation and representation.

The results evidenced that children need to experience problematizing activities involving diverse events so that their observations can lead them to the beginning of the development of probabilistic reasoning. Solving problems to learn the enumeration process is crucial so that they can perceive in an event what is most likely or less likely to happen. In relation to statistics, we observed that children need to go through a research cycle in order to problematize a topic, collect, tabulate and represent data about this problem in order to develop statistical reasoning.

5.2 Combinatorial, Probabilistic and Statistical Thinking in Childhood

We understand that thinking is brought to existence through intellectual activity. We can say that it can arise through rational activities of the intellect or through abstractions of the imagination. Thinking may imply a series of rational operations, such as analysis, synthesis, comparison, generalization and abstraction. On the other hand, we must take into account that thinking is both determined by and reflected in language; as we try to convey concepts or judgments in our mind, those thoughts inevitably change.

There are different types of thinking. Deductive thinking goes from the general to the particular and inductive thinking moves from the particular to the general. Analysis consists of the separation of the whole into parts, which are identified or categorized while synthesis looks across the parts in search of unifying themes. Systemic thinking reveals a complex view of multiple elements with their various interrelationships; and critical thought evaluates created knowledge.

Reasoning is a logical, discursive, and mental operation. It can also be considered as an integral part of the mechanisms of higher cognitive processes for the formation of concepts and for the solution of problems. The human intellect uses one or more propositions to conclude, by mechanisms of comparisons and abstractions, which are the data that lead to true, false, or probable answers. From the premises, we arrive at conclusions. It was through the process of reasoning that the development of the mathematical method occurred, which is considered a purely theoretical and deductive instrument and dispenses with empirical data (Lopes, 2012).

Reasoning is, therefore, a thought process through which one can justify or defend a specific conclusion from a set of premises: if a specific fact causes admiration, one seeks to explain it; if one fears some event, one seeks to infer its consequences; if there are doubts regarding a specific observation, one seeks to verify; if it is necessary to ensure that there is an equivalence, one seeks to show the validity of such a claim. When reasoning, there are no accurate premises that lead to the execution of an action; in this case, reasoning is likely located in the realm of the probability (Leighton, 2004). All these forms of reasoning—explanation, inference, verification, demonstration—enable us to establish relations of consequence among judgments.

This diversity of forms of reasoning is essential for the learning of statistics and probability. Statistics is the science that provides tools to describe variability in data and, based on this description, enable decision making. In statistical thinking, variability is a key concept in statistics because it is at the core of the process of finding relations about the problem investigated, and designing the construction and analysis of data.

Variability that is present in data necessitates a form of reasoning that requires a combination of ideas, which points to an intersection between combinatorial, probabilistic and statistical thinking. It allows data to be collected, displayed, summarized, examined and interpreted to discover patterns and deviations from these patterns. Quantitative data can be described in terms of its main features: measures of shape, centre and dispersion.

Similarly, the randomness is an important phenomenon to the development of statistical and probabilistic reasoning. The uncertainty and the random nature of data distinguish statistical investigation from the more precise and finite nature that characterizes mathematical explorations. Campos, Wodewotzki, and Jacobini (2011) argue that such principles cause statistics to break away from the deterministic aspect of mathematics, and favour reflexive knowledge, which is enabled by the development of statistical thinking and provides "the ability to see the statistical problem in a global way, with its interactions and whys, understand its several relations and the meaning of variations" (p. 61). Therefore, mathematical concepts and procedures are used, in part, to solve statistical problems, but are not limited by them.

Therefore, to reason statistically is to be able to understand and explain statistical processes, as well as interpret their results. This requires understanding of how and why statistical investigations are conducted. It includes recognizing and comprehending the investigative process as a whole, from the elaboration of a query and the selection of tools, to the gathering, analysis, and interpretation of data. This form of reasoning is fundamental for making decisions and predictions, which are usually based on large amounts of data in contexts involving many variables.

All this process still requires the contributions of combinatorial and probabilistic reasoning that are connected to each other, since after enumerating the possibilities one can analyse the odds and make predictions. All this process still requires the contributions of combinatorial and probabilistic reasoning that are connected to each other, since after enumerating the possibilities one can analyse the odds and make predictions. These forms of reasoning are essential to construct data from a problem, which leads to statistical reasoning and allows the understanding of statistical information that involves linking one concept to another (e.g. median and mean) or enables combining ideas about data and facts.

> While probabilistic reasoning aims at structuring our thinking through models, statistical reasoning tries to make sense of observed data by searching for models that may explain the data. Probabilistic reasoning usually starts with models, investigates various scenarios and attempts to predict possible realizations of random variables based on these models. The initial points of statistical reasoning are data and suitable models are fitted to these data as a means to gain insight into the data producing process. (Batanero, Chernoff, Engel, Lee, & Sánchez, 2016, p. 12)

These relationships between combinatorial, probabilistic and statistical reasoning should be considered in early years statistical education through exploration of the intuitive ideas of children; this will bring them closer to the ideas of randomness and variability. The intuitive ideas of children about fundamental concepts of chance enable them "to use probability as a tool to compare likelihood of different events in a world filled with uncertainty" (Batanero et al., 2016, p. 3).

Thus, developing such forms of reasoning in childhood requires an interconnectedness between statistics education and mathematics education. The imbricated approach of mathematical and statistical concepts and procedures, in childhood education, will enable children to have a broader understanding of the situations and problems that they encounter and thereby gain a better understanding of the world in which they live.

5.3 Reasoning in a Culturally Responsive Classroom

Culturally responsive pedagogy allows children to maintain their personal perspective when learning. This includes a child's cultural knowledge, prior experiences, frames of reference and performance styles (Gay, 2010), maintaining student strengths as a foundation for learning. Beginning with the mathematics of a child, a culturally responsive educator will create a nurturing and cooperative environment where power is shared, not held (Morrison, Robbins, & Rose, 2008). This requires reshaping the prescribed curriculum.

We were interested in studying the ways local teachers reshaped the prescribed curriculum in the areas of probability and statistics. Our data stems from activities developed by local teachers in response to observed probabilistic reasoning related to enumeration of elements, analyse possibilities, idea of chance, collection, organization and representation of data.

Before we tell the story of what happened as a result of this reshaping, we must first position the type of reasoning we observed as being both in line with mathematical reasoning and also separate from it. The lessons we documented occurred in the larger context of a mathematics classroom, but were centred around probabilistic and statistical ways of knowing. We will share here our assumptions about early childhood learning and different forms of reasoning.

We begin with the assumption that mathematics education requires an approach that is developmentally appropriate. Childhood is a time of play and spontaneity as well as forms of expression that children engage in that are different from those of adults: their multiple languages and creative terminology, the relationships they establish in the construction and creation of games and playful interactions, and the methods of play and what playing means to them (Prado, 1999).

Mathematical and statistical education during childhood should go beyond the use of algorithms, rules, or conventions. Children are entitled to mathematical knowledge that is present in their imaginary world, and in their real world. As their lives unfold, children are entitled to reason and to establish mathematical habits and patterns based on lived experience. Children construe the world and question what they see. They need educational spaces where they can express their doubts and socialize their hypotheses and solutions. Culturally responsive pedagogy encourages children to bring their whole selves to class and engage publically in these certainties and doubts.

This is a reflexive paradigm, which postulates education as investigation, through which students and teachers question each other. It should be noted that "education for reasoning results in the development of higher thinking". Higher thinking is "a combination of critical thinking and creative thinking" (Lipman, 1995, p. 100).

This form of reasoning can be promoted during childhood education, provided that those involved realise how much a child's world is marked by imagination and creativity. To consider such a notion in childhood education is to help children overcome possible manipulations, for when we begin to explore the paths of childhood activities, we are immediately confronted with the concentration of power in the hands of very few, and increasingly, of large corporations that serve a capitalist market and manipulate children by means of advertising and incentive relating to the consumption of toys, electronics, clothes and shoes.

Children are affected by the economic system under which they live (Kasser & Linn, 2016). The current system, corporate capitalism, has the focus on profit and power not safeguarding the most humane values in the education of children. Marketing to children is a practice known to be associated with a variety of negative outcomes for children as corporations are completely free to produce almost any kind of lucrative childhood cultural entity (Steinberg & Kincheloe, 2001). Schools must promote education that allows children to have experiences that are in opposition to the "aggressions" perpetrated on childhood culture, which rob children of the pleasure of discovery, observation about the movements of nature, of lifetime relationships with peers, of the joy of imagining and creating, of acting differently. In this way, teaching can be critical and can advocate for issues of social justice central to even very young children (Gay, 2013).

This points to a perspective in which

> Problem solving, used as a means to teach mathematics, points towards the design of a plan for mathematics education that encompasses the social experiences of students. Teaching through problem solving starts with an investigation of students' social interactions and

> invites them to formulate problems derived from such situations. The classroom becomes a place for questioning, contextualizing and formulating problems, instead of dealing with ready-made questions and predictable answers. School activities focused on problem solving enable the development of citizens who are equipped to deal with uncertainty, possibilities, and decision-making, thus contributing to their independence and autonomy. (Lopes, Grando, & D'Ambrosio, 2017, p. 254).

While exploring social relations, manipulating objects and interacting with other people, children are able to formulate ideas, test them, and accept or reject what they learn. Children construct meaning from their efforts to discover or invent.

5.4 Methodology

The opportunity to better understand the impact of culturally responsive teaching on statistical and probabilistic learning of elementary children led us to collect data with children at two different years of schooling (second and third year of elementary school). The activities discussed here were prepared by researchers in partnership with teachers from the assumption that the learning of mathematics and statistics is centred on solving problems with aspects of children's culture.

The development these activities with children gave rise to different forms of data. First, activities conducted with children were documented with video and audio recordings. For the construction of this chapter the data from four activities were transcribed and analysed.

Second, field notes from meetings with teachers as well as students were kept by the researchers. These field notes recorded direct observations of activity and conversation, but also contain evidence of the impressions of these observations on the researcher.

Third, in-depth interviews, of a semi-open nature, were carried out with teachers and students. These interviews are characterized by flexibility, enabling a more balanced analysis of the development of students' probabilistic and statistical reasoning.

By working to braid together these three forms of data, we were able to capture how children generate mathematical and statistical meaning and describe the conditions that support the development of probabilistic and statistical reasoning of students during problem-solving.

The current work requires close partnership with the teachers and the children themselves. As D'Ambrosio and Kastberg (2012) have pointed out, listening to students is essential. The authors advocate the importance of listening to the children to understand their different ways of thinking and they warn that

> It should not be confused with "giving kids reasons," meaning to explain things to a learner in the hope that they will absorb the teacher's explanations. In mathematics teaching, giving reason to the learner would mean considering the mathematics learner as a mathematical thinker with system of knowledge that is internally consistent. (p. 22)

In this sense, in order to conduct this research, it was necessary to include the voices of the children and their teachers, thus adopting a collaborative perspective focused on the processes of classroom interviews and observation.

5.5 The Study

The research project that gave rise to the present article stems from the premise that "children have the right to an educational space in which they are stimulated to express their thinking using language; they need to invent and narrate their own stories to colleagues. This education should focus on children's culture" (Lopes, 2012, p. 163).

5.5.1 Goals

The objective of our study was to discuss structural elements and triggers of mathematical and statistical learning from activities, based on probabilistic and statistical content, prepared by the teachers who are responsible for the learners (aged 7–8) in the class. This research also aims to understand and describe indicators of the development of different forms of combinatorial, probabilistic and statistical reasoning that children acquire throughout their second and third year of schooling.

5.5.2 Context and Participants

The research that gives rise to the discussions brought in this chapter is being developed in the city of Valinhos, state of São Paulo, which is located at a distance of 90 km from the state capital, and has a population of approximately 125,000 inhabitants. In 2016, Valinhos stood out in the economic data as the city with the highest income per capita in the interior of the state.

The municipal school *EMEB Cecília Meirelles* was chosen for the study because it is located in a low-income neighbourhood, and it takes only children aged 7–8. The children begin attending *EMEB Cecília Meirelles* in their second year of elementary school, and remain in the same school until the conclusion of the fifth year. There are no significant levels of student drop-out or failure. This institution has a tenured administrative and teaching staff, but needs to improve its rating in the *Índice de Desenvolvimento da Educação Básica—IDEB* (Basic Education Development Index), which is an indicator created by the Brazilian government to assess the quality of education in public schools.

In Brazil, the mathematics curriculum for children (6–8 years old) recommends that reading and interpreting information be expressed through symbols, signs and codes in different situations and in different configurations (graphs, tables, labels, advertisements). Also, it is suggested that children can formulate questions that generate research and observations to collect quantitative and qualitative data. Students should collect, organize and construct their own representations to communicate what is collected. It is recommended that in this phase of childhood, children read and interpret simple tables, double entry tables, and graphs, especially in terms of pictorial representations. It is also proposed that children produce texts from the interpretation of graphs and tables, and also problematize and solve problems from the information contained in tables and graphics. They should recognize and differentiate deterministic and probabilistic situations, as well as, identify the greater or lesser chance of an event occurring.

5.5.3 Methods

In this context, we sought to examine the mathematical and statistical practice of the children while they engaged in problem posing and problem-solving activity, and to infer the conditions considered useful for the development of their probabilistic and statistical reasoning. Data used in the study included recorded interviews with students and class observations, in order to register mannerisms and speech, which are essential to the analysis process. Furthermore, monthly meetings were held with the teachers. At these meetings, the teachers reported on activities that they had developed with the students both orally and in writing. Through discussion, teachers and researchers worked together to analyse the activity and design new curriculum. The aim was to gather data that shows the relationship between problem solving and mathematical and statistical reasoning at each moment of the teaching and learning process, making it possible to detect children's reflections and interactions. The interviews and classroom activities were conducted by researchers and teachers. The children always worked in groups.

5.5.4 The Episodes

Four episodes were selected to describe the ways in which the children in our study reasoned during the conduct of probabilistic and data analysis experiments. We are going to share four episodes. Each episode has three distinct components. First, we will share a catalyst for curricular change. In each episode, there is a moment where students shared a conception, cultural story, or perspective on which curriculum could be built. Second, an activity was built by the local teacher in response that catalyst. We will describe these activities in detail. Third, we will present the result of the activity. The first two episodes occurred in the context of face-to-face interviews between a researcher and the children. The other two episodes were reported by the teachers, recalling the activity they had planned and implemented in the classroom.

5.5.4.1 Episode 1: Guessing Card Colours

Before starting the activity with the children, we asked them:

Researcher: Do you know what chance is?

Suellen: I don't know.

Yasmin: It's when we choose some food.

Researcher: Suellen, do you think it may rain today?

Suellen: Not today.

Maria Eduarda: Rain is impossible now.

(Researcher's record)

This activity involved drawing a card out of a bag in which there were two blue cards and one yellow card. The children were supposed to bet on the colour that would be picked and write down the result. Each child of the group drew cards three times. Then they recorded the colours which were drawn in a column graph and discussed the results. Even after this procedure, the children did not realize that the chance of drawing a blue card was higher. In this first year of work, it could be observed that the children were betting on their favourite colour, or on the colour of the card that had been placed in the bag first, because they believed that this would be the one which would be drawn, according to a determined sequence.

Researcher: Jennifer, why did you bet on blue?

Jennifer: Because that's the sequence.

Researcher: What sequence?

Jennifer: That was the card that was put into the bag first.

Researcher: Then, let's see what happens.

(Researcher's record)

When the card was drawn, it was blue. Jennifer reacted by stating "I said it would be a blue one". From Jennifer's insights, other situations were worked out that would lead children to observe random movements in order to allow them to break their personal preferences. In this way, it was developed during the first-year activities that involved the idea of chance, and the children started to realize that possibilities of occurrence of an event can be higher or lower.

In the second year of work, activities that involved statistics and probability in an articulated way were developed and integrated into the mathematics curriculum planned by the school. After this regular work done by the teachers, the researcher returned to school and we asked Jennifer about the same situation.

Researcher: Why did you bet on blue? Is it your favourite colour?

Jennifer: No, my favourite colour is pink. I bet on blue because blue has a higher chance of being drawn.

Researcher: Why do you think so?

Jennifer: Because you put more blue cards. There is only one yellow card.

(Researcher's record)

We noticed in this exchange that the children, including Jennifer, were not yet used to dealing with the idea of chance and confused it with preferences. That provoked the teachers to consider another activity, the Peek Box.

5.5.4.2 Episode 2: Peek Box

The peek box was another activity carried out during the first year. Seven green marbles and three white marbles were placed in a small box, totalling 10 marbles. The children were asked to carry out the experiment by shaking the box ten times and checking what colour they could see through a hole drilled on the box. They recorded the colours they saw, and the result was that green marbles were seen eight times and white marbles were seen twice. Then we asked the children about the contents of the box.

Researcher: How many marbles do you think there are in the box?

Suellen: 10 marbles.

Researcher: What colour marbles?

Maria Eduarda: Green and white.

Jennifer: But there are more green ones than white ones.

Researcher: Why, Jennifer?

Jennifer: Because we saw eight green marbles and two white ones.

Researcher: Then, how many green marbles must there be in the box?

Maria Eduarda: Eight green marbles

Jennifer: And two white ones.

Researcher: Let's check!

(Researcher's record)

During the discussion, the children presented several possibilities that were not articulated by the results of the withdrawals. The researcher provoked, through questioning, a methodological observation of the colours that appeared. The children began to compare how many times the green marble and white marble had been seen. They realized through repeated and systematic observation that there should be many more greens than white. Thus, when we opened the box and found that there were seven green marbles and three white ones, we discussed the fact that although they saw eight green marbles, that did not necessarily match the contents of the box.

In the second year of work, we carried out the same activity, but this time we changed the number of marbles in the box: we placed four white marbles and sixteen green ones in the box. The children looked 20 times to see the colours that appeared in the hole drilled onto the box, and recorded that green marbles came up thirteen times and white marbles seven times. Then we asked:

Researcher: How many marbles do you think there are in the box?

Maria Eduarda: I think that now there are twenty because the box is heavier.

Suellen: Yes! And the other time we saw the colour ten times.

Rayssa: Now we recorded twenty times.

Researcher: And what colour were the marbles in the box?

Maria Eduarda: They are still green and white.

Kerollyn: But there are still more green marbles than white ones.

Researcher: Why, Kerollyn?

Kerollyn: Because now we saw thirteen green marbles and only 7 white ones.

Researcher: Jennifer, how many green marbles must there be in the box?

Jennifer: I don't know, but there are more green ones, many more.

Maria Eduarda: I think there must be around fifteen green marbles.

Rayssa: Yes! There are few white ones.

Researcher: Suellen, how many white marbles do you think there are in the box?

Suellen: If there are twenty marbles, then there must be fourteen green ones and six white ones, because the other time I remember that there was one more marble than we thought.

(Researcher's record)

We observed that the children no longer related the number of times the colour had been seen and could estimate that there would be more green marbles. However, they still did not relate the results to a numerical perception.

5.5.4.3 Episode 3: Discovering the Favourite Game

In May 2016, the teacher designed an activity whose purpose was to draw a graph of the group's favourite games. The children selected five games. They collected, tabulated and represented the data on a column graph. Then they discussed the results and recorded their conclusions considering the graphical representation.

The children had never conducted an opinion poll or drawn graphs before. This was the first time they developed a procedure of statistical investigation. First of all, the teacher asked the students to select five games and they listed: catch, soccer, tug-of-war, flag-pick and tablet games. Then they recorded the frequencies relating to each selection: one vote for catch, eight votes for soccer, two votes for tug-of-war, five votes for flag-pick and no votes for tablet games.

The teacher needed to use graph paper to keep the proportion when developing the graph. She needed to instruct the children about the entire data entry procedure, since they had no idea how to group the records. They also had difficulties developing the graph, as they had never in the past used this type of representation. Even with punctual guidance by the teacher, we noticed that it was difficult for them to consider the origin in the Cartesian plane.

In May 2017, we asked the teacher of Grade 3, who had this very same group of students in Grade 2, to carry out a similar procedure with the children differing only by topic, which was changed to the children's favourite animal. They selected dogs, cats, turtles, horses and bears. The children signalled that they had already made graphs the year before and made the data entry and representation of the graph with greater skill. This time they had no difficulty in determining the origin of the Cartesian plane, but they still needed to use graph paper to ensure proportionality between columns.

5.5.4.4 Episode 4: Raising Possibilities, Analysing the Least and Most Probable While Casting Dice

Another activity, reported and carried out by the teachers of the same group when they were in Grades 2 and 3, involved casting dice to check the probability of different sums resulting from two dice. The questions proposed to the group were as follows:

1. When I cast two dice, what sums can I get?
2. Bet on the sum that you think will occur most often.
3. Each one of you, cast the dice three times and write down the numbers that appear.
4. Make a graph [the class as a whole] with the number of times that each sum occurred.
5. Who won the bet?
6. What are the sums that might occur when tossing two dice? What are the sums that have the greatest chance of occurring? And, what sums have the least chance?

In the second year of project development, in September 2016, the children needed to project only how numbers from 1 to 6 might occur while throwing dice. There were children who were not familiar with dice. These children had difficulty placing bets because, before betting, they had to list the sums that were possible when casting two dice. After writing down the possible sums, they made their bets but were unable to list the possibilities. The choices took place in several ways: they either chose the greatest or the smallest sum, or the number they liked most. After casting the dice and writing down the results, they were still not able to perceive the other possibilities.

At the end of the second year of our study, the same activity was conducted with the same group of students, who were now in Grade 3. The improvement in children's performance was noticeable. They were able to correctly figure out the greatest and the smallest possible sums. Although they still had no perception of which sums would be the least probable, they were able to notice that the sums 6, 7 and 8 occurred more often than the rest of the possible sums. This perception occurred after the teacher wrote down on the blackboard the winners of the bets who had chosen such numbers.

5.6 Discussion and Conclusions

The exploration of the impact of culturally responsive teaching in the learning process of statistics and probability of primary children revealed through the four episodes the importance of fostering children's understanding about phenomena that are not deterministic. The impact of this study should not be interpreted as a call to incorporate the specific activity documented by this study in the education of all children. Rather, the impact of this study should be a call to respond to students as whole beings with existing intellectual, emotional, social and physical knowledge and forms of reasoning. If we intend to honour students as human, we should give them agency and voice and curriculum should be grounded in their realities rather than our own adult ways of knowing.

Supporting the statistical education of children requires increasing awareness of the presence and impact of randomness. The improvement in children's ability to solve problems involving random events and data analysis situations that were witnessed from one year to the next, points to the need of offering young learners opportunities for conducting systematic probability and statistics studies.

Breaking away from the idea that there are only deterministic factors is critical when understanding the situations the children experience in their social environment. For this reason, it is necessary to promote the study of probabilistic ideas centred on the development of the notion of randomness, so that students can understand that certain events are sure to happen, other events are impossible and other events are probable. It is essential to provide children with opportunities to investigate and draw conclusions about school and family events involving chance; this allows them to build a sample space and analyse possible outcomes of a random phenomenon.

Therefore, we advocate for culturally responsive statistics instruction that takes into consideration students' life context (see the context of activities presented to children in Chap. 3) as an important arena to explore the ideas of probability and statistics. It must also include work with games (such as the tasks and activities in Chaps. 3 and 4), which comprise the origins of probabilistic ideas, and analysis of data about investigations developed by the children themselves, so that they can verify variability within data. Providing children with moments of reflection about random phenomena will enable them to develop their probabilistic and statistical reasoning.

The approach to probabilistic and statistical concepts and procedures proposed by the current study is not tied to the use of complex algorithms and solutions with a significant level of abstraction. We consider that probabilistic reasoning and statistical reasoning have similarities and differences between each other, and therefore are prone to be approached intrinsically.

For significant probabilistic and statistic learning to take place in childhood, a pedagogical environment must be created where the student can be heard and actively participate in an investigation process about authentic topics and situations.

The development of the research reported in this chapter was made possible by the close relationship of the researchers with the teachers and students, and the extended length of time they spent in the school setting in order to be immersed directly in planning for and enacting classroom activity. This chapter has exposed some of the benefits and challenges of this work, as well as the need for future research to investigate more questions about the development of combinatorial, probabilistic and statistical reasoning in childhood. It is still necessary to investigate the aspects of statistical and probabilistic training that should be emphasized in the early childhood mathematics curriculum. This cannot be done from afar or without immediate interrogation of children's mathematical activity and reasoning as well as the teachers' intentional pedagogy and planning. This work can only be accomplished by working directly and empathetically with children and teachers in the context where learning occurs.

References

Batanero, C., Chernoff, E. J., Engel, J., Lee, H. S., & Sánchez, E. (2016). *Research on teaching and learning probability*. London: Springer Open.

Campos, C. R., Wodewotzki, M. L. L., & Jacobini, O. R. (2011). *Educação Estatística - Teoria e prática em ambientes de modelagem matemática*. [*Statistical education—Theory and practice in mathematical modeling environments*]. Belo Horizonte: Autêntica.

D'Ambrosio, B. S., & Kastberg, S. (2012). Giving reason to prospective mathematics teachers. *For the Learning of Mathematics, 32*(3), 22–27.

Gay, G. (2010). *Culturally responsive teaching: Theory, research, and practice* (2nd ed.). New York: Teachers College Press.

Gay, G. (2013). Teaching to and through cultural diversity. *Curriculum Inquiry, 43*(1), 48–70.

Kasser, T., & Linn, S. (2016). Growing up under corporate capitalism: The problem of marketing to children, with suggestions for policy solutions. *Social Issues and Policy Review, 10,* 122–150.

Leighton, J. P. (2004). The assessment of logical reasoning. In J. P. Leighton & R. J. Sternberg (Eds.), *The nature of reasoning* (pp. 291–312). Cambridge: Cambridge University Press.

Lipman, M. (1995). *O Pensar na Educação*. [*Thinking in education*]. Petrópolis: Vozes.

Lopes, C. E. (2012). A educação estocástica na infância. [Stochastic education in childhood]. *Revista Eletrônica de Educação, 6*(1). Retrieved from: http://www.reveduc.ufscar.br/index.php/reveduc/article/viewFile/396/179.

Lopes, C, E. Grando, R. C., & D'Ambrosio, B. S. (2017). Experiences situating mathematical problem solving at the core of early childhood classrooms. *Early Childhood Education Journal, 45*(2), 251–259.

Morrison, K. A., Robbins, H. H., & Rose, D. G. (2008). Operationalizing culturally relevant pedagogy: A synthesis of classroom-based research. *Equity & Excellence in Education, 41*(4), 433–452.

Prado, P. D. (1999). As crianças pequenininhas produzem cultura? Considerações sobre educação e cultura infantil em creche. [Do little children produce culture? Considerations on education and child culture in day care.] *Pro-Posições, 10*(28), 110–118.

Steinberg, S. R., & Kincheloe, J. L. (2001). *Cultura infantil: a construção corporativa da infância*. [*Children's culture: The corporate construction of childhood*]. Rio de Janeiro: Civilização Brasileira.

Chapter 6
Inscriptional Capacities and Representations of Young Children Engaged in Data Collection During a Statistical Investigation

Aisling M. Leavy and Mairéad Hourigan

Abstract Recent research has provided important insights into young children's statistical reasoning when engaged in core components of data modelling, namely attribute selection, data representation and metarepresentational competence. The research described in this chapter, however, explores the stage prior to attribute selection—the collection of data. We describe young children's inscriptions when collecting data within the context of a four-day statistical investigation. The investigation involved 26 children aged 5–6 years in interpreting and investigating a context of interest and relevance to them. The context involved decision making around the design of a zoo. We describe the repertoire of inscriptions that children used to track the appearance of zoo animals and explore their justifications for their invented inscriptions. The rationale for and genesis of inscriptions ranged from aesthetic considerations, ease and simplicity, to contextually driven decision making and approaches motivated by efficiency and by efforts to distinguish between repeated data values and different instances of the same attribute. We argue that when task interest is high the context provides affordances that support authentic data inquiry and data-based reasoning. Moreover, when the focus of the statistical investigation is on having children reason about and understand situations, what emerges are relatively sophisticated approaches to data inscription arising from efforts to make sense of and communicate statistical situations.

A. M. Leavy (✉) · M. Hourigan
Department of STEM Education, Mary Immaculate College-University of Limerick, Limerick, Ireland
e-mail: aisling.leavy@mic.ul.ie

M. Hourigan
e-mail: mairead.hourigan@mic.ul.ie

A. Leavy et al. (eds.), *Statistics in Early Childhood and Primary Education*, Early Mathematics Learning and Development,
https://doi.org/10.1007/978-981-13-1044-7_6

6.1 Introduction

Most educators agree that young children start school with powerful mathematical ideas developed from informal experiences acquired in home and play settings. However, there is growing recognition that in order to maximize the mathematical development of young children we need to recognize the ways in which these preschool experiences contribute to their mathematical development. The field of statistics education has shown great promise in recognizing the nascent abilities of young children to engage in statistical thinking and reasoning through broadening what counts as evidence of statistical thinking in the early years. For example, recent studies of young children engaging in data handling activities have sought ways to create conditions in formal school settings to support and foster children's continued meaningful engagement in and development of statistical thinking (English, 2010, 2012; Kinnear, 2013; Leavy & Hourigan, 2018). Our study builds on previous research by exploring the knowledge and understandings that young children bring when collecting and tracking the appearance of data within the context of a data handling task. Specifically, we identify the different inscriptions or marks children use to record the appearance of data values and explore the justifications children provide for the decisions they make. Thus, we hope to address the concern of van Oers (2010) that 'not paying attention to these events (related to children's graphical marking) means that educators may neglect important and stimulating early events for the promotion of mathematical thinking' (p. 32).

6.2 Theoretical Perspective

6.2.1 Representations in Early Childhood Mathematics

There are many definitions of 'representation'. Most simply stated, a representation is something that stands for something else. In mathematics, representations can be thought of as *internal* or *external.* Internal representations are usually abstractions of mathematical ideas or internal cognitive schema that the learner creates. External representations take many forms such as verbal/gestural, enactive (manipulatives), visual/iconic (pictures, graphs) and symbolic/abstract (equations and formulas); these forms are external manifestations of concepts which communicate meaning and support the development of understanding. The process of concept development involves interaction between both internal and external representations which mutually reinforce, support and influence each other (Pape & Tchoshanov, 2001).

Perkins and Ungers (1994) define representations to mean 'symbols in any symbol system (formal notations, language, picturing, etc.) that serve to denote or exemplify' (p. 2). The important role of representations in the development of mathematical thinking is acknowledged in research and in national curriculum frameworks in many countries. While earlier work in representations focused on college-level learners, there is growing awareness of the richness of young children's representations. Research indicates that the mathematical underpinnings and communicative function of early representations are often overlooked resulting in a missed opportunity for harnessing and promoting young children's emerging mathematical potential (Worthington & Carruthers, 2003; van Oers, 2010). This has led to interest in the representations young children make when moving from informal preschool mathematical experiences into the more formal abstract symbolism of school mathematics.

There is ample evidence that young children can understand one thing as representing another. Prior to entering school, children understand the representational qualities of pictures and video images (DeLoache, 2004) and by the age of four have constructed a wide range of inscriptional techniques (Karmiloff-Smith, 1992; diSessa, 2004). Much of the research on children's representations focuses on the graphic marks of very young children aged 0–3 (Lancaster, 2007) and on children's representations of number (Hughes 1986). A seminal study by Hughes (1986) explored the efforts of ninety-six children (ages 3–7) when asked to represent on paper the number of blocks on a table. Four different categories of responses emerged: idiosyncratic (lacking meaning), pictographic (representing the appearance of the blocks—shape/colour/orientation), iconic (discrete marks to represent blocks) and symbolic (conventional symbols). A larger study by Worthington and Carruthers (2003) of what they term the 'mathematical graphics' of 700 samples of young children's work resulted in the development of a taxonomy of mathematical graphics. This work extended that of Hughes (1986) to include two new categories of 'dynamic' and 'written'. Representations categorized as 'dynamic' capture some form of action and are used by children to represent quantities that are not counted. Representations categorized as 'written' refer to efforts to use words or letter type marks. The presence of a 'transitional period', which refers to representations that combine two categories of response, was also uncovered by Worthington and Carruthers (2003) and this period 'may be important as children move towards the abstract forms of mathematics'. The authors of both these studies argue that even in situations where the meaning of children's marks did not make sense to others, the marks have meaning for the children themselves and serve a communicative function. Indeed van Oers (2009) argues that a process of interactional construction of mathematical meaning between children and educators will 'finally yield meaningful mathematical symbols that may turn out to be more functional for the development of mathematical thinking than conventional symbols imposed onto the child's mind' (p. 33).

Research has uncovered factors which support the construction of representations. A Vygotskian view on early education emphasizes the critical role played by the educator in clarifying and interpreting children's marks and in promoting and stimulating meaning making (Vygotsky, 1978). Other factors include prior knowledge of the learner, the nature of the task, purpose for creating the representations, the learners' own internal representations of concepts (Pape & Tchoshanov, 2001), and the interaction of the learner with the social and material setting of the activity (Meira, 1995). Representations play an important role in the learning of mathematics, and it is important that students 'learn to use multiple forms of representation in communicating with one another' (Greeno & Hall, 1997, p. 363). While representations serve an important communicative role in the early years, their function is greater than that. Representations themselves serve as cognitive tools that help organize thinking, reduce demands on memory and cognitive load and support argumentation and discussion (Greeno & Hall, 1997; Pape & Tchoshanov, 2001).

6.2.2 Use of Representations in Statistics Education

Statistics education, as a discipline, has become more responsive to the challenges and pitfalls of introducing young learners to formal statistical symbolism, conventions and representations. We are now aware that a focus on teaching statistical procedures isolated from the broader view of the statistical inquiry leads to the absence of meaningful and connected understandings for learners.

Thus, what has emerged from recommendations and approaches to teaching school-level statistics is an awareness of the importance of opportunities to engage in a cycle of statistical investigation (Wild & Pfannkuch, 1999), and in particular, a focus on data modelling experiences for young children (Leavy, 2008). Studies have shown that in data modelling situations, when focusing on graphical representations, the task is to create a representation that reveals and displays patterns in the data. In these situations, children often created their own inscriptions or used and modified inscriptions with which they were already familiar such as letters, drawings, diagrams and symbols. In turn, the construction and use of inscriptions by children has lead to the development of metarepresentational competence and knowledge relating to representations (Lehrer & Lesh, 2003; diSessa 2004; diSessa, Hammer, Sherin, & Kolpakowski, 1991). This work has been further facilitated by the development and use of technologies and software which have provided much of the needed support to authentically engage school-age children in the collection, management, representation and analysis of data (Ben-Zvi, Gil, & Apel, 2007; Hancock, Kaput & Goldsmith, 1992; Cobb, McClain & Gravemeijer, 2003; Paparistodemou & Meletiou-Mavrotheris, 2008).

However, engaging very young children in data modelling activities is challenging due to a variety of factors such as their limited ability to use technology, to work with large quantities (numbers) and to reason abstractly. Recent studies, however, which have incorporated design features specifically to support young children in data modelling, have shown great promise. In her study of the data modelling processes of six-year olds, English (2010, 2012) revealed how children's representations and inscriptions changed over time and reported on the metarepresentational competence displayed by children. When representing data, the majority of groups on their first effort constructed pictographs to communicate aspects of the data. When asked to consider whether their attributes and representations should be changed all but two groups, who moved from using pictograms to more formal bar graphs, continued to use pictograms. Interestingly, however, in response to this prompt they made one or more changes to their pictograms involving changes to inscriptions, paper orientation, attributes, orientation of column/row data on the pictogram and used a mix of names and drawings. This and other studies indicate that young children have the ability to communicate data in appropriate representational forms (English 2010, 2012). However, there is evidence that some children tend to show a lack of awareness of the viewer in their selection of novel and esoteric design features and the concomitant neglect of design features that communicate meaning more clearly (Lehrer & Schauble, 2007).

These recent studies have provided critical insights into young children's statistical reasoning when engaged in core components of data modelling (English, 2010, 2012; Leavy & Hourigan, 2018), namely attribute selection, data representation and metarepresentational competence. Efforts to gauge the representational competence of children have focused primarily on children's inscriptions when constructing representations of data rather than the collection and tracking of data. The research reported in this chapter is an effort to address this gap by exploring children's representations in the very early stages of data modelling—when tracking and recording data.

6.3 Methodology

6.3.1 Participants

This study explores young children's (ages 5–6) approaches to collecting and representing data collected as part of a 4-day data modelling investigation. Participants were an intact multi-grade class of 26 primary school children. The 26 children engaged in four 60-min lessons focusing on data generation and collection, identification of attributes, structuring and representation of data and making informal inferences about the results. This chapter focuses on the outcomes of the first lesson which engaged children in generating and collecting data arising from a story context.

6.3.2 Procedure

The children were shown a purpose-made video of a fictitious zookeeper named Zach. Zach stated:

> Hi everyone. My name is Zach and I wonder if you can help me today. I am designing a new zoo and I want to get some friendly animals for my zoo. But I am not sure which animals to pick. I need you to help me. I am going to read you a story 'A walk through the zoo'. It is about some pictures I took one day when I walked through a zoo. I want you to help me figure out which animals are the friendliest. I think the friendly animals are the ones that came out during my walk.

The pages of the story (and the associated images) were projected onto a large interactive whiteboard in the classroom. As the pages on the story changed, new animals appeared on screen. Each child was given responsibility for tracking the appearance of one specific animal during the story. Children were instructed to follow along as the teacher read the story and use their page to 'make a mark' whenever they saw their animal. Children were told that marks could take any form, and the teacher was careful not to provide any examples in case they influenced children's approaches to mark making.

In order to aid memory, each child was given a blank page that had an image of their animal at the top of the page. Each child had the opportunity to track the data twice; this provided the opportunity to assess the stability of representations used by children across two data collection cycles (called cycles 1, 2). Cycle 1 was explained to children as an opportunity to 'practice'; in cycle 2, children were assigned a different animal to track. At the completion of each cycle, in order to gain insight into the meanings attributed to their marks, children were questioned (based on a protocol of pre-designed questions) about their choice of mark(s).

We were interested in the approaches children took when provided with some 'degree of freedom' (van Oers, 2009) in their choices around tools and representations. Hence, rather than teaching the formal convention of tallying, we encouraged children to select their own representations to track the occurrence of data values.

6.3.3 Data Collection and Analysis

The lesson was designed by two teacher educators (authors of this chapter) in conjunction with five pre-service elementary teachers. It was the first of four lessons that engaged children in a cycle of statistical inquiry modelled broadly on the PPDAC cycle (Wild & Pfannkuch, 1999). Each pre-service teacher was assigned to a group of 5–6 children and took the role of 'facilitator' with their group. Their role was primarily to pose a selection of pre-designed questions in an effort to reveal children's reasoning and justification for the selection of marks when collecting data. They did not provide any feedback or support in the selection of marks or in the decision making around the choice of marks made. Conversations within the groups were audio recorded and transcribed.

Data were analysed by both researchers, and the Worthington and Carruthers (2003) taxonomy of mathematical graphics was used to categorize responses as either dynamic, pictographic, iconic, written, symbolic or transitional. Drawing from the taxonomy, responses that captured some form of action were coded 'dynamic' and those that represented the appearance of the animal were identified as 'pictographic'. The use of discrete marks was classified as 'iconic' (Figs. 6.1, 6.2, 6.3, 6.4, and 6.5) and the use of words/letters was considered 'written'. The use of numbers was coded as 'symbolic' (Fig. 6.6), and responses that combined two categories of representations were identified as 'transitional' (Fig. 6.7a).

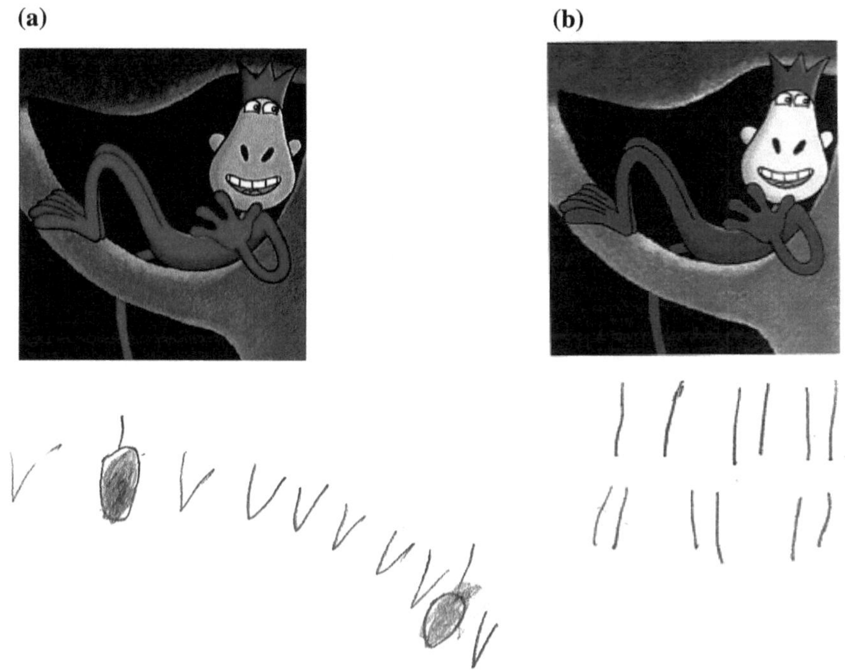

Fig. 6.1 **a** Use of checkmarks to represent patterns in the frequency of occurrence of data. **b** Use of tallies to represent patterns in the frequency of occurrence of data

Fig. 6.2 Barbara's use of pictures of food to represent the occurrence of animals

Fig. 6.3 Paul's use of pictures of food to represent the occurrence of animals

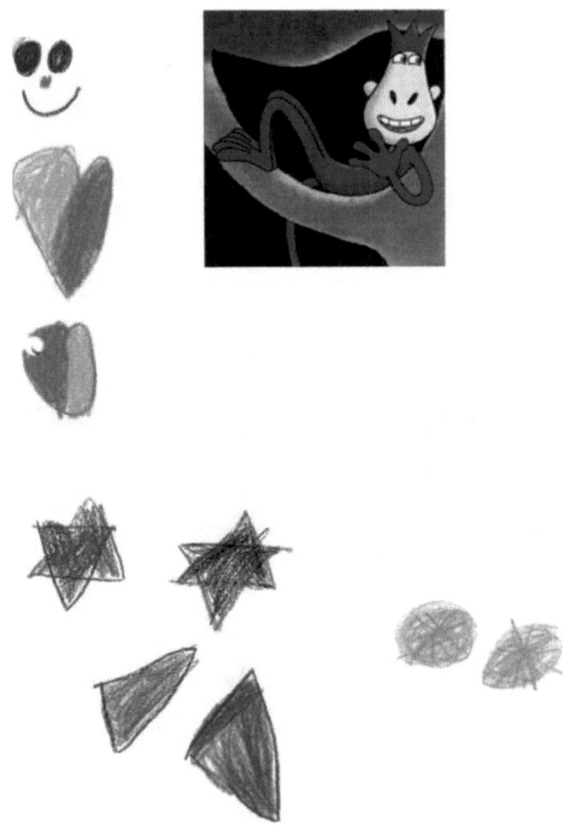

Fig. 6.4 Kate's use of different icons to facilitate accuracy in counting

Fig. 6.5 Polina's use of circles to represent (lack of) variability

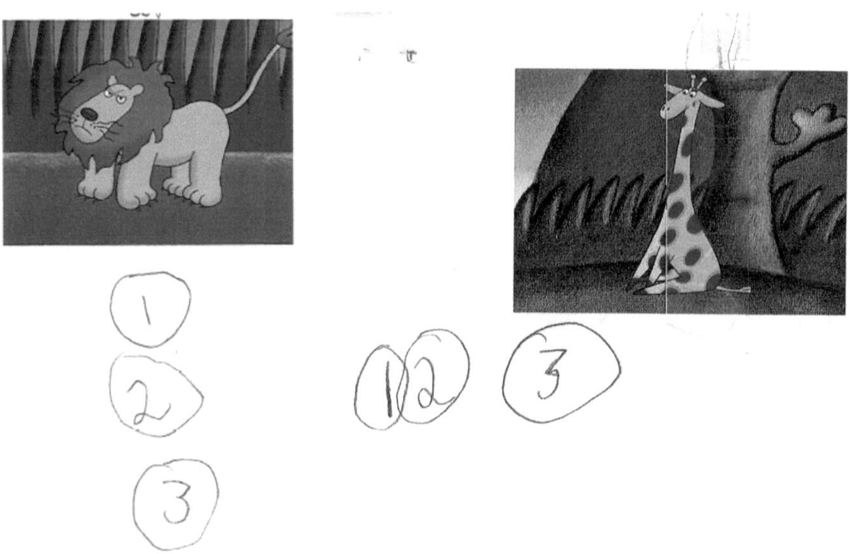

Fig. 6.6 Use of numbers (symbols) to represent the occurrence of data

6.4 Findings and Discussion

The data examined the nature of the representations children used within and between cycles. Table 6.1 summarizes the categorization of responses during the first and second data cycles.

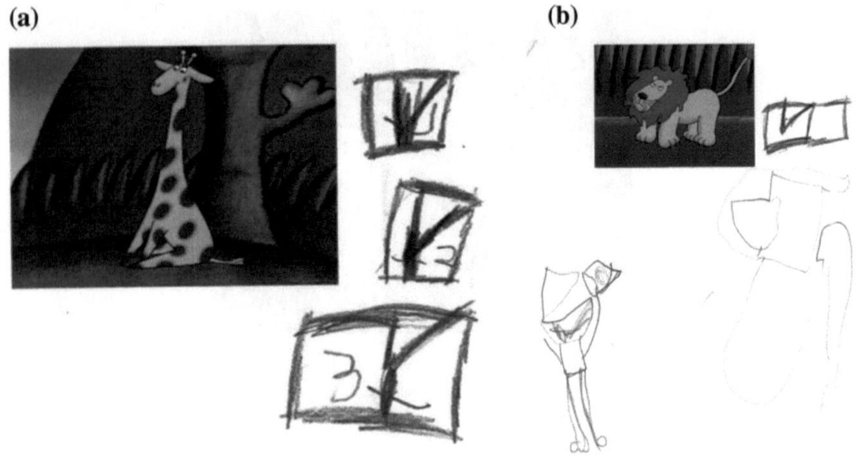

Fig. 6.7 **a** Transitional representation in cycle 1 of data collection. **b** Iconic representation in cycle 2 of data collection

Table 6.1 Classification of representations used to record data

Cycle	Dynamic	Pictographic	Iconic	Written	Symbolic	Transitional
#1			23		2	1
#2			23		3	

None of the approaches was classified as dynamic or pictographic. When asked about the meanings of the marks, analysis of the children's responses indicated they understood that a mark symbolised the appearance of an animal.

Teacher	When you look at that mark you made Eva, what does it tell you?
Eva	It means I saw a lion.

Teacher	How many times did you see the elephant Laura?
Laura	3 times.
Teacher Laura	How do you know? Cause I can count the marks. They are kind of like circles. Look 1, 2, 3. 3 times.

6.4.1 *Iconic Representations*

The tendency of children to use iconic representations (Figs. 6.1, 6.2, 6.3, 6.4, and 6.5) to record data is evident from Table 6.1. During cycle 1, 23 children used marks that were categorized as iconic. These marks demonstrated an understanding of one-to-one correspondence as the child placed one mark each time the animal appeared. While the children used what Hughes (1986) refers to as 'marks of their own devising' (p. 58), there were some general patterns and trends in the types of marks they made. The researchers further classified iconic representations as: tallies, check marks and pictures. The latter of these iconic representations, pictures, were classified as discrete images/illustrations that were drawn to index each occurrence of the appearance of an animal (see Figs. 6.2 and 6.3). While tallying, check marks and pictures were equally prevalent in cycle 1, the second cycle brought about changes in the types of iconic representations children used (Table 6.2).

6.4.1.1 Tallying and Checkmarks

The use of *checkmarks* was very prevalent and accounted for 33% ($n=8$) of responses initially and dropped to 12% ($n=4$) on the second data collection round (Fig. 6.1a). With the exception of two children, checkmarks were recorded in a horizontal line. Responses that resembled the traditional convention of *tallying*, consisting of vertical lines to represent the occurrence of each value, were classified as belonging to this category (Fig. 6.1b). None of the children grouped the tallies in groups of 5 through the conventional use of a diagonal line. In cycle 1, tallying was very common and accounted for 33% of responses. While on almost all occasions tallies were recorded in a horizontal line, Cornelia, organized her tallies vertically down the page during both cycles of data collection. The prevalence of tallying decreased significantly in cycle 2, and it became the least prevalent approach used by 8% ($n=2$) of children.

When asked to explain why they used tallies and checkmarks, many of the responses related to ease and simplicity. Justifications for using tallies were 'they are easy to make and easy to count at the end' (Tomi) and 'they are nice and tidy' (Ayesha). Similarly, checkmarks were described as 'simple to do' (Sheena) and 'easy to count' (Mia). In the following conversation with Matthew he was asked why he used tallies during both cycles of data collection.

Table 6.2 Types of iconic marks used across both data cycles

	Cycle #1	Cycle #2
Tallies	8	2
Checkmarks	8	3
Pictures	7	14

Teacher	Matthew, you used these same marks for when you saw the hippo and the elephant. Can you tell me about the marks?
Matthew	They are lines.
Teacher	Why did you pick lines?
Matthew	Lines make me count properly and help me.

6.4.1.2 Pictures

The pictures drawn by children were categorized as iconic as each image represented the occurrence of one data value/event. Thus, pictures in this study differ from the categorization of marks identified as 'pictographic' used by Hughes (1986) and Worthington and Carruthers (2003). In the aforementioned studies, the picture was merely a rendering of the object in front of the children, whereas in our study, each picture was drawn to represent a reoccurrence of the data value (i.e. the animal being tracked) and is more akin to an 'illustrative tally mark'.

The use of pictures was particularly interesting for a number of reasons. Firstly, the influence of the task (counting the appearance of animals) and the image colour was evident in the pictures. Both Paul and Barbara made marks to represent the food their selected animal ate. In cycle 1, Barbara tracked the appearance of the giraffe and drew bananas to represent the frequencies (Fig. 6.2). When asked why she chose bananas, she stated 'giraffes like bananas as much as monkeys do. And giraffes are yellow like bananas'. During cycle 2, she tracked the elephant and drew peanuts as her marks (Fig. 6.2) and justified it as 'I'll draw nuts to feed the elephant'. Similarly, Paul appeared to be influenced by both the context (animals) and the colour. On both occasions, he drew pears to feed the elephant and hippo. During cycle 1, he had a colour copy and his selection of pear was influenced by the colour as he said 'Well the elephant is green in this picture and pears are green'. However, on the second occasion he had a black and white picture of a hippo and he selected a pencil to draw black pears (Fig. 6.3).

Analysis of conversations with children revealed that in a small number of cases ($n = 5$) pictures were used to distinguish *ordinality* and facilitate *accuracy in counting*. While in most situations, the pictures drawn were the same for each occurrence (a banana or peanut each time, Fig. 6.2), there were five occasions where children chose to represent each occurrence of an animal with a different mark. The following excerpts provide insights into the children's reasoning.

Hence, we can see that for Laura the different marks could also be used to register the *ordinality of the event,* whereas for Kate it facilitated *accuracy in counting* (Fig. 6.4). For some children, the reason was aesthetic as is evident in Mia's response justifying why she drew different pictures for each data value as 'I wanted to make a design'.

Teacher	What marks did you make Kate?
Kate	I used a happy face, love hearts, stars and triangles.
Teacher	And why did you make all different marks?
Kate	So when I count them, I know I counted one and I won't count it again. I count them better.

Teacher	Tell me about your marks Laura. Why are they different?
Laura	They are all different sizes. And they remind me of when I saw the animal.
Teacher Laura	What do you mean? Well that one [pointing to one mark] was the last time I saw him.

6.4.2 Changes in Use of Iconic Representations Across Data Cycles

The use of pictures as a method to record data became more prevalent in cycle 2 (see Table 6.2). Six children who used pictures in cycle 1 continued to use pictures in cycle 2 and two changed to using tallies. However, eleven children who used tallying and check marks in cycle 1 changed to using pictures in cycle 2.

Analysis of the data showed that clusters of children who sat together in groups moved from using tallies and checkmarks to using pictures; hence, there may have been a *social effect*. Some of these children appeared to have observed and been influenced by the pictures a child in their group had drawn in cycle 1. For example, Leah was in the same group as Paul and Barbara (Figs. 6.2, and 6.3); she had used check marks in cycle 1 and moved to using tufts of grass in cycle 2. She explained her use of grass because 'he looks really sad, I wanted to give the lion some grass to eat'. Another group of children who were sitting together (Tomi, Caitlyn, Hannah) had all used tallies in cycle 1 and changed to using pictures (hearts and stars) in cycle 2. This move to the use of pictures may also be attributed to these young children's focus on the *aesthetic* dimensions of their work. Several children were asked to explain why they changed their marks to pictures during cycle 2, and they explained 'to make it look better' (Sian), 'it looked good' (Kate) and 'to make the page look nice' (Darren). It is important to note that while the type of icon changed, the function remained the same.

6.4.3 Efforts to Convey Meaning in Iconic Representations

In five instances, the marks used were intended to convey information beyond frequencies of occurrence of data values. These efforts were quite sophisticated and appeared to arise out of effort to communicate information relevant to the task context.

Polina used a series of small circles to represent the occurrence of her animal. She stated 'they are all circles. They are the same size so we would know we are still counting the same animal'. Hence, the circles represented the frequency of occurrence but also had the potential to *record variability* in data values using size as an indicator (Fig. 6.5).

There were some examples of interesting strategies used when collecting data on monkeys. With the exception of the monkeys who appeared in *pairs on 5 occasions*, there was never more than one of the same animal appearing on any one page of the story. Five children recorded the monkeys, and two of these, Darren and Aidan, made efforts to differentiate between occasions where just one monkey appeared and occasions when two appeared together. Darren used a tally mark when there was just one monkey and an apple when the second monkey occurred in close proximity to the other (Fig. 6.1a). He explained it as 'Every time I saw a monkey I used a tick but I drew an apple when the monkey had a friend'. Darren correctly recorded 2 of the 5 occasions when both monkeys appeared together. Aidan used a different strategy and referred to the fact that monkeys appeared in pairs and tried to have his 'lines' appear in groups of two—hence physical proximity of the tallies was used to communicate pairs. He said 'I always get them in pairs', and 'the animals come in 2 s' (Fig. 6.1b).

6.4.4 Symbolic Representations

A small number of children favoured the use of numerals as a strategy to record data; their justifications were based on ease of use and efficiency. Peter stated that he used numbers because 'it would be easier to see what you have' (Fig. 6.6). Sian made an implicit reference to cardinality when she said she used numbers because 'I don't have to count how much the giraffe came up. I just see the last number and that's the answer' (Fig. 6.6).

6.4.5 Changes in Representations

As there were two data collection cycles, it provided an opportunity to explore the stability of representations used by children. Table 6.3 reveals that 20 children used the same type of representation across cycles 1 and 2. One child used symbolic representations in both cycles, and 19 used iconic representations across both cycles.

Table 6.3 Changes in strategy across events

No change in representation	Iconic -> Iconic	19
	Symbolic -> Symbolic	1
Change in representation	Iconic -> symbolic	2
	Transitional -> Iconic	1

However, as discussed in the previous section, there was movement between the different types of iconic representation used in both cycles. Four children showed movement between representations. Two moved from iconic to symbolic, one from symbolic to iconic and one children who was identified as transitional used an iconic representation on the second cycle.

As mentioned in the previous section, one child was classified as transitional during cycle 1. During the first cycle, he used a combination of check marks and symbols to track the appearance of the giraffe thereby combining aspects of more than one category of response (Fig. 6.7a). This was a relatively complex response, which suggests he was moving from iconic towards more symbolic and abstract representation forms. It was somewhat a surprise then, when during the second cycle he used marks classified as iconic and no longer used symbols. Examination of Fig. 6.7b reveals the use of three very different and somewhat complex images, to represent the appearance of the lion, and no effort to use symbolic forms. One possibility to account for the change in approach is that he may have been influenced by other children in the group, the majority of whom drew pictures in cycle 2.

6.4.6 *Reasoning About the Context*

As children were looking at the data and tracking the appearance of animals, they were engaged in thinking about the driving question which motivated the data collection, i.e. What animals are friendliest? When discussing the friendliest animals, while not required to do so, they frequently referred to factors that may have influenced the frequency of appearance of different animals. Some of these factors were contextual and based on knowledge of the animal kingdom. Knowledge of the contexts within which problems are set was also found to be a support for 6-year old children in the research described by Jane Watson in Chap. 4 of this book. In her chapter, Watson reports that when the data were presented in familiar contexts (such as lollipops and books), as compared to less familiar contexts (weather in this case), that it influenced children's ability to comprehend the questions asked. To take one example from our research in this chapter, children were very eager to discuss why the lion appeared only 3 times; the following statements point to the importance of the context in supporting children's reasoning.

Other suggestions to account for the frequency of appearance of animals were based on patterns that were observed within the data. For example, in reference to

Peter	The lion must be minding her babies, that's why we only saw her 3 times.
Melios	He must be hibernating.
Cornelia	The lion wasn't here much. He is afraid of people.
Kate	I only saw the lion 3 times. He's asleep.

the monkeys, Polina explained the high frequency of appearance as 'because there's two monkeys in each picture—when it shows one monkey there's another monkey next to it'.

6.5 Conclusion

This study reveals insights into the inscriptions young children make when collecting data in a data-modelling environment. Conversations with children around their choice of inscriptions uncovered the multiple meanings they incorporate into the marks they make. Analysis of the data confirmed that once the child embarks on making marks, these marks become both a record of and an abstraction for the real event. For these children, the marks stimulate an image of and support recall of the event that motivated the collection of data. In this way, this study supports the finding of van Oers (2010) that for young children inscriptions serve a communicative function and represent the beginnings of abstract thought. Thus, we argue that when the focus of the statistical investigation is on having children reason about and understand situations, what emerges are relatively sophisticated approaches to data inscription arising from efforts to make sense of and communicate statistical situations.

Children's justifications for inscriptions ranged from aesthetic considerations, ease and simplicity, to contextually driven decision-making and approaches motivated by efficiency and by efforts to distinguish between repeated data values and different instances of the same attribute. These explanations indicate that their representations were more than a record of frequencies and served in some cases as cognitive tools (Greeno & Hall, 1997; Pape & Tchoshanov, 2001) to help organize thinking. This was evident in Laura's efforts, which she justified as keeping an account of ordinality, in Polina's use of same-sized circles to communicate variability and in the efforts of both Darren and Aidan to convey the structure of the data. The variety of approaches used to index and describe the data may be an indicator of the meaningfulness of the task; this influence of the task context on the creation of meaningful representations has also been a finding of other studies (Vellom & Pape, 2000). The levels of interest displayed by the children in the task (recording the appearance of animals) and the larger context (the zoo), we believe, provided affordances that supported these young learners in authentic data inquiry and in data-based reasoning. Children were motivated and on-task throughout the activity and eagerly discussed reasons for the different frequency of occurrences of animals. Thus, the context both maintained interest and in some cases informed the choice of representations to use when collecting data. This critical role played by context was also a feature in data modelling studies with young children using story (English, 2010, 2012; Hourigan & Leavy, 2016; Kinnear, 2013) and technology (Paparistodemou & Meletiou-Mavrotheris, 2008).

The large repertoire of inscriptions used by children were similar to those identified by Hughes (1986) and Worthington and Carruthers (2003). The absence of responses categorized as dynamic or pictographic attests to the understanding on the part of the children that the inscription was an abstraction of the real event, i.e. represented the occurrence of a specified data value. Iconic inscriptions were by far the most popular approach to record data. Understandings of one-to-one correspondence and the consideration of frequencies were apparent in inscriptions classified as iconic. The stability of iconic responses across both tasks was indicative of children of this age and also resulted in accurate recording of the occurrence of data values. While the absence of formal tallying (in groups of 5) was not surprising as it is a school mathematics practice, the abundance of informal tallies ties in with other studies where tally marks appear very frequently in the spontaneous representations of children as young as 3 and 4 years old (Hughes, 1986; Lehrer & Schauble, 2000). The relationship between tallies and finger counting, evidenced in the accounts reported by historians of mathematics, may account for the prevalence of informal tallies. Indeed, Hughes (1986) following an overview of the written number systems of other cultures refers to the 'extremely fundamental nature of tallying' (p. 83) over thousands of years.

An interesting, and somewhat unexpected finding, was the fluidity in the types of iconic responses used across both cycles of data collection. Children moved between tallies, check marks and pictures to record events and were influenced by the efforts of others. In particular, there was a move to using pictures to record data in the second cycle of data collection; this did not reduce the accuracy of children's responses. This change in representation may have been motivated by the possibilities to use colour and the aesthetic dimensions of the work. Alternatively, as this was their second effort at collecting data, the children may have had more cognitive resources available to them during the second cycle thus enabling them to concentrate on the construction of pictures. The absence of a transitional period wherein children's inscriptions combine aspects from more than one category of response distinguishes these findings from those of Worthington and Carruthers (2003). This may be due to the similarity in ages of the children in this study as compared to the research by Worthington and Carruthers (2003) which looked at children across a wide span of ages.

While the constructions of representations within the context of data collection serves a more confined purpose than representation use in later years' mathematics; these representations are important nonetheless. They form a communicative role (Greeno & Hall, 1997) in conveying the frequency of occurrence of a data value. Moreover, they serve as antecedents to more formal and standard representational forms for which children will later have to negotiate meaning (diSessa et al., 1991). When we open up opportunities for young children to participate in statistical activities in ways that make sense to them, we provide opportunities for children themselves to construct meaning and promote genuine developmentally appropriate learning. The representations produced by the young children in this study attest to the quality of their participation in the data modelling environment.

The research has a number of limitations. Firstly, this is a case study of one class of 5–6-year olds; hence, the results of this study cannot be generalized to all children of this age. However, there is potential for further study to examine additional age groups in a variety of educational settings. The second limitation relates to the relatively short time the children engaged with the task. The extension of the study over a longer duration would facilitate more thorough analysis and reap interesting findings where the nature of children's inscriptions could be tracked over a number of years.

Acknowledgements The authors wish to acknowledge the funding received from Mary Immaculate College Faculty Seed Funding.

References

Ben-Zvi, D., Gil, E., & Apel, N. (2007). What is hidden beyond the data? Helping young students to reason and argue about some wider universe. In D. Pratt & J. Ainley (Eds.), *Reasoning about informal inferential statistical reasoning. Proceedings of the fifth international research forum on statistical reasoning, thinking, and literacy*, University of Warwick, UK.

Cobb, P., McClain, K., & Gravemeijer, K. (2003). Learning about statistical covariation. *Cognition and Instruction, 21*(1), 1–78.

DeLoache, J. S. (2004). Becoming symbol-minded. *Trends in Cognitive Sciences, 8*(2), 66–70.

diSessa, A. A. (2004). Metarepresentation: Native competence and targets for instruction. *Cognition and Instruction, 22*(3), 291–292.

diSessa, A. A., Hammer, D., Sherin, B., & Kolpakowski, T. (1991). Inventing graphing: Metarepresentational expertise in children. *The Journal of Mathematical Behavior, 10,* 117–160.

English, L. D. (2010). Young children's early modelling with data. *Mathematics Education Research Journal, 22*(2), 24–47.

English, L. D. (2012). Data modelling with first-grade students. *Educational Studies in Mathematics, 81*(1), 15–30.

Greeno, J. G., & Hall, R. P. (1997). Practising representation: Learning with and without representational forms. *Phi Delta Kappan, 78*(5), 361–367.

Hancock, C., Kaput, J. J., & Goldsmith, L. T. (1992). Authentic inquiry with data: Critical barriers to classroom implementation. *Educational Psychologist, 27*(3), 337–364.

Hourigan, M., & Leavy, A. M. (2016). Practical problems: Introducing statistics to kindergarteners. *Teaching Children Mathematics, 22*(5), 283–291.

Hughes, M. (1986). *Children and number: Difficulties in learning mathematics*. Buckingham: Open University Press.

Karmiloff-Smith, A. (1992). *Beyond modularity: A developmental perspective on cognitive science*. Cambridge, MA: MIT Press.

Kinnear, V. (2013). *Young children's statistical reasoning: A tale of two contexts*. Unpublished Ph.D. dissertation. Queensland University of Technology.

Lancaster, L. (2007). Representing the ways of the world: How children under three start to use syntax in graphic signs. *Journal of Early Childhood Literacy, 7*(2), 123–154.

Leavy, A. M. (2008). An examination of the role of statistical investigation in supporting the development of young children's statistical reasoning. In O. Saracho & B. Spodek (Eds.) *Contemporary perspectives on mathematics education in early childhood* (pp. 215–232). Information Age Publishing.

Leavy, A., & Hourigan, M. (2018). The role of perceptual similarity, data context and task context when selecting attributes: Examination of considerations made by 5–6 year olds in data modelling environments. *Educational Studies in Mathematics, 97*(2), 163–183. https://doi.org/10.1007/s10649-017-9791-2.

Lehrer, R., & Lesh, R. (2003). Mathematical learning. In W. Reynolds & G. Miller (Eds.), *Comprehensive handbook of psychology* (Vol. 7, pp. 357–390). New York: Wiley.

Lehrer, R., & Schauble, L. (2000). Inventing data structures for representational purposes: Elementary grade students' classification models. *Mathematical Thinking and Learning, 2*(1&2), 51–74.

Lehrer, R., & Schauble, L. (2007). Contrasting emerging conceptions of distribution in contexts of error and natural variation. In M. C. Lovett & P. Shah (Eds.), *Thinking with data* (pp. 149–176). New York, NY: Taylor & Francis.

Meira, L. (1995). The microevolution of mathematical representations in children's activity. *Cognition and Instruction, 13*(2), 269–313.

Paparistodemou, E., & Meletiou-Mavrotheris, M. (2008). Enhancing reasoning about statistical inference in 8 year-old students. *Statistics Education Research Journal, 7*(2), 83–106.

Pape, S., & Tchoshanov, M. (2001). The role of representations in developing mathematical understanding. *Theory into Practice, 40*(2), 118–127.

Perkins, D. N., & Unger, C. (1994). A new look in representations for mathematics and science learning. *Instructional Science, 22*(1), 1–37.

van Oers, B. (2009). Developmental education: Improving participation in cultural practices. In M. Fleer, M. Hedegaard, & J. Tudge (Eds.), *Childhood studies and the impact of globalization: Policies and practices at global and local levels—World yearbook of education 2009* (pp. 293–317). New York: Routledge.

van Oers, B. (2010). Emergent mathematical thinking in the context of play. *Educational Studies in Mathematics, 74*(1), 23–37.

Vygotsky, L. S. (1978). *Mind in society: The development of higher psychological processes*. Cambridge, MA: Harvard University Press.

Vellom, R. P., & Pape, S. J. (2000). EarthVision 2000: Examining students' representations of complex data. *School Science and Mathematics, 100*, 426–439.

Wild, C., & Pfannkuch, M. (1999). Statistical thinking in empirical enquiry. *International Statistical Review, 67*(3), 223–265.

Worthington, M., & Carruthers, E. (2003). *Children's mathematics: Making marks, making meaning*. London: Paul Chapman Publishing.

Leavy, A. M. (2008). An examination of the role of statistical investigation in supporting the development of young children's statistical reasoning. In O. Saracho & B. Spodek (Eds.), *Contemporary perspectives on mathematics in early childhood education* (pp. 215–232). Information Age Publishing.

Chapter 7
Scaffolding Statistical Inquiries for Young Children

Jill Fielding-Wells

Abstract Statistics in the early years is often limited to the construction and 'reading' of simple data representations as distinct from employing statistical inquiries that engage students with data in more authentic and meaningful contexts. One of the challenges of engaging with data inquiries is the extent to which students struggle with the lack of structure and direction, thus requiring additional support, or scaffolding. This chapter details the framework used for introducing statistical inquiry to young students and then provides insights that emerged from observation and analysis of a class of 5–6 year olds engaged in their own data investigation to illustrate. The findings suggest that considerable teacher scaffolding is required to progress students through inquiries and this was largely achieved through questioning employed to focus students on both the inquiry process and the statistical content to be addressed.

7.1 Introduction

Statistics is most commonly taught at the early childhood level in a surface and procedural fashion. A characteristic example is the teacher asking the students what their favourite fruit is. Students are provided with fruit cutouts to place on a pre-prepared picture graph as modelled by the teacher. If the students are slightly older, there may be some modelling of tallying first as each child selects their favourite fruit from a pre-populated, limited list of common fruits. In both examples, any questions asked of students will be simple, literal comprehension questions about the graph: *What is the most popular fruit? How many more people like oranges than apples?*

J. Fielding-Wells (✉)
Institute for Learning Sciences and Teacher Education, Australian Catholic University, Brisbane, Australia
e-mail: jill.wells@acu.edu.au

A. Leavy et al. (eds.), *Statistics in Early Childhood and Primary Education*, Early Mathematics Learning and Development,
https://doi.org/10.1007/978-981-13-1044-7_7

This commentary is not intended to diminish the requirement for children to learn to construct and read data representations, but to do so in this way removes the reality of statistics from the classroom: gone is decision making about problems, consideration of the data that is needed, planning for obtaining data and so on. Activities of this nature also remove the inherent messiness of data by artificially restricting the data to be 'workable'.

In real life, problems requiring the application of statistics are not neatly packaged in this way. Students need, from an early age, to develop an appreciation of the dynamics of statistical inquiry—such as the formulation of problems and appreciation of issues associated with planning and measurement—rather than a focus only on the collection, organization and conclusions that could be drawn from data sets (Wild & Pfannkuch, 1999).

To provide a contrast to the example, consider what might have happened had the students been asked to engage with an inquiry, '*What are the best fruits for a class fruit platter?*'. Left to their own planning, students may not have pre-defined potential responses (apple, banana and so on) and then had to deal with the potential for 20 different fruits to be named as 'favourite'. Such messiness would have led to many opportunities to enhance statistical thinking and decision-making, as well as providing a valuable lesson in the planning of data collection/surveys. Another potential benefit of an investigation is in considering the role that context plays. Students may have further considered the authentic purpose of such an investigation: '*We will have to narrow the choices to fruits that are readily available and in season now*'.

While an inquiry approach clearly supports more authentic engagement with statistical understandings, the ill-structured nature of inquiry problems means that students do not immediately or intuitively see how to address them. As inquiry is not widespread in classrooms, it isn't surprising that students struggle with the unaccustomed lack of structure and direction, and require additional structured support, or *scaffolding* to engage with them. So, what supports work?

The aim of the research described in this chapter was to provide insight into the ways in which a statistical inquiry could be facilitated with very young children. These insights emerged from observation and analysis of teacher-student interactions as an experienced inquiry teacher immersed a class of 5–6-year-old students in their first data inquiry. Sufficient detail of the classroom context has been provided to enable the reader to envisage the learning. Implications and suggestions for educators have been addressed.

7.2 Statistical Inquiries and Investigations

Simplistically, statistical investigations can be considered activities in which students engage with a genuine, contextualized problem they can apply statistical methods to, in order to lead to a data-based solution. As distinct from approaches often seen in schools—in which students are given neat, organized, *convenient* problems—investigations address the complexities and difficulties inherent in more genuine problems; thus, apprenticing students into the discipline of statistics.

Statistical inquiries and statistical investigations can be viewed along a continua of problem structure from well-defined to ambiguous, with inquiry problems lying further towards the ill-structured end of the continua. Ill-structured problems are those which have multiple potential solutions and solution paths: they 'contain uncertainty about which concepts, rules and principles are necessary for the solution' (Chin & Chia, 2006, p. 47). One of the essential components of addressing ill-structured problems is that students must engage in discussion to establish the elements of the problem; that is, they must first interpret the problem in its context and consider how data could address it (Allmond & Makar, 2010). Thus, investigative and inquiry problems develop a need for additional stages of statistical investigation to be addressed—refining of problems and planning of approaches (Shaughnessy, 2007; Wild and Pfannkuch, 1999)—in turn providing genuine opportunities for addressing these complexities and engaging in authentic statistical decision-making and reasoning. Shaughnessy argues:

> If students are given only pre-packaged statistics problems, in which the tough decisions of problem formulation, design and data production have already been made for them, they will encounter an impoverished, three-phase investigative cycle and will be ill-equipped to deal with statistics problems in their early formulation stages. (Shaughnessy, 2007, p. 963).

One commonly used model for statistical investigation is the investigative cycle, or 'PPDAC' due to the model acronym, as adapted by Wild and Pfannkuch (1999):

Problem	The deconstruction, negotiation and refining of the problem in conjunction with *context familiarization*
Plan	The identification of the data needed to address the problem and consideration of effective collection, recording and analysis of that data
Data	Data collection, recording and cleaning
Analysis	Organizing, manipulating, representing, and interpreting data to identify trends or patterns and provide evidence with which to address the problem; and
Conclusion	Reflecting upon the evidence identified in the analysis stage and linking it back to the initial problem to provide a response to that problem

This cycle is useful in identifying and describing the stages of investigation, in enabling teachers to plan investigations, and in providing students with structure by displaying this cycle for them.

Existing research supports the capacity of young children to engage in statistical inquiries (Fielding-Wells, 2010; Fielding-Wells & Makar, 2013; McPhee & Makar, 2014), but little research has been undertaken into *how* teachers scaffold young students in their initial inquiries. Fielding-Wells (2016) suggests there are three domains of knowledge that are drawn upon in inquiry: knowledge of the context; knowledge of [mathematical and statistical] content; and, knowledge of [statistical] inquiry.

The shift from a view of early childhood statistics practice from data collection, display and literal interpretation, to a rich inquiry approach is complex and requires a significant shift in teaching and learning practices from one which is teacher-led, to one in which the children are following a complex, 'messy' practice. Previous research has demonstrated that many teachers struggle with such a shift (Makar & Fielding-Wells, 2011).

7.3 Scaffolding

Scaffolding was first introduced by Wood et al. (1976) as a building metaphor to describe the expert support that could enable learners to reach goals considered beyond their reach (1976, p. 90). Scaffolding extends from Vygotsky's Zone of Proximal Development (ZPD),

> the distance between the actual developmental level as determined by independent problem-solving and the level of potential development as determined through problem solving under adult guidance, or in collaboration with more capable peers. (1978, p. 86)

The scaffolds used to support the learner are then gradually withdrawn until the goals can be achieved unaided. To be considered scaffolding, supports are required to meet the criteria of contingency (the type, timing and strength of supports must be responsively adapted to the student's current level of performance), fading (supports should be gradually withdrawn, or faded, as the student develops increased confidence and competence), and transfer of responsibility (accountability for performance should be progressively shifted to the learner) (van de Pol, Volman, & Beishuizen, 2010). This definition of scaffolding has been adopted in this chapter, as distinct from the informal notion of general 'support'.

Research into scaffolding has predominantly focussed on one-to-one or small group scaffolding due to the ZPD underpinning, as whole class scaffolding can be complex due to the nature of the numerous student ZPD's involved (van de Pol, et al., 2010). However, the practicalities of classrooms require whole class instruction; therefore, in practice, a teacher must work with group ZPD as well as consider individual learner's ZPD (Smit, van Eerde, & Bakker, 2013).

7.3.1 *Scaffolding Framework*

Wood et al. (1976, p. 98) identified six *functions* or *intentions* of the expert scaffolder from observation of young children being taught to perform a repetitive, mechanical task. These categories are provided below, but it is worth noting that modelling is often regarded in the literature as a means, rather than function, of scaffolding (van de Pol, et al., 2010):

1. Recruiting the learner: engaging the problem solver's interest in the task requirements.
2. Reducing the degrees of freedom: reducing the number of component processes or performances required to achieve a solution.
3. Maintaining student direction: maintaining the learner's focus and motivation.
4. Marking critical features: noting and drawing attention to the specific and relevant features of the task to identify discrepancies in production.
5. Frustration control: dealing with the affective state of the learner.
6. Demonstration: modelling a solution to a task for imitation.

The mechanistic task was chosen to enhance opportunities for the researchers to identify mastery but potentially reduces the applicability of the framework to tasks requiring higher cognitive and metacognitive regulation—tasks such as the solving of ill-structured problems (Shin, Jonassen, & McGee, 2003).

While Wood et al. (1976) focused on the functions or intentions of scaffolding, Tharp and Gallimore (1988) focused their attention on ‘how’ the expert assists the performance of the learner. They identified six ways that learners were supported in one-to-one interactions between an expert and the learner: modelling, contingent rewards and punishments, feeding back, instructing, questioning, and cognitive structuring. Again, this framework has reduced applicability for whole-class research as children in a classroom setting are less likely to work on an individual basis with a teacher and more likely to engage in group or whole class activities. To address the need to work with entire classes, van de Pol et al. (2010) built upon the work of these two seminal studies to tackle the issue of analysing scaffolding intentions and means in a natural classroom setting, deriving categories of feedback, hints, instructing, modelling, questioning, explaining and miscellaneous. These categories were used as a starting point for the research reported in this chapter.

7.3.2 *Research Question*

The question specifically addressed by this exploratory research was: What insight can be gleaned into scaffolding of statistical inquiry through observing a teacher as she worked to support young learners new to addressing ill-structured problems?

7.4 Method

The aim of the research described in this chapter was to ascertain how an experienced teacher of inquiry supported young learners to engage with an ill-structured problem. The classroom and lesson sequence, that is the focus of this chapter, was a single iteration of a larger design-based research (DBR) study. DBR was adopted as a methodology as DBR suits the intent of this research: to develop theory [about the process and scaffolding of learning during the teaching of statistical inquiry], to undertake highly interventionist research [teacher and researcher seeking together to implement and study classroom interactions during inquiry], and to have a practical focus and application [the support of young students engaging in statistical inquiry]. These principles underpin design-based research (Cobb, Confrey, diSessa, Lehrer, & Schauble, 2003).

The class participating in this research was a Year 1 class (5–6 years old) from a large suburban, mid-range socio-economic school in Australia. The teacher, Miss O, was an early career teacher with two years of experience in using inquiry pedagogies to teach mathematics. The unit was implemented in the third quarter of the school year, with a focus on data gathering, representation and analysis based on informal measurement of area. This was the students' first experience with statistical inquiry.

The ill-structured question that drove this inquiry was, '*How big are most Year 1 feet*?'. This problem lacked structure in that both 'big' and 'most' are ambiguous words. The former may refer to length, height, volume, area, mass and so on, whereas the latter might mean the most in our class, school, state, country or world. The teacher deliberately chose a context that would be familiar to all students in the class (their feet), and the measurement concepts included had been addressed previously with these students. Thus, the use of measurement for data collection served to reinforce previous work these children had undertaken.

Each lesson was videotaped and transcribed in full. All non-relevant speech was removed prior to coding (e.g. requests for bathroom breaks). The remaining teacher discourse was then coded using the thematic framework described below. Coding was undertaken concurrently with the viewing of the videotape to ensure alterations to meaning, for example because of an interrogative inflection, were not overlooked as may have been the case if relying on transcript alone.

Thematic codes were taken from van de Pol et al. (2010) as described in the scaffolding framework section above. These originally included *feedback, hints, instructing, modelling, questioning, explaining* and *miscellaneous*. After an initial pass, the need for additional codes became apparent to distinguish purpose or intent of questioning as most teacher interactions were of that nature. To address this, the categories were re-coded to reflect both the scaffolding means/purpose and the linguistic sentence type. The sentence types identified were as follows: interrogative [Q]—those which aim to elicit information; declarative [D]—those that provide information; or imperative [I]—those that convey a command or request (Brinton & Brinton, 2010). The codes are provided in Table 7.1 with the added categories marked with *. To illustrate, a code of Q-AMS is an interrogative [question] that serves to draw attention to a mathematical or statistical concept: similar to the marking of critical features as identified by Wood et al. (1976), with the difference that there was no intent to highlight discrepancies in performance, but rather, to highlight important concepts or understandings. Two category codes drawn from van de Pol et al. (2010) were not noted during the period of the inquiry: explaining and modelling. By the definitions of these category codes, these means of scaffolding address more explicit interactions—explicit showing or explicit telling. Explaining might include an explanation of the purpose of a column graph, whereas modelling might include step-by-step demonstration of how to create one.

Table 7.1 Code guide for analysing classroom discourse

Means	Code	Description	Example
Feedback	F	Direct evaluation of the work of the student	Yes, we could definitely use something to measure it
Hints	H	Provision of a hint but without providing whole solution	… we have been learning a little bit about the space in between things, haven't we?
Instructing	I	Provision of information so the student knows what to do/how to do it	So, here's an idea, why don't you go back over and have as little gaps as you can and for the space where it doesn't fit
Explaining	E	Provision of information on why to do something or content information	No instance
Modelling	M	Demonstration of behaviour for imitation	No instance
Extending	EX*	Prompting of student to think of/provide greater depth of reaction/response	Does anybody have another idea why tracing our feet is a good idea?
Clarifying	C*	Prompting students to explain their thinking or ideas more coherently.	Why do you think that is a good idea?
Revisit	R*	Reiterating/revisiting a question that did not elicit an adequate response after additional input (rewording, hint, etc.)	Alright so now I have got to think about how am I going to represent that? (Q) How am I going to put that on the board so we can see it and work out how big most year one feet are? (QR)
Attending inquiry	AI*	Drawing students' attention to important aspects of the inquiry process—explicitly or implicitly	So, what we're going to do today, how are we going to find this out? How big are year one feet?
Attending mathematical or statistical knowledge	AMS*	Drawing students' attention to important aspects of the mathematical of statistical content—explicitly or implicitly	There's a little space here, though isn't it? And that's not quite enough for one is it and a bit would be overhanging
Attending to the context	AC*	Drawing students' attention to important aspects of the context—explicitly or implicitly	What are we going to have to do in order to show so we know Katie's foot is probably one of the smallest in the class?
Miscellaneous	O	Other	OK hands on heads. Hands in laps

7.5 Illustrations from the Classroom

The final pass coding count from the entire unit is provided in Table 7.2 (interrogative interactions), Table 7.3 (declarative interactions) and Table 7.4 (imperative interactions). These have been included for demonstrating the highly disproportionate profile of the interactions: over two-thirds of all teacher dialogue being questioning and an almost complete absence of directives (imperatives).

There was also a significant distribution of questions towards the latter end of the cycle. This was only of interest in that the students engaged in less overall discussion during the early parts of the inquiry. This was likely because the problem was posed by the teacher as she was aware of the difficulties novice students have in question posing (Allmond & Makar, 2010).

More generally, there is a consistent pattern of interaction in this classroom. The teacher, Miss O, spent time with the children, talking through what students had done

Table 7.2 Total number of interrogative [question] interactions from teacher by stage of inquiry

Code	Stage of statistical cycle					Total
	P	P	D	A	C	
QAC			2			2
QAI	2	1	3	8	3	17
QAMS	2	1	9	7	8	27
QC		1		2		3
QEX			3		1	4
QH				1		1
QR				2		2
Total Q	4	3	17	20	12	56

Table 7.3 Total number of declarative interactions from teacher by stage of inquiry

Code	Stage of statistical cycle					Total
	P	P	D	A	C	
DAI					1	1
DE	1		1	1		3
DF	3	6	2	3	1	15
DI					1	1
DM					1	1
Total D	4	6	3	4	4	21

Table 7.4 Total number of imperative interactions from teacher by stage of inquiry

Code	Stage of statistical cycle					Total
	P	P	D	A	C	
II		1	1		1	3
IO	1					1
Total I	1	1	1		1	4

and then discussing ways to move forward. The children carried out the next stage of the inquiry in their group or pairs before the teacher drew them back to the whole group for discussion. While the students were working in groups, the teacher spent time circulating and working with students as needed: challenging, providing hints, clarifying and so forth.

The following section will guide the reader through the inquiry from start to finish. While the PPDAC cycle has been used as the organizer, and the unit thus presented chronologically, only sample excerpts have been included to illustrate the activities and interactions. Whole class discussions are prominent to highlight scaffolding means. The sequential numbering is provided only to enable discussion, with dialogue missing between sections but not within a section. Elipses (…) represent short pauses with long pauses indicated as [pause].

7.5.1 Problem

The teacher commenced by posing thc question, drawing the students into a discussion that quickly engaged them. The excerpt below shows the initial whole-class conversation, demonstrating the students becoming excited about the topic in a relatively brief time. The teacher involved the students in narrowing the focus of the inquiry to aspects of measurement and data, both making the inquiry manageable and directing the students towards desired curriculum foci of measurement and area.

In this first interaction with the students, the teacher can be seen to be *marking* those features which are critical to statistical inquiry by focusing attention on important aspects: the problem posed [1] and the need for evidence [3]. Once these ideas were planted, she quickly moved to the planning phase.

1.	Miss O:	OK. How big are most Year 1 feet? How big do you think they are?	Q-AI
2.	Katie:	About this big? [holding hands close together]	
3.	Miss O:	Seems about right	D-F
		but we've got to prove that right?	Q-AI
4.	Kylie:	I think this big. [holding hands an extended distance apart]	
5.	Miss O:	You think this big? I think Miss O's feet aren't as big as that! [smiling]	D-F
6.	Students:	[calling out] This big, this big	
7.	Miss O:	OK hands on heads. Hands in laps. Still waiting….	I-O

7.5.2 *Planning*

The teacher intentionally limited the student involvement in the posing of the question; however, she wanted students to become more involved in the planning phase to begin to envisage the data collection process. By using predominantly questioning and extensive feedback, the teacher guided the conversation and language use and thus the planning.

8.	Miss O:	So, what we're going to do today, how are we going to find this out? How big are year one feet? How are we going to find out what the space of the bottom of our feet are?	Q-AI
			Q-AMS
9.	Students:	Draw around them.	
10.	Miss O:	Draw around our feet? Yep, we definitely could do that.	D-F
		Why do you think that's a good idea?	Q-EX
11.	George:	Because it's the shape of our feet.	
12.	Miss O:	Yes, because it gets the whole shape of our feet, that's pretty good. Good explanation!	D-F
		Does anybody have another idea why tracing our feet is a good idea? James what do you reckon?	Q-AMS
13.	James:	Because then we can get some stuff to move around our feet.	
14.	Miss O:	Yes, we definitely could use something to measure our feet. Yep?	D-F
15.	Harry:	Tells us how big our feet are?	
16.	Miss O:	Yes, it could tell us about how big our foot is if we trace it. Yes, that's another good reason.	D-F
17.	Tanya:	We can use something to measure it.	
18.	Miss O:	Yes, we could definitely use something to measure it. Yes Jessica?	D-F
19.	Jessica:	You could write the number inside if you draw around it like the number inside it.	
20.	Miss O:	Do you mean once you've traced the foot you could actually write the measurement is that what you mean?	Q-C
		Alright well let's find out a little more about it. We're going to get on with it because that's the fun part.	D-I

The students could now see a way forward and required little assistance to envision data collection. They proceeded to collect materials and trace around their feet before measuring using whatever informal unit of measure they wished (these included dominoes, unifix cubes, tangram pieces and so on). The teacher did not lead the students to consider uniform units as she wished this to arise naturally through observation.

7.5.3 *Data Collection*

During data collection, the teacher initially focused on accuracy [as is valued by the discipline] before discussing other aspects of 'fair' data collection more broadly. This was a precursor to understanding sources of variability as due to either natural variation or error. The teacher worked individually with students to discuss their approaches at a level appropriate for the child. In the excerpt below, the teacher has the students consider the spaces on the foot outline (the 'yellow' in [21]) that are not covered by the unit of measurement, that is, 'gaps':

21.	Miss O	I'm looking and I'm thinking that's pretty good [the student's tiling] but I am also looking and I can see a lot of your yellow.	D-F
		What do you think that might mean for your measurement?	Q-AMS
22.	Alex	There's gaps	
23.	Miss O	Yes, there's gaps, so if there's gaps on your foot is that going to give you a correct measurement?	Q-AMS
24.	Alex	No	

The second opportunity the teacher orchestrated was to establish a need for accurate measurement to enable fair comparison. By allowing students to choose their own unit of measure, the students were led to see that this did not enable the data to be meaningfully compared. In the excerpt below, the students had measured their foot outline, recorded their data on the outline and were then discussing the results. One student, Peter, tried two different units, and the teacher took the opportunity to draw out the issue of 'fairness' to address variation that would be caused by measurement discrepancy (error), using the context of the inquiry to develop appreciation of the need for similar units [33].

25.	Miss O:	You used dominoes and unifix did you? Which one did you use first Peter?	Q-EX
26.	Peter:	Dominoes.	
27.	Miss O:	And how did you find that? Did it work? And what was the measurement of your dominoes?	Q-EX
28.	Peter:	10	
29.	Miss O:	It was 10 dominoes and what was the measurement with your unifix cubes?	Q-EX
30.	Peter:	24	
31.	Miss O:	24. Um, why was the unifix cubes a bigger number?	Q-AMS
32.	Peter:	Because they [unifix] are smaller.	

33.	Miss O:	So, while Seth has a bigger foot, his measurement shows if we looked at that, that his foot is actually smaller but what I am saying, yeah, but the units told us differently didn't it? What are we going to do then?	Q-AC Q-AMS
		Let's go back to our question, 'How big are year one feet?'. What do we need to make it a fair test? What are we going to have to do in order to show so we know Katie's foot is probably one of the smallest in the class? *Probably* [stressed]. And Seth probably has one of the bigger feet?	Q-AI
			Q-AC
		What are we going to do? Amanda? What could we do?	Q-AMS
34.	Amanda:	Use um the same units.	

The students elected to use unifix cubes as a comparative unit. These data were collected, and results are recorded by students, along with their name, on the outline of their feet. Giving students opportunities to plan and collect the data established connections between context and data: students could see the natural variation that was inherent as they made connections between what they observed in each other's foot sizes and the data as they were recorded. The questioning by the teacher [33] suggested the familiarity of context enabled her to facilitate appreciation of variation resulting from erroneous data collection while the students were recognizing that natural variation in data also occurs.

7.5.4 Analysis of Data

During data analysis, the teacher's purpose was to have students focus on the type and range of the data to organize them meaningfully. The teacher was supporting the students to experience the decision making inherent in dealing with raw data. Harry suggested focusing on the value of the tens in the measurements taken [35–39], and the teacher effectively privileged that suggestion. Binning the data, as Harry suggested, may somewhat mask the shape of the data; however, it was a valid and, for his age, sophisticated means of organization. From [39] on, the teacher tried persistently to establish, through questioning, how Harry's idea could be represented. It would have been far easier to create a bar graph of 'bins' at this point for the students; however, this would not have given them the experience of grappling with the issue themselves and resolving it. Throughout this discussion, the teacher preferenced responses of order and organization: characteristics of effective graphing.

35.	Miss O:	What are we all looking for to answer this question 'How big are most Year one feet?'.	Q-AI
		Now we know that we had to collect all our information that we have right now.	Q-AI
		What's the information I am looking for?	Q-AMS
36.	Harry:	The first number of the two numbers. [Harry is referring to the digit in the tens place on the measured feet, that is, if the foot was 24 blocks in area then the first of the two numbers is a 2—representing two tens].	
37.	Miss O:	What do you mean by that Harry?	Q-C
38.	Harry:	Like if the person had two and a number or if they had something different… if most have 2 and something then … then 20's is the most …more than 30 …	
39.	Miss O:	So, let me just clarify what you're saying. Are you saying that we need to collect the numbers and we're finding out if it's a teens number, something in the 20s, something in the 30s, something in the 40s, is that what you're saying? So, collecting the measurements? [Henry indicates assent]	Q-C
		Alright so now I have got to think about how am I going to represent that?	Q-AI
		How am I going to put that on the board so we can see it and work out how big most year one feet are? [waits]	Q-R
		So, we need some ideas now and you're going to help me to display the data and find out what we are going to do with all that information we have collected over the last couple of days. How are we going to figure this out? [waits]	Q-R
		If you just call numbers out at me we're probably not going to get a good idea, are we?	Q-H
		Because they're just numbers going all over the place inside our brains, outside our brains and we're going to lose all of that information so we need to figure out a way to display it on our whiteboard.	D-E
		So, does anyone have an idea?	Q-AMS, Q-AI
40.	Matt:	We could [stick] um this onto the whiteboard so we know what the number is and who it is.	
41.	Miss O:	OK we could do that ok that's not a bad idea.	D-F
		Well what will I do though will I just get all of these feet and just [stick] them wherever?	Q-AMS, Q-AI
42.	Students:	No.	
43.	Miss O:	Why isn't that going to work?	Q-AMS
44.	Peter:	Because they wouldn't be in order.	

45.	Miss O:	OK you're saying then that we need some kind of order to put the…	Q-AMS, Q-AI
46.	Peter:	[interrupting] Columns.	
47.	Miss O:	Columns? That's an interesting word. Alright so you're thinking, 'how about we put them in columns'.	D-F
		What am I going to do? Put all the blue ones in a column, all the green ones in a column, the orange ones in a column, the yellow ones in a column.	Q-AMS
48.	Jenny:	All the ones that are 20 in columns	
49.	Miss O:	All the ones that are 20 in a column. Ok is everybody agreeing with that? Does everybody think that might be a good idea to display our ideas?	Q-AMS, Q-AI
50.	Students:	Yes.	

The teacher proceeded to work with the students, through questioning, to determine what each column should represent and how these columns could be labelled. There were several instances [as in 47] where she provided the students with an incorrect process to ascertain understanding. This setting up of 'adversary' statements was a popular approach with this teacher. If the students argued, and corrected her, she would move on. This dialogic movement served (and this was confirmed by the teacher) to assess student knowledge 'on the fly' to monitor understanding and enabled her to adapt contingently.

7.5.5 Conclusions

The final stage of the statistical investigation was to provide a conclusion derived from data. The students were seated on the floor in front of the whiteboard looking at the image seen in Fig. 7.1. In the exchange below, the teacher supported and guided student appreciation of the investigative cycle by drawing attention to the need for a conclusion [51] and for that conclusion to be linked to evidence and drawn from such [53, 55]. In doing so, the teacher privileged the key features of the students' data representation [66] to enhance their interpretation. Finally, the teacher used questioning heavily to draw the students to a conclusion.

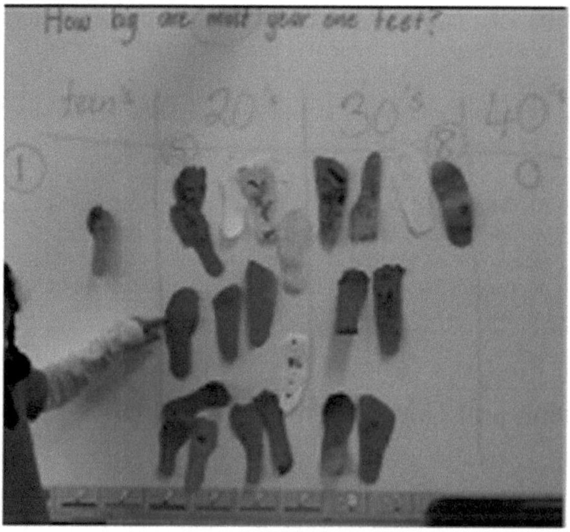

Fig. 7.1 'Binned' representation of the students' feet

51.	Miss O:	Can we answer our question from the data we've collected and represented on the board? Can we answer it?	Q-AI
		How big are most year one feet or at least in this classroom?	
		How can we answer it? Who can tell me?	
52.	Bethany:	The 20s, the most biggest because lots of people have… [tails off]	
53.	Miss O:	Well, how do you know? [pause]	Q-AI
		Have you checked? [drawing focus to data]	Q-AI
54.	Students:	Yeah	
55.	Miss O:	How am I going to tell exactly? Because you are telling me there's lots of feet in there but there's lots of feet in the 30s too. But how are you going to tell me? How do I know, how do you know that there is the most in the 20s?	Q-AMS
56.	Bethany:	There's lots of space there and there's not much space in the 20s. [meaning absence of data]	
57.	Miss O	Well that's true but I could make less space by doing this. [Moving all the feet together]. But that's not answering it really because now I've got lots of space in here too. How do you know for sure?	D-F Q-EX
		It's not about space in those columns, [waits]	D-H
		but how do you know for sure? That the answer is the 20s James?	Q-AMS
58.	James:	The 20s have 14 because I counted and the 30s have 8. [the students continue to count all items in each column]	

59.	Miss O:	So, can I answer my question? Who can answer my question? Who can answer my question that we have been trying to answer for the past two days?	Q-AI
		I'm hoping everybody has their hands up? How big are year one feet? William?	Q-AMS
60.	William:	20's	
61.	Miss O:	Ah in the 20s's. So when I say How big are most year one feet? We could all say because the data shows us, all the measurements we took over the last couple of days and all the information we collected tells that—Year one feet are mostly in the 20s.	D-M D-AI

7.6 Discussion and Conclusion

The aim of this research was to develop insight into how a teacher familiar with inquiry scaffolded her students to engage with the complexities of addressing ill-structured statistical problems. To begin, the first question was whether the teacher was in fact providing scaffolded support. Returning to van de Pol et al.'s (2010) requirement that scaffolding meet the requirements of contingency, fading and transfer of responsibility offers a structure to explore this.

Contingency addresses the need for the type, timing and strength of the supports to be responsive to the learner. The strength of the response is important as the teacher must have knowledge of the students' current level of understanding to respond. A teacher needs to assess and consider both class and individual ZPD in order to make contingent responses, and this is quite complex and difficult a requirement. Here, we saw the teacher exploring the degree of support required: in exchanges with students [eg. 39], she provided a question prompt, and then worked to reformulate the question, and eventually provided a minor hint to establish what the students needed in order to continue. The teacher neither 'told' students nor 'rescued' them but rather provided the minimum of progressive assistance. The practice of offering minimal support and building on it served to address class ZPD but may be problematic in terms of building up future capability. The teacher was experienced at inquiry and had some experience in teaching early childhood: this poses the question of how we can 'bottle' this approach to assist novice teachers. The contrasting argument is that should the teacher have chosen to provide stronger scaffolds, or more direction, earlier, the task challenge may have been overly reduced and the benefits of the statistical inquiry lost. Obtaining a balance is a challenge.

The second requirement of scaffolding is the fading of supports (van de Pol et al., 2010). Unfortunately, the examination of a single unit of work to develop a complex understanding of the statistical inquiry process does not provide a realistic measure of fading as students have only had the opportunity to cycle through the process once. It is thus impossible to demonstrate fading sufficiently without a longitudinal study.

Finally, transfer of responsibility deemed a necessary component of scaffolding (van de Pol et al., 2010). Again, a longitudinal study would be of more benefit in ascertaining the effectiveness of transfer. The purpose of implementing statistical investigation is to develop students capable of engaging as apprentice statisticians, and therefore, the aim of the process is to use the PPDAC cycle (Wild & Pfannkuch, 1999) as a scaffold via a learnt process. With this goal in mind, increased adoption of responsibility and decreased support would be target outcomes, at least in terms of students developing the *process* of statistical inquiry. It is likely that support for the content being developed would alter and deepen as students progressed through schooling, and therefore, the scaffolding specific to the content may be of an ongoing, changing nature.

A second aim of this study was to ascertain the means of support or scaffolding. In observing the overall pattern of class interaction, there were several strategies or means of scaffolding that were employed by this teacher for specific and clear purposes. The predominant means of scaffolding was the use of questioning, so much so that multiple sub-categories were needed to assist with the classification of interrogative interactions. The purpose of these questions was largely to draw students' attention to one of three identified knowledge categories: knowledge of the statistical inquiry process; knowledge of the mathematics and statistics content areas; and knowledge of context (Fielding-Wells, 2016). While the latter was familiar to students, the teacher's specific linking of results to context served to allow the context to support statistical ideas.

In supporting the statistical inquiry process, the teacher focused on the broader objectives: the need to address a question; the need to plan, gather and organize the data; and the need to provide a data-based conclusion which addressed the initial question. By leading the students through this cycle, she explicitly and implicitly demonstrated the valuing of the process to the students. Simultaneously, she guided students through the 'discovery' of statistical understandings essential to students' appreciation of the 'big ideas' of statistics (e.g. Watson, 2006): variation (error and natural, within group), distribution and centre (implicitly).

7.7 Implications for Teaching and Research

The results of this study have implications for both teaching practice and future research studies. In terms of classroom teaching practice, the findings suggest that considerable teacher scaffolding is required to guide young students with little experience in engaging in statistical inquiries. To enact a shift from more traditional data activities that utilize neat, organized data sets to learning that involves the 'messiness' of more realistic data investigations is complex. Such a shift requires that teachers have a solid grounding in the process of statistical inquiry, have the conceptual knowledge of statistics to guide and draw out key statistical concepts and be willing to relinquish the control they might have otherwise had of the learning process. Beyond this, a shift to statistical inquiry also requires facility with scaffolding students through the process. This research suggests that the predominant supports that were beneficial to young children came in the form of questioning and feedback.

Questioning supports that draw students' attention to both the process of statistical inquiry and the underpinning knowledge of statistics/mathematics was required, and this reinforces the importance of teachers having underpinning understanding of these statistical and inquiry knowledge domains.

In consideration of future research, further studies that provide insight into the scaffolding of statistical inquiry are needed; in particular, studies that address engaging children of varied ages and levels of experience with statistical inquiry and also longitudinal studies that focus on the extent to which the crucial scaffolding requirements of *fading* and *transfer of responsibility* can be more adequately addressed. A second area of future focus needs to be the identification of mechanisms for assisting teachers to identify class ZPD accurately so as to provide the least amount of support necessary to progress students. In this way, students can be engaged as authentically as possible with the inquiries, while not exceeding the class ZPD and therefore requiring explicit instruction or direction which may undermine their ability or interest in making their own judgements.

Acknowledgements The author wishes to acknowledge the contributions of the participating students and teacher. This work was funded by Australian Research Council Grant DP170101993 and an Australian Postgraduate Award.

References

Allmond, S., & Makar, K. (2010). Developing primary students' ability to pose questions in statistical investigations. In C. Reading (Ed.), *Proceedings of the 8th international conference on teaching statistics*. Voorburg, The Netherlands: International Statistical Institute.

Brinton, L. J., & Brinton, D. M. (2010). *The linguistic structure of Modern English* (2nd ed.). Amsterdam, The Netherlands: John Benjamins.

Chin, C., & Chia, L. (2006). Problem-based learning: Using ill-structured problems in biology project work. *Science Education, 90*(1), 44–67.

Cobb, P., Confrey, J., diSessa, A., Lehrer, R., & Schauble, L. (2003). Design experiments in educational research. *Educational Researcher, 32*(1), 9–13.

Fielding-Wells, J. (2010). Linking problems, conclusions and evidence: Primary students' early experiences of planning statistical investigations. In C. Reading (Ed.), *Proceedings of the 8th international conference on teaching statistics*. Voorburg, The Netherlands: International Statistical Institute.

Fielding-Wells, J. (2016). "Mathematics is just 1 + 1 = 2, what is there to argue about?": Developing a framework for Argument-Based Mathematical Inquiry. In B. White, M. Chinnappan, & S. Trenholm (Eds.), *Opening up mathematics education research (Proceedings of the 39th annual conference of the Mathematics Education Research Group of Australasia)* (pp. 214–221). Adelaide: MERGA.

Fielding-Wells, J., & Makar, K. (2013). *Inferring to a model: Using inquiry-based argumentation to challenge young children's expectations of equally likely outcomes*. Paper presented at The Ninth International Conference on Statistical Reasoning, Thinking and Literacy Superior Shores, Minnesota.

Makar, K., & Fielding-Wells, J. (2011). Teaching teachers to teach statistical investigations. In C. Batanero, G. Burrill, & C. Reading (Eds.), *Teaching statistics in school mathematics-challenges*

for teaching and teacher education: A joint ICMI/IASE study (pp. 347–358). Voorburg, The Netherlands: Springer.
McPhee, D., & Makar, K. (2014). Exposing young children to activities that develop emergent inferential practices in statistics. In K. Makar, R. Gould, & B. daSousa (Eds.), *Sustainability in statistics education: Proceedings of the ninth international conference on teaching statistics.* Voorburg, The Netherlands: International Statistical Institute.
Shaughnessy, J. M. (2007). Research on statistics learning and reasoning. In F. K. Lester (Ed.), *Second handbook of research on mathematics teaching and learning* (pp. 957–1009). Charlotte, NC: NCTM.
Shin, N., Jonassen, D. H., & McGee, S. (2003). Predictors of well-structured and ill-structured problem solving in an astronomy simulation. *Journal of Research in Science Teaching, 40*(1), 6–33. https://doi.org/10.1002/tea.10058.
Smit, J., van Eerde, H. A. A., & Bakker, A. (2013). A conceptualisation of whole-class scaffolding. *British Educational Research Journal, 39,* 817–834.
Tharp, R. G., & Gallimore, R. (1988). *Rousing minds to life: Teaching, learning, and schooling in social context.* New York: Cambridge University Press.
van de Pol, J., Volman, M., & Beishuizen, J. (2010). Scaffolding in teacher–student interaction: A decade of research. *Educational Psychology Review, 22*(3), 271–296. https://doi.org/10.1007/s10648-010-9127-6.
Vygotsky, L. S. (1978). *Mind in society.* Cambridge, MA: Harvard University Press.
Watson, J. (2006). *Statistical literacy at school.* Mahwah, NJ: Lawrence Erlbaum Associates.
Wild, C. J., & Pfannkuch, M. (1999). Statistical thinking in empirical enquiry. *International Statistical Review/Revue Internationale de Statistique, 67*(3), 223–248. https://doi.org/10.2307/1403699.
Wood, D., Bruner, J. S., & Ross, G. (1976). The role of tutoring in problem solving. *Journal of Child Psychology and Psychiatry, 17*(2), 89–100. https://doi.org/10.1111/j.1469-7610.1976.tb00381.x.

Chapter 8
How Kindergarten and Elementary School Students Understand the Concept of Classification

Gilda Guimarães and Izabella Oliveira

Abstract Teaching statistics has been increasingly valued in recent years. To understand the physical and social world that surrounds us, knowing how to systematize information and/or understand the information matched is fundamental. Thus, knowing how to classify data is a fundamental skill. This article aims to analyse what students between 5 and 9 years old and teachers who teach those grades know and can learn about activities involving classification. To this end, we present the results of three different studies conducted with elementary school children and teachers. The results reveal that children are able to classify from a previously defined criterion and to discover a classification criterion, but experience more difficulties when creating criteria to classify. We believe this may be explained, partially, by the lack of familiarity with this type of activity both in everyday life and at school, as they are generally asked to classify from pre-defined criteria instead of producing their own. However, since kindergarten children are already able to classify in different situations and, most importantly, they are able to learn easily the skills needed to classify, we believe that if they have instruction that leads them to reflect about classification, they learn easily, thereby evidencing the important role of the school.

Keywords Classifying · Kindergarten · Elementary school · Statistics

One of the phases of an investigative cycle is data classification. Only with classified data is it possible to interpret the situation and obtain conclusions. We consider it fundamental that people are able to create criteria so that they can organize the information from the goals defined by them.

G. Guimarães (✉)
Universidade Federal de Pernambuco, Recife, Brazil
e-mail: gilda.lguimaraes@gmail.com

I. Oliveira
Université Laval, Quebec City, Canada
e-mail: Izabella.Oliveira@fse.ulaval.ca

A. Leavy et al. (eds.), *Statistics in Early Childhood and Primary Education*,
Early Mathematics Learning and Development,
https://doi.org/10.1007/978-981-13-1044-7_8

Fig. 8.1 Example of a student using two criteria (colour and means of locomotion)

8.1 What Is Classification?

Classification is a natural activity that begins very early for humans. From a very early age, children classify objects according to analyses based on similarities and differences (Vergnaud, 1991). Piaget and Inhelder (1983) defined classification as a procedure that enables the individual to assign all the elements of a certain collection to a category, according to certain criteria. For them, a classification is correct when the exhaustiveness and exclusivity criteria are met. Exhaustiveness implies that all elements are classified, and exclusivity implies that each element can only be part of one of the classes or groups. In other words, categories must be able to exhaust and, at the same time, be mutually exclusive. In the following example, the child used two different criteria (colour and means of locomotion) when trying to classify the nine cartoon figures. The child verbally explains that "they are the yellow ones" (SpongeBob, Tweety, Garfield), "they are the ones who swim" (Shrek, Nemo, and Little Mermaid), "are the ones who fly" (OddParents, Monica, Superman, and Spider-Man). In this case (Fig. 8.1), it did not meet the exclusivity criterion.

On the other hand, it is fundamental to point out that there is more than one way to classify. The same elements may be classified in different manners one at a time or in hierarchical classifications.

For example, animals may be classified according to their origin: in cooking, we might divide them into seafood and red meat. We can classify animals as carnivorous or herbivorous, according to an ecological classification in their natural habitats. These same elements may also be classified as echinoderms and mollusks, according to a zoological classification based on their biological evolution (Lecointre, 2004). Is there a hierarchy among these criteria? No. So, which of them should be used for classification? The choice depends on the purpose of who classifies. Different objectives suggest different classifications. Within a zoology laboratory, the term "seafood" is meaningless. Similarly, referring to echinoderms and mollusks in a kitchen makes no sense. These differences go beyond the vocabulary used; it is the specific objectives of the context of a kitchen or a zoology laboratory that determine the coherence and pertinence of a classification. Therefore, it is of fundamental importance to use a classification system that is related to the established goals and the context in which they are stipulated. Classifications are not neutral operations. Thus, it is important to point out that the objective of a classification is to understand a set of data and/or information. The one who classifies chooses the criteria according to his/her needs.

We would further point out that the classification criteria adopted depend on the context in which the classifications take place, including the historic moment and the person's needs. Thus, it is possible to find as many classification systems as there are classifiers.

Classifying is part of people's daily lives, and activities that involve classification can provide children and adults with models for organizing things in the real world, such as putting blocks away or setting the table for dinner.

Ware (2017) and Kalénine and Bonthoux (2006) state that categorization is a fundamental aspect of cognition and a critical task of child development that helps children to organize experience and understand relations between entities. Classifications may be based on thematic, causal, functional, temporal, perceptual and relational criteria, among others. Bonthoux and Kalénine (2007) state that children classify based on different aspects and that their choices depend on development, context, and individual factors.

8.2 The Investigative Cycle as a Means of Developing Statistical Reasoning

One way to develop statistical literacy is to carry out an investigative cycle. The investigative cycle relates to the way an individual thinks and acts during an investigation (Wild & Pfannkuch, 1999). Statistics can be considered an important tool for carrying out projects and investigations in various areas, used for planning, data collection, and analysis, and in inferences for decision-making with the intent to support statements in various areas, such as health, education, science, and politics. When students carry out an investigation, they may reflect autonomously and consequently be capable of interpreting reality through their own data systems or of interpreting the data systems of others critically (see this chapter). Therefore, statistics plays a fundamental role in education for citizenship.

An investigative attitude must necessarily include a concern for developing questions; elaborating hypotheses; defining samples; collecting, classifying, and organizing information in graphical or table representations; analysing and reaching possible conclusions to solve the proposed problem; and making decisions based on said information (Gattuso, 2011).

In this sense, we believe that statistical investigation must be the main axis in students' and teachers' statistical education from the earliest grades through every educational level. Research must be an essential element in teachers' education and practice, since it allows a reflective attitude in teaching and requires teachers to master the procedures of scientific investigation (Guimarães & Borba, 2007).

However, to understand fully how statistical investigation is developed, students must participate in it from its very beginning to its conclusion, experiencing every phase (Gal & Garfield, 1997; Ponte, Brocardo, & Oliveira, 2003; Batanero & Diaz, 2005; Ben-Zvi & Amir, 2005; Makar & Rubin, 2009; Fielding-Wells, 2010; Leavy & Sloane, 2017; among others). It is in these situations that students are able to perceive the function of statistical concepts.

Having statistical investigation as a structural axis of statistical learning and teaching, it is fundamental to consider that this can happen throughout a whole investigative research cycle, as well as reflect on each of its phases. So, one of the phases of the investigative cycle is data classification.

More precisely, in this chapter, we are interested in discussing the development of understanding about creating criteria to classify.

8.3 What Is the Importance of Classification in the Statistical Investigative Cycle?

One way to organize statistical thinking is the realization of the research cycle. Wild and Pfannkuch (1999) argue that the investigative cycle concerns the way the individual acts and thinks during an investigation. In this sense, Silva and Guimarães (2013) propose different phases of an investigative cycle (Fig. 8.2).

This diagram presents the different stages of the investigative cycle. Each of these elements contributes to the development of students' investigative skills. Even though they are all important, this chapter will deal with classification.

To reflect on this relationship, we begin by presenting an example. Luanna, a teacher, asked her fourth-grade students (aged 9) to build a bar graph based on a list of items bought at a supermarket. To build the graph, students first had to create criteria to organize the products. One student created the criteria "storage place", with the categories freezer, fridge, cupboard, open air. Accordingly, he built a database with which he could check if all the products had been classified and that none of them was in more than one category, meeting the criteria of exhaustivity and exclusivity.

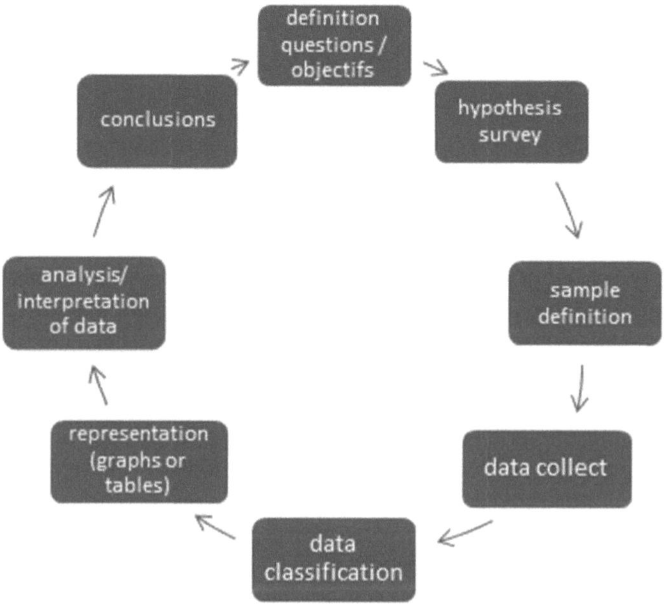

Fig. 8.2 Investigative cycle proposed by Silva and Guimarães (2013)

Graphs and tables are ways of representing these classifications when we deal with quantitative data.

The ability to build and interpret a graph also depends on understanding the categories that are represented. In a bar graph, for instance, each bar represents one category of a variable or the criterion analysed. Thus, classifying data is an important ability in the development of statistical literacy.

As previously mentioned, the ability to interpret a graph depends on understanding the categories involved. In this sense, categorizing data is an important ability for the development of statistical literacy. If interpreting a graph or table requires a significant level of understanding, representing statistical data is even more complex. According to the type of data obtained from a classification, students may choose various types of representation: a hierarchical classifications tree, sets, tables, and graphs.

8.4 Errors in Classification Carried Out by Children

Piaget and Inhelder (1983) identify that people present different understandings in their attempts to learn the ability to classify: figure collections, grouping elements in pairs, not classifying every element, dichotomous forms of classification, classifying without specifying the criteria, and adequate classification. We point out that these performances do not take place in any given order, since the same subject may try to organize the elements in pairs and later classify them dichotomously.

Collections of figures are when the student puts together a triangle and a square to draw a house. When a student groups elements in pairs, he/she is seeking a relation of resemblance, but is not relating these elements to the whole group.

Another type of strategy used by the student is not exhausting all the elements to classify, since they used criteria that did not involve all the elements. In a dichotomous classification, the subject chooses a property and analyses whether the element has that property or not. For example, he/she may classify the ones that are blue and those that are not. This type of classification is also called a binary classification. Some subjects are able to classify, but not to specify the criteria. They classify their toys as wooden, plastic, or metal, but are unable to explain that the criterion chosen was the type of material the toys are made of. Lastly, subjects classify by anticipation, explicitly stating a criterion (such as size) and using said criterion to classify.

According to Gitirana (2014), classifying must follow well-defined criteria, as well as keep in mind that every object can belong to one or various classes. Every concept, in and of itself, is a class, and once we define it, we have one of the necessary characteristics for an element to be part of that class (concept). It is exactly the use of these characteristics that allows us to decide whether a given object is part of a given class or not.

Thus, the importance of classification work reinforces the need for systematic work, with interesting, challenging activities that encourage students to think for themselves. In this sense, it is essential to discuss which abilities related to classification children need to develop.

8.5 Research on the Ability to Classify

In general, we have found several studies that investigate how adults and children build concepts based on classifications (Deák & Bauer, 1995; Nguyen & Murphy, 2003; among others), as well as studies with very young children or babies on the relation between thought and language (Mareschal & French 2000; Vieillard & Guidetti 2009; among others).

Besides these, Clements (2003) and Rodrigues (2016), among others, state that most of the studies of children's mathematical classification concern geometric shapes, progressively making possible their access to the process of shape classification based on the characteristics and properties of the simple geometric shapes, which allows them to identify and recognize inclusive types of classifications.

In this same sense, Amorim and Guimarães (2017), upon analysing Brazilian math textbooks for students aged 6–8, observed an almost complete lack of activities that involve classification, and when such activities were proposed, they also involved teaching geometric concepts. In these situations, students are not led to create classification criteria. The objective is to see whether students are able to classify the shapes based on defined criteria for existing categories. According to Gitirana

(2014), although essential to the development of concepts, schools have given little importance to the logical procedure of building criteria.

Other research suggests that reasoning about data classification is encouraged when students are invited to invent and revise models (Hancock, Kaput, & Goldsmith, 1992). Thus, teachers need to know how to propose situations that develop creation, critique, and revision of data classification.

The objective of this chapter is to analyse what students between 5 and 9 years old and teachers of those grades know and can learn about activities involving classification.

In the following section, we present, in detail, three studies carried out in Brazil involving students and/or teachers of the earliest school years. The first study was done with 20 kindergarten children (age 5); the second, with 48 third-grade children (age 8) and 16 early grade teachers; and the third, with 72 fourth-grade children (age 8–9).

8.6 Method and Data Analysis

This chapter presents the outcomes of three different studies. The results were obtained from the three studies conducted by the GREF team—*Grupo de Estudo em Educação Estatística no Ensino Fundamental* (Study Group on Statistical Education in Elementary Education). The first two (diagnostic) studies were carried out based on an individual Piagetian clinical interview, in which the researcher has a script that is modified depending on the student's answers, in order to allow the researcher to investigate how the student is thinking about the question. The third study is an experimental study, in which a pretest, two different intervention situations, and a post-test were performed to verify the learning from the interventions.

From the students' answers during the interviews, a qualitative and quantitative analysis was carried out in order to identify the students' understanding of classification, considering the types of answers identified by Piaget and Inhelder (1983). Thus, a hit-and-error analysis was initially performed. Then, a qualitative analysis of the types of errors and hits was performed. The categorization of strategies was performed according to the types of responses observed by Piaget and Inhelder (1983) and in the studies previously presented in this chapter.

In the following sections, we will present each of the studies that have contributed to our reflection on the understanding that kindergarten and elementary school students have about classification.

Study 1—What do kindergarten students understand about classification?

In this study, Barreto and Guimarães (2016), upon identifying the scarcity of activities that request that children create classification criteria, both in textbooks and in classrooms, decided to investigate what students understand about classification. To this end, they carried out Piagetian clinical interviews with 20 children (age 5) from three schools in Recife, Brazil. The authors chose to work with different schools to

METAL	PLASTIC	GLASS	PAPER

Fig. 8.3 Activity based on a given criterion

avoid similar results due to teaching methods or curricular organization of a given school.

Each child was asked, individually, to respond to three types of activities involving different classification situations. The order in which the activities were presented was established based on how familiar the children were with the activities. They began with the most common activities in school and in textbooks, followed by an activity that was done rarely, and finally an activity which requested them to create criteria. The interviews were conducted in a support room in the school and lasted approximately half an hour. The activities were:

Based on a given criterion

In this type of activity, the criterion is presented and the student must identify which objects belong in each group. This activity used 11 cutout pictures and a sheet of paper (Fig. 8.3).

Each child was told: "*I'd like you to organize these pictures based on what they're made of. Here are objects made of metal, plastic, glass, and paper*." The researcher always asked the student to justify their answer.

Identifying the classification criterion

In this activity (Fig. 8.4), the children had to do the inverse operation of the previous activity, since in this one the groups are already formed. Students had to discover which criterion was used for the classification. The researcher reads the statement presented in the activity.

When a child did not understand the command, the researcher would explain, "*A boy organized his books in these two baskets because he thought not all the books were the same. So I want to know what he might have thought to organize them this way; why did he put these here and these here?*" (pointing while giving this explanation).

Fig. 8.4 Activity for identifying the classification criterion. (Observe the images below, and discover the criteria used to organize children's books in each basket.)

Fig. 8.5 Activity for creating a classification criterion. (It is time to pack the toys. Cut out the toys from the next page. Stick on the same shelf the toys that you think should stick together)

Creating a classification criterion

In this type of activity (Fig. 8.5), the objects are presented and the child is asked to create a criterion to classify the elements in the way they think best. For this activity, the children received eight cutout pictures to come up with an organization criterion to divide the pictures into three groups. The children would then glue the pictures on the three shelves. "*These objects need to be organized. You will see which ones are similar and that belong together. You are going to organize these toys in three groups and put them on the shelves*".

Once again, as the child was doing the activity, the researchers asked questions, trying to understand what the child was thinking upon gluing each picture. The children were asked: "*Why are you putting these pictures together? Could this picture go somewhere else?*"

Table 8.1 Percentage of correct answers in each activity

Activity	Quantitative of successful students (%)
Classify from a given criterion	95
Identify classification criteria	20
Create classification criteria	35

8.7 Results Study 1

Table 8.1 represents the percentage of correct answers the children gave in each activity. We see great variety in the percentage of correct answers. This result shows that the activity of classifying based on a given criterion had a high percentage of correct answers, as was expected, since this activity is frequently found in textbooks and classrooms. On the other hand, few children were able to identify or create classification criteria, activities that are less frequent in textbooks.

In the activity of identifying classification criteria, we found students gave different types of answers: they either identified correctly the classification criteria; or described the themes of the books; or used more than one criterion in their attempts to classify. In these cases, the children would group the books as this example: giraffe and dragon books/family, dolls, and fairy tale or horror books (theme)/“girl” books (gender). We point out the difficulty, but not impossibility, of early schoolchildren doing this task, since four children were able to answer adequately. They classified the books as animal books/fairy tale books, dinosaur books/princess books.

In the activity to create classification criteria, seven children were able to classify correctly, but only three were able to explain the criteria (dolls, musical instruments, games). The others distributed the pictures randomly among the shelves. We always asked the students to explain how they had made their classifications, just as Lehrer and Schauble (2000) did in their study. We considered it important to investigate students’ metacognitive ability to explain how they classified.

However, based on this study, we can conclude that children from early childhood education have shown that they understand classification activities, and in the first place, they are capable of creating certain sorts of classification. Thus, if they are able to create criteria to classify, they will be able to conduct research, collect data, and classify responses so that conclusions can be drawn.

Study 2—How elementary school students and their teachers created classification criteria.

Guimarães, Luz, and Ruesga (2011) investigated how elementary school students’ and their teachers created classification criteria. Forty-eight students (aged 8) in the third grade of elementary school and 16 teachers of this same level of education participated in the research.

The author carried out Piagetian clinical interviews, during which they handed subjects nine cutout pictures and asked the students/teachers to create a criterion to classify them. Half of the participants were asked to create two groups, and the other half three groups.

Since context has been shown to be a determining factor in classification, the authors used two groups of pictures (toys and cartoons) which were familiar to the participants. Although the pictures were familiar, there is no type of classification commonly used neither in everyday life nor in school.

8.8 Results Study 2

The results show that most of participants performed poorly on this task (only 33% of the students and 44% of the teachers succeeded in the task). This result is very important because it shows that the teachers present a difficulty similar to that of their students. In this way, two points should be taken into account: how can teachers lead their students' learning about creating classification criteria if they themselves do not know how to do it? Second, it is evident that the life experience of adult teachers participating in this study has not been enough to lead all of them to learn how to classify in some situations.

In face of the importance of knowing how to classify, it is fundamental that teachers learn to create classification criteria and that they provide their students with activities that will guide them to learn it also.

Once again, in their attempts to classify, the participants (56% of students and teachers) would use more than one criteria when asked to create a criterion to classify a group of figures (Fig. 8.6). In this example, the teacher does not seem to realize that all the pictures have shapes (which can be the same or different) and that a game always leads to a competition.

However, it is important to point out that some children carried out this activity successfully, both in two groups and in three, showing once again that children can create classification criteria correctly.

In Fig. 8.7, we present an example of a student who appropriately classifies into three groups. Although he wrote the name of the first figure in each group, when he was asked how he had classified it, he explained that "*these* (referring to the Shrek group) *I watch a lot, these* (referring to the Garfield group) *I sometimes watch, and these* (referring to the Monica group) *I don't watch*".

Few participants wrote the name of the criterion (29% of students and 44% of teachers). Furthermore, in several cases where the participants wrote the name of the criterion, they do not fit into the classification used. Vieillard and Guidetti (2009) had already observed that adults and children would name the groups and not the criterion they had used.

Fig. 8.6 Response of a teacher upon request to classify into three groups (shapes/games/competition)

Fig. 8.7 Example of a correct classification of a child in three groups

The results obtained on this type of activity are similar to those found by other authors (Leite, Cabral, Guimarães & Luz, 2013; Lehrer & Schauble, 2000; Guimarães, Gitirana, & Roazzi, 2003), in which they identified the difficulty experienced by children and adults present when classifying.

For example, Lehrer and Schauble (2000) researched classrooms of students and their teachers (first, second, fourth, and fifth grades) in the task of classification. The authors observed that the youngest children evolved systems of attributes that described their categories in a post hoc fashion, but failed to regard those rules as a model to guide classification. In contrast, fourth and fifth graders considered their category systems as models that logically constrained the members admitted into categories, although many continued to include redundant or foreign information. They incorporated and discussed a variety of kinds of decision rules, and they had the opportunity to see the intellectual work performed by practices of data modelling. Also, Guimarães and Oliveira (2014) investigated how 113 future teachers in Recife, Brazil; Quebec, Canada; and Burgos, Spain, created criteria to classify and used these criteria in a free classification activity. Although most students managed to reach a correct classification in two groups, when the activity required three groups, the performance was significantly weaker ($x^2 = 13.717$, gl 1, $p \leq .000$). Only those who defined a descriptor were successful in their classification. Thus, knowing how to classify appears not to be an ability learned solely through life experience. This difficulty faced by both students and teachers can be partially explained by the absence of any systematic schoolwork on classification.

In this way, we can conclude that some elementary school students and their teachers have difficulties in creating classification criterion. This conclusion needs to be drawn very carefully, since, despite the difficulty of some, others are able to create criteria to classify. More than that, some children are able to create classification criteria in an appropriate way since kindergarten, as evidenced by Barreto and Guimarães (2016). Thus, we believe that the difficulties we met can be explained by an absence of reflections from teachers and students on how to create classification criteria. Thus, we believe it is essential that teachers understand the importance of knowing how to classify and how to propose that kind of learning for their students.

Study 3—Learning about classification

Since studies have found that children can create classification criteria since kindergarten, we present a study in which the researchers sought to work with classes of students that were expected to learn to create classification criteria.

Cabral (2016) investigated 72 students (9 years) of six classes of fourth-grade students from three different schools. This study involved three phases: a pretest, sequences of activities, and a post-test. During the pretest phase, each student was asked, individually, to classify nine pictures, with the purpose of determining if they were capable of creating an adequate criterion for classification.

To carry out the intervention processes, different types of activities were used involving the ability to classify proposals in textbooks and previous research: (1) classifying based on a given criterion, (2) discovering the criterion used in a classification, (3) presenting the criterion and asking students to analyse the pertinence of the classes, (4) listing properties of the elements, (5) analysing whether elements in a class belong or not, (6) identifying classes based on a criterion, and (7) creating a classification criterion. For the accomplishment of the sequence of activities, the students were divided into three groups. Two classes (G1) participated in a sequence of teaching activities that involved understanding only one criterion, based on situations 1, 2, and 3. Another two classes (G2) participated in a sequence of teaching activities that involved an understanding of element, class, and criteria, based on situations 4, 5, and 6. Two classes (G3) had no teaching activities, thus serving as control group. The sequences of activities were carried out in class during regular class time, involving all the students, over 2 days, for approximately two hours each

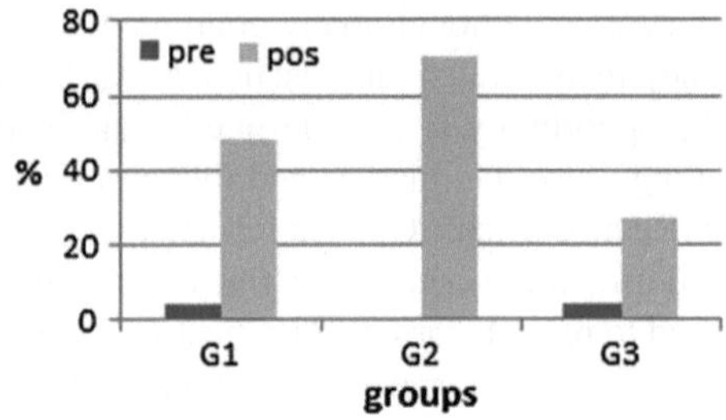

Fig. 8.8 Percentage of correct answers in each group in the pre- and post-tests

day. After students had done the activities, they were organized in pairs, and together with the teacher, the whole groups would once again analyse their answers, reflecting on how appropriate they were.

During the post-test, the students were asked once again to classify another nine pictures that did not present an explicit or common criterion. The purpose was to identify whether the students had learned to create a criterion and classify correctly.

8.9 Results

In the pretest, the performance of students in all three groups when creating a classification criterion was very weak, and no group presented significantly different performance from the others $F[(2.71)=0.419, p=0.659]$ through an analysis of variance.

In Fig. 8.8, we present the percentage of correct answers in each group in the pre- and post-tests. We can see that the performance of groups G1 and G2, which took part in the sequences of learning activities, improved quite a bit between the pre- and post-tests. Albeit much lower, G3 also presented improvement $F[(2.71)=4.702, p=\leq 0.012]$. We believe these results to be quite important, since they show how easy it seems for children to learn how to classify when they are systematically stimulated to do so.

In this way, the two sequences of activities allowed learning, emphasizing that systematic work with students on what it means to classify is possible and fundamental. It is also possible to affirm that the two intervention proposals led to learning and that there is no significant difference between them; that is, both types of activity allowed an advancement in understanding what it is to create classification criteria.

The variety of criteria used by students in the post-test is also interesting (Fig. 8.9). We stress that it is our intent that students create criteria and not use the habitual classifications. Why is this so important? If we want students to be able to research, collecting data to answer their questions, it is fundamental that they know how to organize the information they collect. Knowing how to create classification criteria will allow them to group answers according to their objectives and thus develop answers.

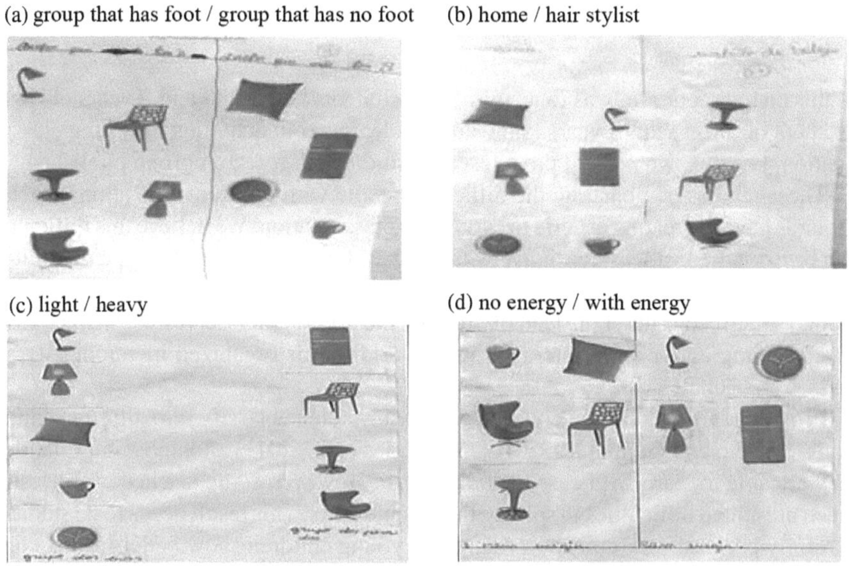

Fig. 8.9 Variety of criteria used by students in the same task (**a** group that has foot/group that has no foot; **b** home/hair stylist; **c** light/heavy; **d** no energy/with energy)

In this type of activity, the majority of the students (83%) presented an excellent performance.

In another study, Leite et al. (2013) investigated the knowledge of 30 third (aged 8) and fifth (aged 10) grade elementary school students creating categories to classify, and a potential improvement in performance based on a teaching intervention. This intervention was carried out over 2 days, during which it was proposed that students, in pairs, classify groups of pictures, then present these to the class, and, together with the teacher, reflect on the criteria used. The results show that students in both grades had difficulty classifying; however, the reflections in class allowed significant improvement in their understanding of classifying, showing the possibility of quick learning.

Thus, we observe the importance of proposing trajectory of work to promote the understanding of classification through different activities. We can also say that the activities imply different abilities, since the performance was different for the students. However, it is the set of them that supports the development of students' knowledge of classification. For the construction of knowledge, a variety of situations involving different skills is necessary. Finally, it is important to mention that the work of the teacher plays a fundamental role in understanding the classification of students.

8.10 Conclusions

In this chapter, our aim was to analyse what students between 5 and 9 years old and teachers of those school years know and can learn about activities involving classification. For this, we present three research studies our research group published.

These studies demonstrate the difficulty people with different educational backgrounds have in creating criteria to carry out a classification. We believe this difficulty can be explained, at least partially, by the lack of familiarity with this type of ability, both in everyday life and at school, since students are generally asked to classify from pre-defined criteria, instead of creating the criteria themselves. However, from a very young age, people interact with a world that is organized hierarchically in classes and subclasses.

From kindergarten, some children are already able to classify into different situations (Barreto and Guimarães, 2016; Guimarães et al., 2011) and, most importantly, they are able to learn with ease (Cabral, 2016). However, some teachers we investigated presented difficulties in solving these same tasks. This result needs to be studied carefully because it is rather surprising that young children show better results than adults do. Perhaps, children are more likely to create criteria because they feel freer to create categories that are not standardized. Adults, in some situations, use common categories in schools and textbooks. Guimarães and Gitirana (unpublished) observed that some graduate students in mathematics education also presented difficulties in creating classification criteria. Thus, studies on how to promote adult learning (teachers or others) need to be carried out so that this issue can be better analysed. Probably, if children and adults have opportunities that lead them to think about the classification, they learn easily, thereby showing the important role of the school.

Thus, the results provide evidence of the possibility of stimulating the learning of students when encouraged to reflect on classification. In addition, we believe that the different types of activities presented here will allow teachers to diversify classroom work by seeking to develop their students' ability to classify.

If current teachers have difficulty classifying, how will they be able to teach their students? The process of training teachers needs to lead them to systematic reflection that allows them to learn to create criteria to classify a given group of objects, respecting both exhaustiveness and exclusiveness. It needs to go beyond activities in which classes are already defined and where the student is only expected to distribute the elements. Developing students' independence in creating classifications will allow them to classify and analyse whatever data they wish, be it in school or in their daily lives, in a relevant manner.

It is fundamental to citizenship that everybody knows how to analyse the criteria chosen for a classification and how to create criteria to classify a set of data they wish to analyse. Knowing to create classification criteria will allow the children to participate in the universe of research, making it possible for them to become autonomous decision makers.

References

Amorim, N., & Guimarães (2017). Statistics education in textbooks: Brazil's National Textbook Program and the Teachers' Manuals. In *Proceedings of International Conference on Mathematics Textbook research and development—ICMT*, Rio de Janeiro.

Barreto, M., & Guimarães, G. (2016). Estratégias utilizadas por crianças da educação infantil para classificar. *Revista Iberoamericana de Educação Matemática e Tecnológica EM TEIA, v7*(1).

Batanero, C., & Diaz, C. (2005). El papel de los proyectos en la enseñanza y aprendizaje de la estadística. *Anais do I Congresso de Estatística e Investigação Operacional da Galiza e Norte de Portugal.*

Ben-Zvi, D., & Amir, Y. (2005). How do primary school students begin to reason about distributions? In K. Makar (Ed.), Reasoning about distribution: A collection of current research studies. *Proceedings of the Fourth International Research Forum on Statistical Reasoning, Thinking, and Literacy (SRTL-4)*, University of Auckland, New Zealand, 2–7 July, 2005. Brisbane, University of Queensland.

Bonthoux, F., & Kalénine, S. (2007). Preschoolers' superordinate taxonomic categorization as a function of individual processing of visual vs. contextual/functional information and object domain. *Cognitie Creier Comportament, 11,* 713–731.

Cabral, P. (2016). *A classificação nos anos iniciais do ensino fundamental.* Dissertação (mestrado) - Programa de Pós-graduação em Educação Matemática e Tecnológica - Universidade Federal de Pernambuco.

Clements, D. H. (2003). Teaching and learning geometry. In J. Kilpatrick, W. G. Martin, & D. Schifter (Eds.), *Research companion to principles and standards for school mathematics.* NCTM: Reston, VA.

Deák, G., & Bauer, P. J. (1995). The effects of task comprehension on preschoolers' and adults' categorization choices. *Journal of Experimental Child Psychology, 60,* 393–427.

Fielding-Wells, J. (2010). Linking problems, conclusions and evidence: Primary students' early experiences of planning statistical investigations. In *Proceedings of the Seventh International Conference on Teachings Statistics—ICOTS 8*, Slovenia.

Gal, I., & Garfield, J. (Eds.). (1997). *The assessment challenge in statistics education.* International Statistical Institute: Amsterdam.

Gattuso, L. (2011). L'enseignement de la statistique: où, quand, comment, pourquoi pas? *Statistique et Enseignement, 2*(1), 5–30.

Gitirana, V. (2014). A pesquisa como eixo estruturador da educação estatística. In: *Pacto Nacional pela Alfabetização na Idade Certa: Educação Estatística*/Ministério da Educação, Secretaria de Educação Básica, Diretoria de Apoio à Gestão Educacional. Brasília: MEC, SEB.

Guimarães, G., & Borba, R. (2007). Professores e graduandos de pedagogia valorizam e vivenciam processos investigativos? *Revista Tópicos Educacionais, 17,* 61–90.

Guimarães, G., & Oliveira, I. (2014). Does future teachers of primary school know how to classify? In *Proceeding of 38th Psychology of Mathematics Education (PME 38)*, Vancouver, Canada from July 15 to July 20.

Guimarães, G., Gitirana, V., & Roazzi, A. (2003). Interpretar e construir gráficos de barras: o que sabem os alunos de 3ª série do ensino fundamental. *Anais do XI Congresso Interamericano de Educação Matemática—CIAEM*, Blumenau.

Guimarães, G., Luz, P., & Ruesga, P. (2011). Classificar: uma atividade difícil para alunos e professores dos anos iniciais do Ensino Fundamental? *Anais do XIII Conferência Interamericana de Educação Matemática—CIAEM*, Recife/ Brasil.

Hancock, C., Kaput, J., & Goldsmith, L. (1992). Authentic inquiry with data: critical barriers to classroom implementation. *Educational Psychologist, 27*(3), 337–364. (Lawrence Erlbaum associates, Inc.).

Kalénine, S., & Bonthoux, F. (2006). The formation of living and non-living superordinate concepts as a function of individual differences. *Current Psychology Letters: Behaviour, Brain, & Cognition, 19*(2).

Leavy, A. & Sloane, F. (2017). Insights into the approaches of young children when making informal inferences about data. In *Proceedings of 10th Congress of European research in Mathematics Education—CERME 10*, Dublin.

Lecointre, G. (Ed.). (2004). *Comprendre et enseigner la classification du vivant*. Paris: Belin.

Lehrer, R., & Schauble, L. (2000). Inventing data structures for representational purposes: Elementary grade students' classification models. *Mathematical Thinking and Learning, 2*(1&2), 51–74.

Leite, M., Cabral, P., Guimarães, G., & Luz, P. (2013). *O ensino de classificações e o uso de tabelas*. Trabalho de Conclusão do Curso de Pedagogia na Universidade Federal de Pernambuco.

Makar, K., & Rubin, A. (2009). A framework for thinking about informal statistical inference. *Statistics Education Research Journal, 8*(1), 82–105.

Mareschal, D., & French, R. (2000). Mechanisms of categorization in infancy. *Infancy, 1*(1993), 59–76.

Nguyen, S. P., & Murphy, G. L. (2003). An apple is more than a fruit: Cross-classification in children's concepts. *Child Development, 74,* 1–24.

Piaget, J., & Inhelder, B. (1983). *Gênese das Estruturas Lógicas Elementares.* 3ª Ed. Rio de Janeiro: Zahar Editores.

Ponte, J. P., Brocardo, J., & Oliveira, H. (2003). *Investigações Matemáticas na sala de aula*. Autêntica: Belo Horizonte.

Rodrigues, M. P. (2016). *Identificar propriedades em quadriláteros – um caminho para a classificação inclusiva*. Porto, Portugal: Anais do Seminário de Investigação em Educação Matemática - XXVII SIEM.

Silva, E., & Guimarães, G. (2013). *Perspectivas para o ensino da Educação Estatística*. Curitiba: XI ENEM—Encontro Nacional de Educação Matemática—Anais.

Vergnaud, G. (1991). *El Niño, Las Matemáticas y La Realidad: Problemas de La enseñanza de Las Matemáticas en La Escuela Primaria*. México Trillas.

Vieillard, S. & Guidetti, M. (2009). Children's perception and understanding of (dis)similarities among dynamic bodily/facial expressions of happiness, pleasure, anger, and irritation. *Journal of Experimental Child Psychology*, 102, 78–95.

Ware, E. (2017). Individual and developmental differences in preschoolers' categorization biases and vocabulary across tasks. *Journal of Experimental Child Psychology, 153,* 35–56.

Wild, C., & Pfannkuch, M. (1999). Statistical thinking in empirical enquiry. *International Statistical Review, 67*(3), 223–265.

Part III
Teaching Statistics and Probability: Curriculum Issues

Chapter 9
Unpacking Implicit Disagreements Among Early Childhood Standards for Statistics and Probability

Randall E. Groth

Abstract Numerous recent curriculum documents around the world recommend that children begin to develop understanding of probability and statistics during early childhood and primary school. Although there is widespread agreement that such learning should occur, standards documents are not uniform in their specific recommendations. In particular, there are implicit disagreements about the roles of student-posed statistical questions, probability language, and variability in children's learning. Unpacking these implicit disagreements is in the interest of teachers, researchers, and curriculum developers because it can stimulate thought and debate about the proper emphasis for the concepts in standards documents. This chapter will help define the space for such thought and debate by summarizing how some key concepts are addressed differently in various early learning standards for probability and statistics. Defensible interpretations of the research literature are considered. Strategies teachers and curriculum developers can use to cope with situations in which standards documents conflict with desirable learning goals for children are also described. Boundary objects, which allow related communities of practice to operate jointly in absence of consensus, are discussed as a means for advancing teaching and research despite the existence of disagreement. Suggestions for working toward a greater degree of consensus across early childhood standards for statistics and probability are also offered.

9.1 Introduction

Statistics is relatively new to early childhood and primary curricula. The first recommendations to include statistics in school mathematics appeared in the first half of the twentieth century and focused on secondary school; it was not until the second half of the twentieth century that statistics appeared in curriculum recommendations for early childhood and primary level students as well (Jones & Tarr, 2010). The

R. E. Groth (✉)
Salisbury University, Salisbury, MD, USA
e-mail: regroth@salisbury.edu

A. Leavy et al. (eds.), *Statistics in Early Childhood and Primary Education*, Early Mathematics Learning and Development,
https://doi.org/10.1007/978-981-13-1044-7_9

standards documents of the National Council of Teachers of Mathematics (NCTM, 1989, 2000) were groundbreaking in recommending specific statistical concepts for study during the early years of school. At the outset of the twenty-first century, it is now unusual for school mathematics standards documents *not* to include statistics in the course of study for the youngest students.

With the proliferation of standards documents that include statistics, implicit disagreements have arisen in the concepts recommended for study and their sequencing. These disagreements can be seen as one compares standards from various nations (Jones, Langrall, & Mooney, 2007) or even those within a single nation such as the USA (Dingman, Teuscher, Newton, & Kasmer, 2013). Such disagreements can be counterproductive because they may make research and development efforts in statistics curriculum and instruction more difficult. Research and curricula from one setting may have limited use in another if the learning standards governing each setting differ. Hence, disagreements among standards documents have some negative aspects.

Although disagreements are often seen in a negative light, they do not have to be. Discourse devoted to respectfully expressing and unpacking disagreements can lead to deeper examination of different positions even if consensus is not obtained (Matusov, 1996). Since standards documents are usually written by separate groups that do not always have direct contact with one another, discourse about curricular disagreements can be limited. In this chapter, I aim to create a space in which disagreements about curriculum standards for early childhood and primary statistics are made explicit and then respectfully analyzed. I also consider steps that can be taken to support early statistics education in the absence of consensus on curriculum standards.

9.2 Scope of the Chapter

This chapter is not an exhaustive treatment of all points of disagreement among all curriculum documents. Instead, it deals with three salient issues at the foundation of early statistics education: the posing of statistical questions, the development of probability language, and the study of variation. In order to illustrate the nature of disagreement in regard to each issue, I compare recommendations from several standards documents (Table 9.1), including: *Principles and Standards for School Mathematics* (PSSM, NCTM, 2000), *Guidelines for Assessment and Instruction in Statistics Education* (GAISE, Franklin et al., 2007), *Common Core State Standards for School Mathematics* (CCSSM, Common Core State Standards Initiative (CCSSI), 2010b), Turnonccmath.net bridging standards (Confrey et al., 2012), the New Zealand Curriculum (Ministry of Education, 2014), the Australian Curriculum (Australian Curriculum, Assessment, & Reporting Authority (ACARA), 2015), and the National Curriculum in England (Department for Education (DfE), 2013). Although this sample is not inclusive of all curriculum standards around the world, these documents collectively bring to light the important differences in approaches to early childhood and primary statistics. Several of these differences are summarized in Table 9.2 and subsequently explored in this chapter.

9.3 Statistical Questions

All of the curriculum standards documents shown in Tables 9.1 and 9.2 recommend the posing of statistical questions during the early years of school. However, they differ in regard to the types of questions students are to pose. Some documents prescribe specific types of questions to be posed, whereas others are more open. The documents also differ in their portrayal of the nature and purpose of statistical questioning.

Some standards documents espouse a relatively narrow perspective on posing statistical questions in the early grades. The CCSSM include just one explicit mention of student-posed statistical questions in the measurement and data strand for Grades

Table 9.1 Standards documents considered in this chapter and their sections for early statistics

	Document sections for early statistics education	Explanation of section content
PSSM (NCTM, 2000)	Grades Pre-K-2 data analysis and probability standard	Recommendations for children younger than 5 (Pre-K) through approximately age 8 (Grade 2)
GAISE (Franklin et al., 2007)	Level A	Beginning level in GAISE Pre-K-12 report; precise grade levels/ages not specified
CCSSM (CCSSI, 2010b)	Grades K-5 measurement and data strand	Recommendations for children approximately 5 years old (K) through approximately age 11 (Grade 5)
Turnonccmath.net bridging standards (Confrey et al., 2012)	Grades K-5 elementary data and modeling	Written to enhance the teaching of the CCSSM Grades K-5 Measurement and Data strand
New Zealand Curriculum (Ministry of Education, 2014)	Years 1–3 statistics strand of mathematics and statistics learning area	Learning objectives to be accomplished after 1, 2, and 3 years in school
Australian Curriculum (ACARA, 2015)	Foundation Year—year 2 statistics and probability strand of mathematics learning area	Proficiencies to be attained by children approximately 5–8 years of age
National Curriculum in England (DfE, 2013)	Key Stage 1—years 1–2 statistics portion (starting year 2) of mathematics program of study	Prescribed program of study for children approximately 5–7 years of age

Table 9.2 Summary of recommendations across standards documents

	Statistical questions	Probability language	Variability
PSSM (NCTM, 2000)	Teachers encourage children to ask questions about their experiences and develop ways to gather data (data analysis and probability standard)	Earliest emphasis on distinguishing among *likely, unlikely, more likely*, and *less likely* (Pre-K-2 data analysis and probability standard)	No explicit recommendations for the types of variability to be encountered during first years of school
GAISE (Franklin et al., 2007)	Children pose questions about contexts of interest with teachers' help (Level A)	Development of a continuum from *impossible* to *certain*, with *less likely, equally likely*, and *more likely* lying in between (Level A)	Children's early experiences should encompass measurement, natural, and induced variability (Level A)
CCSSM (CCSSI, 2010b)	Children ask questions about the total number of points in a data set, how many in each category, and how numbers in categories compare (Grade 1 measurement and data standard)	No explicit recommendations for development of probability language	Variability is not explicitly mentioned until Grade 6, where the focus is on quantifying it with formal statistical measures
Turnonccmath.net bridging standards (Confrey et al., 2012)	Children pose questions of interest to launch statistical investigations (Grade 1)	No explicit recommendations until Grade 7	In Grades 2 and 3, children should work with natural, experimental, and measurement-related variation
New Zealand Curriculum (Ministry of Education, 2014)	Children engage in statistical enquiry cycle, which includes statistical questions, with support in years 1–3	Children describe likelihoods with everyday language in Year 2, language such as *most likely* and *least likely* used in Year 3	No explicit recommendations for the types of variability to be encountered during first years of school
Australian Curriculum (ACARA, 2015)	Children pose simple questions of interest; emphasis on categorical variables through year 3	In Year 1, children use *will happen, won't happen*, and *might happen*; In Year 2, likely, *unlikely, certain*, and *impossible* are used	No explicit recommendations for the types of variability to be encountered during first years of school
National Curriculum in England (DfE, Department for Education 2013)	Children ask questions that require counting and comparing the numbers of objects in different categories (Key Stage 1, Years 1–2)	No explicit recommendations at Key Stages 1 and 2	No explicit recommendations for types of variability to be encountered during first years of school

K-5. It appears in Grade 1, where students are to, "Organize, represent, and interpret data with up to three categories; ask and answer questions about the total number of data points, how many in each category, and how many more or less are in one category than in another" (CCSSI, 2010b, p. 16). The National Curriculum in England takes a similar approach, stating that children should "ask and answer questions about totaling and comparing categorical data" (DfE, 2013, p. 16) during Key Stage 1.

Other standards documents put less limitations on the types of statistical questions young children are to pose. The Australian Curriculum, for example, is more restrictive only during the Foundation Year, stating that children should "answer yes/no questions to collect information and make simple inferences" (ACARA, 2015, p. 9). It prescribes work with categorical variables during Years 2 and 3, but it does not specify the types of questions children are to ask about the data after the Foundation Year. PSSM, GAISE, and the Turnonccmath.net bridging standards are even more open in their recommendations. These three documents emphasize the importance of having children choose questions of interest. Although children's questions may often involve categorical variables, there is no recommendation to limit their questions to those types of variables. GAISE does, however, emphasize the importance of teacher guidance in helping students select appropriate questions for investigation during level A.

In some cases, standards documents portray the posing of statistical questions as part of an iterative investigative cycle that includes activities such as gathering data, constructing representations, drawing inferences, and perhaps revising the original question. All of the recommendations in the GAISE document are situated within this type of cycle. The Turnonccmath.net bridging standards and New Zealand Curriculum situate questioning within a statistical inquiry cycle as well. These types of documents characterize question-posing as part of an overall process for conducting statistical investigations. Participating in this process brings the work students do in the classroom closer to what statisticians do during the course of professional practice.

Reviewing the range of recommendations for statistical question-posing, some readers might be inclined to criticize overly restrictive curriculum standards that do not prioritize students' interests and do not situate question-posing within an investigative cycle. Before making such judgments, however, it is important to consider what is known about children's tendencies in posing questions and teachers' abilities to support them. When prompted to pose questions about a given situation, children's initial questions may be too ambiguous to yield useful data; in such cases, they often need help to make the questions more precise (Russell, Schifter, & Bastable, 2002). It takes considerable skill for teachers to help students transform such questions into manageable ones. Research suggests that such skill may be elusive for teachers, as they can exhibit many of the same statistical difficulties as younger students (Groth, 2007) and may have trouble formulating interesting statistical questions of their own (Heaton & Mickelson, 2002). Given these circumstances, one might argue that it is appropriate to constrain the types of classroom questions to those which teachers and students are likely to be able to manage.

On the other hand, if one believes that standards documents should set forth aspirational goals rather than those that may be more immediately achievable, then setting ambitious goals for the range of questions to be asked and situating question-posing within a close approximation of statistical practice are advisable. In general, it is best to avoid the tendency to impose ceilings on young children's thinking, as they frequently show the ability to exceed adults' expectations (Moss, Bruce, & Bobis, 2016). Additionally, prompting students to pose enticing questions about a context of interest can encourage them to persevere through all phases of a statistical investigation (Konold & Higgins, 2003). As children pose questions, teachers need to be ready to help them form questions that can be addressed with data and avoid those that are too broad, inadequate, or produce too much data (English, 2014). Preparing teachers to support statistical question-posing in such a manner is a non-trivial task (Franklin et al., 2015), so standards for children's learning should not be developed without considering teacher preparation. Teachers must learn to focus children's attention on both the process of inquiry and the statistical content to be addressed to scaffold children's abilities to pose statistical questions (Fielding-Wells, this volume). Ideally, teacher preparation and standards documents for students would be developed and supported in tandem so that the written curriculum set forth in ambitious standards documents has a greater chance of becoming the enacted curriculum (Stein, Remillard, & Smith, 2007) in classrooms.

9.4 Probability Language

Standards documents can be separated into two broad groups in their recommendations for young children's development of probability language. One group makes no explicit recommendations for probability language development, and the other group does. In the latter group, there is not uniform agreement about how probability language should initially be developed. The lack of agreement across documents raises a number of issues to consider.

The CCSSM, the National Curriculum in England, and the Turonccmath.net bridging standards exemplify documents that contain no explicit recommendations for young children's development of probability vocabulary. The Turnonccmath.net bridging standards do address the development of probability vocabulary starting in Grade 7, but do not speak to the issue in the earlier grades. In the Turnonccmath.net bridging standard for Grade 7, students are to learn language to express certain events, impossible ones, and those whose probabilities lie in between. In the same grade level, they are to learn to quantify probabilities. The bridging standard is intended to help students prepare to succeed in their work with the CCSSM probability content standards, which also first appear in Grade 7.

PSSM and the New Zealand Curriculum take the approach of focusing on initially developing young children's use of everyday language to describe probabilities. In the Pre-K-2 portion of PSSM, this includes developing children's use of *likely* as a standalone word to describe the chance of an event and adding qualifiers to form everyday phrases such as *more likely* and *less likely*. Children are also to encounter *impossible* events and describe them as such. In PSSM, the use of everyday terms is to precede quantification of probabilities. PSSM recommend that calculations of exact probabilities occur in the later grades.

GAISE and the Australian curriculum also seek to leverage children's experiences with everyday language to describe probabilities, but take a slightly different approach. The earliest levels in both of these documents explicitly refer to everyday language that can be used to describe both ends of the probability scale (0 and 1) as well as terms in between. According to GAISE, at Level A, "Events should be seen as lying on a continuum from impossible to certain, with less likely, equally likely, and more likely lying in between" (Franklin et al., 2007, p. 33). Level A students are also to informally assign numerical probabilities to events corresponding to these terms. The Australian curriculum is also designed to help young children begin to think about both ends of the probability scale, but includes a scaffold not present in GAISE. During Year 1, Australian children are to use *will happen, won't happen*, and *might happen* to describe events. During Year 2, these terms *certain, impossible, likely*, and *unlikely* come to the forefront, presumably under the assumption that *will happen* is more understandable than *certain*, and *won't happen* more so than *impossible*. Terms associated with these key points on the probability scale continuum are included in other curriculum documents as well, but not necessarily at the earliest levels.

Examining the disparate recommendations for probability language development in light of existing literature is informative. One of the robust findings of research is that vocabulary learning tends to take place over multiple word encounters (Leung, 2005) and even several high-quality encounters with words do not ensure learning (McKeown, Beck, Omanson, & Pople, 1985). Everyday encounters with probability vocabulary are not necessarily of high quality. The word *certain*, for example, has a colloquial use of describing something that is very likely to occur (Certain, n.d., n.p.). This use in the everyday register differs from the more precise meaning in the mathematical register. One might conjecture that the more encounters one has with the word in the everyday register, the more difficult it becomes to incorporate the meaning from the mathematical register into existing cognitive structures. If this is, in fact, the case, then postponing explicit attention to probability vocabulary until the later grades is not advisable. Young students need multiple opportunities to distinguish colloquial meanings of probability words from mathematical ones.

The literature contains multiple examples of how students at times struggle to use probability vocabulary in a manner resonant with conventional mathematical discourse. In one study, Fischbein, Nello, and Marino (1991) found that some students used the word *possible* to describe events that were certain to occur, and that many used *rare* to describe impossible events. Nacarato and Grando (2014) reported students' use of *less probable* to describe events that cannot occur, and their use of *improbable* to describe events that occur frequently. In some studies, students have used the phrase *50–50 chance* to describe events they believe to be possible (Watson, 2005) or to describe outcomes that do not have the same probability of occurring (Tarr, 2002). Given well-documented examples of this nature, teachers may find themselves trying to counteract meanings students bring to school for probability terms rather than leveraging them as an intermediary step toward quantifying probabilities. Early attention to the mathematical meanings of vocabulary heard in everyday language could help counteract this problem, making it difficult to defend standards documents that include no explicit early attention to probability vocabulary development.

Even among standards documents that do include attention to early probability language development, there are issues to resolve. The most pressing of these appears to be deciding on the most appropriate way to scaffold children's learning of the probability scale continuum. Different sections of the continuum may require different amounts and types of scaffolding. For example, Fischbein et al. (Fischbein et al. 1991) found that students were more successful using *impossible* than *certain* in a mathematical manner. Such a finding seems to support the PSSM approach of drawing students' attention to *impossible* events in Grades Pre-K-2 and delaying formal work with *certain* until Grades 3–5. The Australian Curriculum presents an interesting alternative, however, having students first work with *will happen* and *won't happen* rather than *impossible* and *certain* immediately. Under the Australian progression, children may come to see *certain* as a synonym for *will happen* and *impossible* as a synonym for *won't happen* with teachers' guidance. The extent to which the progression achieves this goal, and in the process counteracts students' difficulties dealing with *certain*, is an empirical question awaiting investigation.

In general, the discrepancies in recommendations for supporting children's learning of probability vocabulary suggest an array of questions in need of systematic research attention, such as:

- To what extent does explicit attention to probability vocabulary in the early grades contribute to probability learning in later years?
- How does children's learning of words and phrases representing the parts of the probability continuum generally progress?
- How might informal outside-of-school encounters with probability vocabulary interfere with or support formal learning?
- Which approaches to introducing probability vocabulary, reflected in different curriculum documents, are the most effective?

As investigations of such questions take place, perhaps a greater degree of uniformity will be achieved across documents.

9.5 Variability

Standards documents can be grouped in two categories in regard to their treatment of statistical variability: those that explicitly identify variability as an object of study for young children and those where experiences with variability may be incidental to work with other standards. Attending to the manner in which variability is treated is a core consideration in teaching statistics. A primary reason statistics exists as a discipline is to study the variability we see in everyday life. Snee (1999) stated,

"If there was no variation, there would be no need for statistics and statisticians" (p. 257). Likewise, Cobb and Moore (1997) argued that the need for statistics "arises from the omnipresence of variability" (p. 801). Variability is pervasive in data and distinguishes the study of statistics from the study of mathematics.

GAISE and the Turnonccmath.net bridging standards identify variability as a core object of study for young children and specify the types of variability they should encounter. GAISE recommends structuring curricula so that children encounter three types of variability at Level A: natural, measurement, and induced. Natural variability is encountered as children measure the same quantity across individuals, such as height, weight, or arm length, and observe that these measurements vary from one individual to the next. Measurement variability occurs when repeated measurements of the same thing vary due to characteristics of the measuring device or because of the system being measured. Induced variability occurs when intentional changes are made to a system to observe their effects. For example, planting one crop in a sunny area and another in a shady area may produce variability in yield from each one. The Turnonccmath.net bridging standards closely mirror GAISE in their treatment of variability, recommending the same three types of experiences starting in Grade 2. In both GAISE and the Turnonccmath.net bridging standards, priority is placed on having children work with the recommended types of variability but not necessarily learning the names for each type immediately.

GAISE and the Turnonccmath.net bridging standards are unusual in identifying specific types of variability students should encounter. The other standards documents shown in Table 9.1 do not explicitly identify types of variability to be studied by young children. For example, the word "variability" is not mentioned at all in the Grades K-5 CCSSM. Instead, the K-5 CCSSM data and measurement standards focus on constructing line plots, picture graphs, and bar graphs. After constructing these displays, students are to perform tasks such as finding the difference between the highest and lowest observation and determining how many more or less one category may contain than another. Although such activities may allow students to encounter different types of variability incidentally, the systematic treatment of different types of statistical variability is not prioritized.

Because variability is a core concept in statistics, it is worth considering how children whose curricula are not guided by documents such as GAISE might still have rich experiences with different types of variability. Although explicit mention of statistical variability might ultimately occur across more curriculum documents in the future, it is likely a long-term aspirational goal rather than something more readily attainable. Since standards documents are often criticized for containing too much content (Schmidt, McKnight, & Raizen, 1997), there is a natural tendency to resist adding to prescribed curricula. The content that ultimately does make its way into a standards document will also reflect the values and beliefs of the document writers. In some cases, standards document writers are driven by the desire to emphasize number and operation to a greater extent during the early years. This is sometimes done at the expense of de-emphasizing statistics. The first page of the CCSSM, for example, cites Ginsburg and Leinwand's (2009) argument that mathematics curriculum standards in higher achieving countries include less emphasis on data analysis in the early grades in favor of more attention to number, measurement, and geometry. This citation helps explain why statistics is de-emphasized in the early portions of the document, shifting much of the load to a compressed timeframe in the middle grades.

9.6 Boundary Objects

Given the lack of agreement about the content important for young children to study, creative ways to ensure rich variability experiences are needed. One theoretical construct that can be of assistance in this endeavor is the notion of *boundary object* (Star & Greismer, 1989). Boundary objects help different communities of practice operate collectively in the absence of consensus. In some cases, instructional plans can serve as boundary objects. For example, when teaching elementary students whose curriculum was driven by CCSSM, I created lessons that included the different types of variability identified in GAISE as they addressed the statistical graphing standards included in CCSSM (Groth, 2015). Careful attention to the contexts in which students produced graphs required in CCSSM helped ensure they would experience the different types of variability described in GAISE without adding requirements or extra time to the curriculum. Proponents of both CCSSM-like curricula and GAISE-like curricula can be satisfied with such a lesson sequence, even though their standards for early statistics differ substantially. As proponents of these different types of curricula discuss boundary objects like this lesson sequence, the prospects for greater consensus about early experiences in statistical variability may improve.

Boundary objects can also play roles in addressing the earlier-identified standards-related dilemmas for probability language and statistical questions. One possible approach to resolving these dilemmas is to look to content areas other than mathematics. For example, the teaching of probability vocabulary has natural connections to the language arts. Although it is not reasonable to expect language arts instruction to be guided by statistics standards, statistics educators can position themselves to collaborate on the design of lessons that meet elementary language arts standards such as, "Use precise language and domain-specific vocabulary to inform about or explain the topic" (CCSSI, 2010a, p. 20). Lessons that use words such as *certain* and *likely* as examples of domain-specific vocabulary could promote probability proficiency without adding extra language arts standards to the curriculum. In science, it is natural to pose questions that generate investigative cycles. Statistics educators can collaborate with science teachers to design classroom investigations that satisfy many existing science standards and simultaneously are motivated by rich statistical questions. Such collaborations with language arts and science teachers could ultimately provide avenues to expand the teaching of statistics and probability significantly beyond mathematics classes.

9.7 Conclusion

Disagreements about statistics standards for young children have a variety of sources. To conclude, I consider several sources of disagreement discussed in this chapter: beliefs about students' abilities, beliefs about teachers' abilities, robustness and influence of the research literature, and priorities for mathematics education in the early grades. In considering these sources, I also propose directions the field might take in order to provide high-quality statistics education for all young children even in a climate of disparate curricular recommendations.

Knowledge of students' and teachers' abilities should, to an extent, drive curriculum recommendations. For instance, recommendations to have students focus on categorical data when first posing statistical questions are reasonable from the standpoint that these may be among the most accessible types of questions for students in the early grades. However, including language only about categorical variables in a standard can have the effect of putting a ceiling on children's activities, even if the standard is meant only to be a minimum expectation. Because high-stakes assessments are often attached to standards, and there are many standards to address over the course of a school year, teachers tend to limit instruction to what is prescribed in the required standards (Breault, 2014). The highest priority for professional development then becomes learning to help students attain what is in the text of the standards, essentially putting a ceiling on teachers' growth as well.

To avoid putting ceilings on students' and teachers' growth, writers of standards documents can take a number of steps. One step would be to carefully phrase standards in a manner that identifies essential content but also encourages deeper study as opportunities arise. The PSSM document does so in its recommendations for children's posing of statistical questions, saying that such questions should arise from students' curiosity about the world around them. Of course, open recommendations of this nature put a greater burden on the teacher, who must skillfully handle unusual or unwieldy questions that students may pose. The presence of this greater burden suggests the desirability of forming standards for students and standards for teacher preparation in tandem, so that teachers might be better prepared to handle challenges that may come about as a result of more ambitious standards for students. At present, the two types of standards documents are usually written by separate groups and/or at different points in time; greater success might be realized by writing the two simultaneously.

Although more ambitious standards documents are desirable, simply having the goal of writing ambitious standards is not enough. The research community has considerable work to do to help guide the process. This is vividly illustrated by the current situation with standards for learning probability language. The disagreements in this area suggest a number of research questions in need of investigation, including: (i) To what extent does early instruction focused on probability language help improve students' probabilistic thinking and discourse throughout their years of school? (ii) How should children's experiences with probability language be sequenced? (iii) What kinds of scaffolding should teachers be prepared to furnish as children begin to distinguish between colloquial and technical meanings of probability vocabulary words and phrases? Answers to questions of this nature can help guide the formation of learning expectations for students and professional development goals for teachers. Along with conducting such studies, researchers need to be conscious of presenting their findings in venues and formats likely to be accessed and understood by writers of standards documents.

Of course, research studies will not resolve all disagreements because research is invariably interpreted in different ways by different individuals (Sierpinska & Kilpatrick, 1998). Some differences in interpretation and use of research stem from different priorities and beliefs about what is important in early childhood mathematics education. Overcoming such philosophical differences may ultimately prove to be the greatest challenge in providing quality early childhood experiences in regard to statistical questions, probability language, and variability. Achieving uniform consensus is not likely. However, boundary objects (Star & Greismer, 1989) allow groups with different beliefs to operate collectively even in absence of consensus. As noted earlier, one promising direction for the creation of boundary objects is designing lesson sequences and tasks that satisfy multiple standards documents simultaneously without over-burdening the curriculum. Designing, implementing, and analyzing such lessons can provide space for collective work among those holding different beliefs about the appropriate focus for the early study of statistics. Even if this collective work does not move individuals toward complete consensus, it can prompt deeper consideration of the beliefs and positions they hold (Matusov, 1996). As beliefs and positions are re-examined, a foundation is formed for well-constructed recommendations for children's and teachers' statistical learning.

References

Australian Curriculum, Assessment, and Reporting Authority. (2015). *Mathematics: Sequence of content F-6*. Retrieved from https://acaraweb.blob.core.windows.net/resources/Mathematics_-_Sequence_of_content.pdf.

Breault, R. (2014). Mining the potential in your state standards. *Kappa Delta Pi Record, 50*(1), 4–8.

Certain. (n.d.). In *Oxford's living dictionary*. Retrieved from https://en.oxforddictionaries.com/definition/certain.

Cobb, G., & Moore, D. (1997). Mathematics, statistics, and teaching. *The American Mathematical Monthly, 104*(9), 801–823.

Common Core State Standards Initiative. (2010a). *Common core state standards for English language arts*. Retrieved from http://www.corestandards.org.

Common Core State Standards Initiative. (2010b). *Common Core state standards for mathematics*. Retrieved from http://www.corestandards.org.

Confrey, J. Nguyen, K. H., Lee, K., Panorkou, N., Corley, A. K., & Maloney, A. P. (2012). *TurnOnCCMath.net: Learning trajectories for the K-8 Common Core State Math Standards*. Retrieved from https://www.turnonccmath.net.

Department for Education. (2013). *Mathematics programmes of study: Key stages 1 and 2*. Retrieved from https://www.gov.uk/government/uploads/system/uploads/attachment_data/file/335158/PRIMARY_national_curriculum_-_Mathematics_220714.pdf.

Dingman, S., Teuscher, D., Newton, J. A., & Kasmer, L. (2013). Common mathematics standards in the United States: A comparison of K-8 and Common Core standards. *The Elementary School Journal, 113*(4), 541–564.

English, L. (2014). Statistics at play. *Teaching Children Mathematics, 21*(1), 36–44.

Fischbein, E., Nello, M. S., & Marino, M. S. (1991). Factors affecting probabilistic judgements in children and adolescents. *Educational Studies in Mathematics, 22*(6), 523–549.

Franklin, C., Bargagliotti, A. E., Case, C. A., Kader, G. D., Scheaffer, R. L., & Spangler, D. A. (2015). *Statistical education of teachers*. Alexandria, VA: American Statistical Association.

Franklin, C., Kader, G., Mewborn, D., Moreno, J., Peck, R., Perry, M., et al. (2007). *Guidelines for assessment and instruction in statistics education (GAISE) report: A PreK-12 curriculum framework*. Alexandria, VA: American Statistical Association.

Ginsburg, A., & Leinwand, S. (with Decker, K.). (2009, December). *Informing grades 1–6 mathematics standards development: What can be learned from high-performing Hong Kong, Korea, and Singapore?* Washington, DC: American Institutes for Research. Retrieved from http://www.air.org/sites/default/files/downloads/report/MathStandards_0.pdf.

Groth, R. E. (2007). Toward a conceptualization of statistical knowledge for teaching. *Journal for Research in Mathematics Education, 38*(5), 427–437.

Groth, R. E. (2015). Royalty, racing, rolling pigs, and statistical variability. *Teaching Children Mathematics, 22*(4), 218–228.

Heaton, R. M., & Mickelson, W. T. (2002). The learning and teaching of statistical investigation in teaching and teacher education. *Journal of Mathematics Teacher Education, 5*(1), 35–59.

Jones, D., & Tarr, J. E. (2010). Recommendations for statistics and probability in school mathematics over the past century. In B. J. Reys & R. E. Reys (Eds.), *Mathematics curriculum issues, trends, and future directions: Seventy-second yearbook of the National Council of Teachers of Mathematics* (pp. 65–75). Reston, VA: National Council of Teachers of Mathematics.

Jones, G. A., Langrall, C. W., & Mooney, E. S. (2007). Research in probability: Responding to classroom realities. In F. K. Lester Jr. (Ed.), *Second handbook of research on mathematics teaching and learning* (pp. 909–955). Charlotte, NC: Information Age & National Council of Teachers of Mathematics.

Konold, C., & Higgins, T. (2003). Reasoning about data. In J. Kilpatrick, W. G. Martin, & D. Schifter (Eds.), *A research companion to principles and standards for school mathematics* (pp. 193–215). Reston, VA: National Council of Teachers of Mathematics.

Leung, C. (2005). Mathematical vocabulary: Fixers of knowledge or points of exploration? *Language and Education, 19*(2), 127–135.

Matusov, E. (1996). Intersubjectivity without agreement. *Mind, Culture, and Activity, 3*(1), 25–45.

McKeown, M., Beck, I., Omanson, R., & Pople, M. (1985). Some effects of the nature and frequency of vocabulary instruction on the knowledge and use of words. *Reading Research Quarterly, 20*(5), 522–535.

Ministry of Education. (2014). *The New Zealand Curriculum online: Mathematics and statistics*. Retrieved from http://nzcurriculum.tki.org.nz/The-New-Zealand-Curriculum/Mathematics-and-statistics/Achievement-objectives.

Moss, J., Bruce, C. D., & Bobis, J. (2016). Young children's access to powerful mathematics ideas: A review of current challenges and new developments in the early years. In L. D. English & D. Kirshner (Eds.), *Handbook of international research in mathematics education* (3rd ed., pp. 153–190). New York: Routledge.

Nacarato, A. M., & Grando, R. C. (2014). The role of language in building probabilistic thinking. *Statistics Education Research Journal, 13*(2), 93–103. Retrieved from http://iase-web.org/documents/SERJ/SERJ13(2)_Nacarato.pdf.

National Council of Teachers of Mathematics. (1989). *Curriculum and evaluation standards for school mathematics*. Reston, VA: Author.

National Council of Teachers of Mathematics. (2000). *Principles and standards for school mathematics*. Reston, VA: Author.

Russell, S. J., Schifter, D., & Bastable, V. (2002). *Working with data casebook*. Parsippany, NJ: Dale Seymour Publications.

Schmidt, W. H., McKnight, C. C., & Raizen, S. A. (1997). *A splintered vision: An investigation of U.S. science and mathematics education*. Boston, MA: Kluwer Academic Press.

Sierpinska, A., & Kilpatrick, J. (Eds.). (1998). *Mathematics education as a research domain: A search for identity*. Dordrecht, The Netherlands: Kluwer.

Snee, R. (1999). Discussion: Development and use of statistical thinking: A new era. *International Statistical Review, 67*(3), 255–258.

Star, S. L., & Greismer, J. R. (1989). Institutional ecology, 'translations', and boundary objects: Amateurs and professionals in Berkeley's Museum of Vertebrate Zoology, 1907–39. *Social Studies of Science, 19*(3), 387–420. https://doi.org/10.1177/030631289019003001.

Stein, M. K., Remillard, J., & Smith, M. S. (2007). How curriculum influences student learning. In F. K. Lester, Jr. (Ed.), *Second handbook of research on mathematics teaching and learning* (pp. 319–369). Reston, VA: National Council of Teachers of Mathematics; Charlotte, NC: Information Age.

Tarr, J. E. (2002). The confounding effects of "50-50 chance" in making conditional probability judgments. *Focus on Learning Problems in Mathematics, 24*(4), 35–53.

Watson, J. M. (2005). The probabilistic reasoning of middle school students. In G. A. Jones (Ed.), *Exploring probability in school: Challenges for teaching and learning* (pp. 145–168). New York: Springer.

Chapter 10
Statistical Graphs in Spanish Textbooks and Diagnostic Tests for 6–8-Year-Old Children

Carmen Batanero, Pedro Arteaga and María M. Gea

Abstract Statistical graphs are complex semiotic tools requiring different interpretative processes of the graph components, in addition to the entire graph itself. Taking this assumption and hierarchies proposed in previous research as a starting point, in this chapter we analyse the content related to statistical graphs of the Spanish curricula, textbooks and external compulsory tests taken by 6–9-year-old children. We examine the types of graphs presented to the children, the activity demanded, the reading levels required from them, as well as the graph semiotic complexity and the task context. The examples and analysis will help understand the expected progression of children's learning of statistical graphs.

10.1 Introduction

Statistical graphs play an important role in summarising and communicating information and are widely used in the media and different curricular topics. Being able to read, interpret and detect biases in these graphs, as well as accurately constructing elementary graphs is considered a part of contemporary statistical literacy (Watson, 2006), which is the union of two related competences: (a) interpreting and critically evaluating statistically based information from a wide range of sources, and (b) formulating and communicating a reasoned opinion about such information (Gal, 2002).

In order to develop these skills in children, curricular guidelines include this content from the lower levels of primary school in many countries (e.g. National Governors Association Center for Best Practices & Council of Chief State School Officers, 2010; MECD, 2014; NCTM, 2000). One important step in the transformation from the intended curriculum—as stated in curricular guidelines—to the curriculum implemented in the classrooms is the written curriculum reflected in the textbooks, which often guides teachers in their final decisions about the activities carried out in the classroom. Another important factor influencing teaching is exter-

C. Batanero (✉) · P. Arteaga · M. M. Gea
University of Granada, Granada, Spain
e-mail: batanero@ugr.es

A. Leavy et al. (eds.), *Statistics in Early Childhood and Primary Education*, Early Mathematics Learning and Development,
https://doi.org/10.1007/978-981-13-1044-7_10

nal assessment, which is compulsory for Spanish children at specific school levels. Consequently, in this chapter we analyse the content related to statistical graphs of the Spanish curricula, textbooks and external compulsory tests in order to inform us about the expected progression in the learning of graphs by 6–8-year-old children.

10.2 Theoretical Framework

Bertin (1967) described a graph as a complex semiotic object: The graph itself and every component of the same are made up of signs that require a semiotic activity by those who interpret them. When reading the graphs, several translation activities between the graph as a whole, and each part of the graph should be performed. Each piece of information (numerical, pictorial, verbal or statistical information) obtained from a graph requires us to establish a correspondence between elements, subsets or sets of this graph. A reader has to perform three successive operations to extract the information in the graph:

- *External identification*, to find the conceptual and real-world referents that support the information contained in the graph (which components are being represented).
- *Internal identification* of relevant dimensions of variation in the graph pictorial content and the correspondence between the visual and conceptual dimensions and scales (which components are mapped to which visual variable).
- *Perception of the correspondence*, by which the reader uses the levels of each visual dimension to draw conclusions about the levels of each conceptual dimension.

Cleveland and McGill (1984) argue that the information in a graph is encoded (e.g. using position or size) and the person that reads the graph should decode that information, through a process of graphical perception, consisting of "visually decoding the information encoded in a graph" (p. 531). The authors identified the following basic graphic perception tasks that are carried out during the visual information decoding process: (a) determining the position of a point or element along a common scale; (b) determining the position when using two scales, for example, in the scatter plot; (c) determining length, direction and angle; (d) estimating an area; and (e) estimating a volume or curvature. In addition to the above competences, Bertin (1967) described the following levels of understanding in the critical reading of graphs:

- *Extracting data* or direct reading of the data represented on the graph. For example, in a bar graph, reading the frequency associated with a value of the variable.
- *Extracting trends*: being able to perceive a relationship between two subsets of data that can be defined a priori or visually in the graph. For example, visually determining the mode of a distribution in a bar graph.
- *Analysing the data structure*: comparing trends or clusters and making predictions. For example, in a grouped bar graph, analysing the differences in mean and range of two distributions.

A related classification was presented by Curcio (1989), who termed the three levels defined by Bertin as: *reading the data* (literal reading of the graph without interpreting the information contained in it), *reading between the data* (interpreting and integrating the data in the graph) and *reading beyond the data* (making predictions and inferences from the data to information that is not directly reflected in the graph). Shaughnessy, Garfield and Greer (1996) expanded the above classification by defining a new level of *reading behind the data,* which consists of judging the method of data collection, and assessing data validity and reliability, as well as the possible generalization of findings (see also Shaughnessy, 2007).

10.2.1 Graph Semiotic Complexity

When producing a graph, we need to perform a series of actions (such as deciding the particular type of graph or, fixing the scale), and therefore, we implicitly use some concepts (such as variable, value, frequency, range) and properties (e.g. proportionality between frequencies and length of bars in the bar graph) that vary in different graphs. We therefore should not consider the different graphs as equivalent representations of one same mathematical concept (the data distribution) but as different configurations of interrelated mathematical objects that interact with that distribution. Bertin (1981, pg. 15) suggested that, "the efficacy of a graphic construction is revealed by the level of question that receives an immediate response" and, therefore, considers a graph to be more effective when more complex questions can be answered from the same. Inspired by these ideas, Batanero, Arteaga, and Ruiz (2010) defined different levels in graphs semiotic complexity, as follows:

L1. Representing only individual results. When, in spite of having a complete data set, only a few cases are represented while other data are excluded from the graph. For example, the teacher provides a data set with information of all the children in the classroom and a child only displays his/her own data. In these graphs, we do not use the ideas of statistical variable or distribution.
L2. Representing all the individual values for one or several variables, without forming the distribution. When data on a graph are represented individually, without an attempt to order the data or to combine identical values. Consequently, on reading the graph these students neither need to interpret the frequency of the different values nor explicitly use the idea of distribution.
L3. Only one distribution in a graph. The graph represents a frequency distribution for only one variable; then, the ideas of frequency and distribution are used.
L4. Producing a joint graph for both distributions. This level corresponds to graphs representing the distributions for two or more variables. These graphs are the most complex, since they represent two different variables in the same frame.

10.3 Curricular Guidelines Related to Statistical Graphs and External Compulsory Assessment

In Spain, the study of statistics starts in the second cycle of kindergarten (4–5 year olds) in relation to the area "Knowledge of your environment", where mathematical abilities are included (MEC, 2007). For example in Andalusia, curricular guidelines at this level recommend selecting problem situations that interest children, in an effort to motivate the collection and organisation of data and reflect on the results of their analysis in order to come to a solution (Consejería de Educación de la Junta de Andalucía, p. 33, 2008).

Statistical graphs are included from the first grade of primary education in Spain (MECD, 2014), which propose compulsory content, assessment criteria and learning standards, for the entire period (grades 1–6, 6–11-year-old children), with no specification of the particular grade in which they should be applied. According to the learning standards related to statistical graphs in these documents, children should be able to:

- Identify and collect qualitative and quantitative data in everyday situations.
- Build and interpret simple graphs: bar graphs, line graphs and pie chart with familiar data.
- Carry out a critical analysis of the information presented in statistical graphs (p. 19393).

One main goal is that children value the benefits provided by statistical knowledge in decision-making and discover the usefulness of mathematics to solve everyday problems. This decree is interpreted by the different Spanish regional governments. For example, in Andalusia (Consejería de Educación, Cultura y Deporte de la Junta de Andalucía, 2015), the following content related to graphs is suggested for 6–9-year-old children:

- *First cycle* (6–7 year olds): Building and interpreting simple graphs: bar graphs; Using elementary techniques to collect and order data from everyday contexts. Oral description of the procedures used to collect data, interpret graphs and solve the problems. Attention and care in recording information and in graphical representation (p. 300).
- *Second cycle* (8–9 year olds): Statistical graphs: bar graph, line graph. Collecting and classifying quantitative information using surveys, observation and measurement. Use and interpretation of bar graphs and line graphs. Verbal description of elements in simple graphs in family contexts Attention and care in recording information and graphical representation (p. 303).

10.3.1 External Compulsory Assessment

An important approach to quality control within the educational system is provided by the external compulsory diagnostic tests in mathematics, science and language that are given to every child at specific educational levels.

Firstly, an initial test is administered (since 2010) to children in the second semester of grade 2 with the aim of evaluating their learning and correct any possible major problems. The test (scale) is a global questionnaire that is completed individually by each child in a standard form where the children write their responses. This test is not specific for mathematics, but assesses simultaneously written language, reading comprehension and knowledge of the social and natural environment. A few familiar contextual situations (e.g. shopping, going to the zoo, sports) are proposed to the children who read a short story and give some answers regarding the situation. After examining the scale tests proposed in Andalusia in the period 2011–2016, we observed that the whole questionnaire comprises of 15 questions, each of them with 4 items; only six of the questions are aimed at assessing the child's mathematics competence and only one of these six is related to statistical graphs.

The task consists of completing a bar graph in which a first bar is drawn (see Fig. 10.1, an adaptation of the task presented in 2012). The data set includes the frequencies for 4–5 different categories of a familiar variable and the vertical scale needed to represent the frequencies is given in the graph using either lines (like in the example) or small squares; generally, all the values in the data set correspond to one of the labels in the vertical scale. The categories for the variable are also represented on the horizontal scale. Therefore, the first competence requested from the child is understanding the problem and establishing a correspondence between each category in the table and the corresponding category on the horizontal scale (establishing an external correspondence, according to Bertin, 1967); then, the child should establish a second correspondence between the frequency corresponding to each category and the corresponding position on the vertical scale (internal correspondence, according to Bertin). Finally, the child should draw a bar of adequate length for each category (correspondence between the contextual and visual dimensions, according to Bertin).

A second external test (diagnostic test) for each of the topics' language, mathematics and science is given to all the children since 2007 following completion of grade 3. The exact points where the tests are given have slightly changed in the past years: until 2013, the children took the tests in the first semester of grade 4 (9 years old), while starting from 2014 they take the tests in the second semester of grade 3 (8 years old). The diagnostic tests for mathematics take into account the mathematical content, the reasoning processes and the mathematical competence needed to solve the different items proposed. Statistical graphs appear linked to the mathematical content *dealing with information, chance and probability,* as well in other mathematical content, such as geometry or number sense, where data in some items are provided in graphical format.

A detailed analysis of the mathematics diagnostic tests proposed in Andalusia over six years (2007–2012) provided the following results: Globally, 24 items (22.6% of

These are the data of books read in the library in May:

Adventure	Fairy tales	Plants	Animals
20	25	15	10

Represent in this bar graph all the data. Use a different colour for each bar.

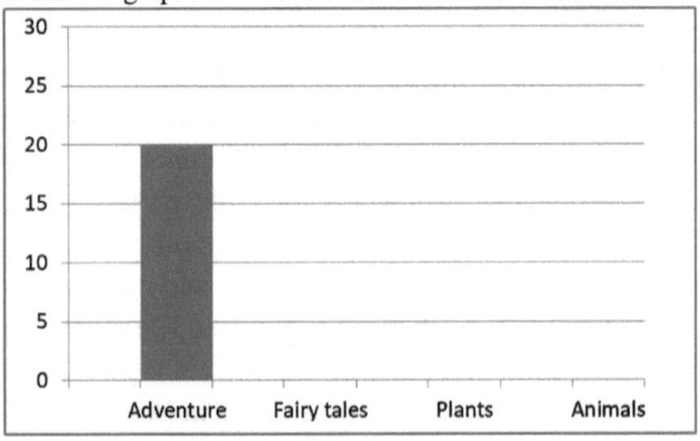

Fig. 10.1 Adapted example of a task similar to one proposed to grade 2 children in the scale test (2012)

the items proposed in all the tests in the period) involved some statistical content; more specifically, the number of items containing statistical information or questions varied between 2 and 6 from a total of 18 items in each test, and, therefore, the weight given to the topic is not homogeneous in the different years. The graphical content of the 20 items (18.8%) that assess graphical competence will be analysed in the following sections.

10.4 Graphs in the Spanish Textbooks and External Assessment Tests

Usually, each school selects the textbooks from a given editor for all the different grades. The statistical graphs' curricular content has been interpreted in different ways by the authors of the textbooks and the type of tasks proposed in relation to these graphs is quite varied.

In a previous comparative study of the textbooks produced for all the primary education grades (1st to 6th grades) in Chile and Andalusia in the period 2008–2011 (Díaz-Levicoy, Batanero, Arteaga, & Gea, 2016), we studied the distribution of the following variables in the graphs activities included in these textbooks: type of graph, reading levels, graph complexity and activity requested to the children. Below, we analyse the same variables, as well as the task context of all the activities related to statistical graphs in some recent textbooks published for grades 1–3 by four editors, as well as in the diagnostic tests given to children at the end of this school period.

The aim is to identify the expected progression in children's learning, as well as the correspondence between the curricular guidelines, textbooks and external assessment. The list of textbooks analysed is presented in the Appendix; these editors were selected because of their long tradition and good reputation in Spain.

10.4.1 Method

We performed a content analysis for each task and the expected solution to the same in order to identify the categories for each variable. This technique helps divide a text in analysis units that could be classified into a reduced number of categories, using underlying variables that help in making inferences about its content (Krippendorff, 2013). Starting from the categories identified in Díaz-Levicoy et al., (2016) and using an inductive and cyclical procedure, we coded and analysed the data to produce frequency tables that helped us to draw conclusions about the distribution of these categories in each variable and to compare results by grade and in the different editor series. We also selected specific examples of tasks translated from the textbooks or diagnostic test that are presented alongside the chapter in order to clarify the definition of the categories, the activity description and questions. Reliability of coding was ensured through independent coding of the data by two researchers; in the few cases of disagreement, the situations were discussed and consensus reached.

10.4.2 Type of Graph

The books usually start with schematic simple bar graphs in grade 1 representing small data sets related to a qualitative variable. In these graphs, the bars are not continuous, but consist of a grid constituted by identical rectangles, where each coloured rectangle represents a case. This facilitated children in interpreting the scale (they only need to count the number of coloured rectangles for each category to find the frequency for that category).

These schematic graphs are progressively turning into more abstract bar graphs with frequencies represented by the bar length in grade 2 (Fig. 10.2), where the variable is still qualitative, but the bars are continuous, the squares in the vertical axe do not appear and the rectangles in the graph framework are substituted by parallel lines.

In grade 2, we found some rudimentary pie charts, where a circle is divided into a small number of equal area sectors each of them indicating a case. Pictograms where the icon size (Fig. 10.3) or number of icons (Fig. 10.4) represent the frequencies of different categories are also introduced in this grade. Notice that the icons in Fig. 10.3 are not correctly displayed, since they do not only vary in length but also in area, which can be misleading for the children. In Fig. 10.4, each icon represents three units, which increases the difficulty for the children in reading the graph. In

Look at the bar graph of favourite meal of a group of children.

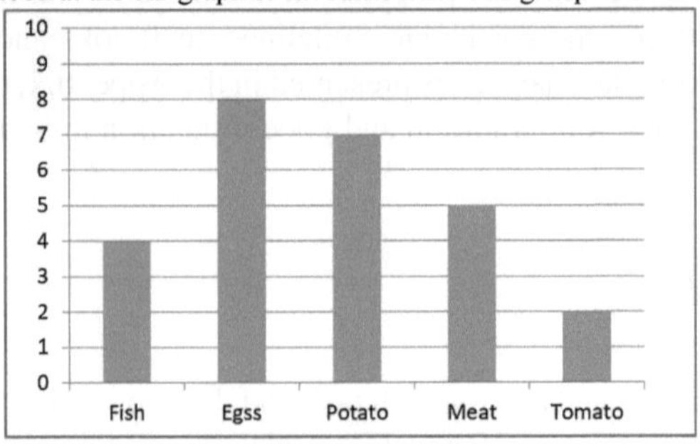

1. How many children prefer potatoes--- fish-- meat –?
2. Order the meals from most to less preferred.

Fig. 10.2 Interpretation of a bar graph (adapted from a similar task in EDEBE, grade 2, p. 102)

In the school we were asked what activity we like the best (adding, painting, reading and solving problems).

1. How many children prefer to add?
2. How many children prefer to paint?
3. How many prefer solving problems?
4. How many prefer to read?
5. Which is the activity preferred by more children?

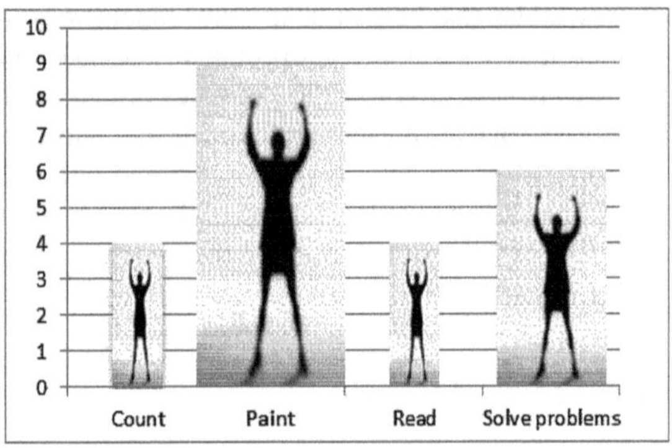

Fig. 10.3 Interpretation of a pictogram (adapted from EDEBE, grade 1 p. 172)

this grade, we also found some grouped bar graphs (Fig. 10.5) where children are requested to compare two or more distributions. Finally, in grade 3 line graphs are also introduced, although the majority of graphs are still bar graphs. See an example in Fig. 10.6 representing a list of values of a quantitative variable.

In Table 10.1, we classify the activities found in the different grades (collapsing the activities from different editors), and the 20 items containing statistical graphs in the external assessment, by the type of graph. In Table 10.2, we classify the type of graphs presented in the three grades presented by the different editors.

Usually, the activities are based on bar graphs, which appear in 60% of the items in the external assessment and are the most common graph in the textbooks, while line graphs appear in a quarter of the activities in grade 3. Pictograms (not recommended in the curricular guidelines) are also found from grade 1 in the different grades and pie charts (included in the curricular guidelines) sporadically appear. Occasionally, we found a few maps, which children use in their social studies lessons, in the external

These pictograms informs on the time a group of children get up on different days ⏲ = 3 children

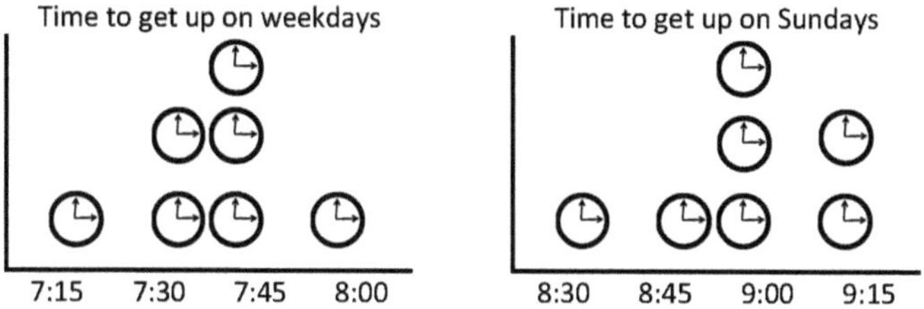

Discuss if you think the following sentences are true or false:

a. Only one child gets up at 7:15 on weekdays
b. Six children get up at 7:30 on weekdays
c. Most children gets up at 9:00 on Sundays
d. Most children get up earlier on weekdays

Fig. 10.4 Comparing two pictograms (EDEBE, grade 2, p. 159)

Look at the following grouped bar graphs that show which hand, right (R) or left (L) a group of children use for different activities. Can you figure out a question that can be answered with these diagrams?

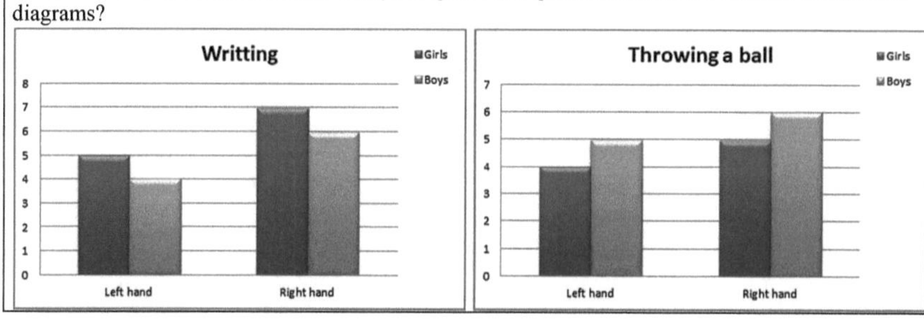

Fig. 10.5 Creative activity on a grouped bar graph (adapted from EDEBE, grade 2, p. 127)

tests, where colours or icons are used to represent a qualitative variable. Therefore, the graphs suggested in the curricular guidelines bar graphs for grades 1 and 2 and line graphs for grade 3 appear in the textbooks and external tests, as well as some additional graphs that are not considered in the mathematical curriculum. Finally, we analysed the distributions of types of graphs presented by the four editors and most notably Anaya presented only bar graphs, while all the other series include a variety of graphs.

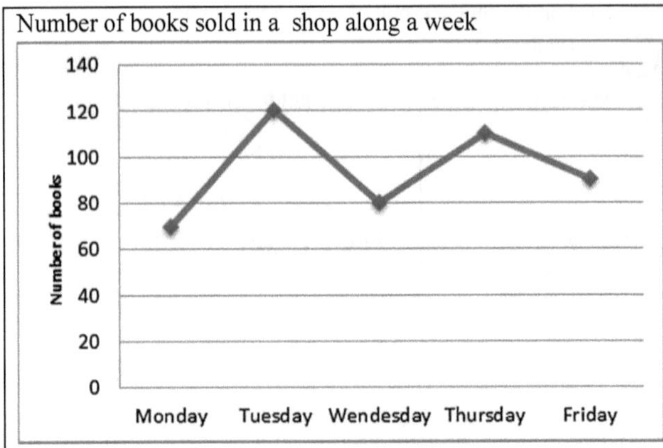

- How many books were sold each day? Can you tell from the graph?
- On which days did the bookshop sell more than 80 books?
- On which days did the number of books sold increase in relation to the previous day?

Fig. 10.6 Interpretation of a line graph (adapted from Santillana, grade 3, p. 200)

Table 10.1 Percentage of activities according to the type of graph in the different grades and external tests

Graph	Grade 1 (n = 22)	Grade 2 (n = 39)	Grade 3 (n = 41)	External tests (n = 20)
Bar chart	86.4	74.4	53.7	60.0
Pictogram	9.1	25.6	19.5	15.0
Maps				10.0
Line graph			26.8	5.0
Pie chart	4.5			10.0

Table 10.2 Percentage of activities according type of graph by editor

Graph	Anaya (n = 12)	Edebé (n = 41)	Santillana (n = 24)	SM (n = 25)
Bar chart	100	65.9	66.7	60.0
Pictogram		24.4	12.5	28.0
Line graph		7.3	20.8	12.0
Pie chart		2.4		

10.4.2.1 Reading Levels

In the three grades, the textbooks pose questions at different reading levels of Curcio's (1989) classification. For example, the first question in the activity reproduced in Fig. 10.2 requests the child to read from the graph the number of children with different preferences for meal. The child only has to recognise on the horizontal axis the label that corresponds to each meal (e.g. that corresponding to meat) and then

Table 10.3 Percentage of activities according to reading levels in the different grades and external tests

Reading level	Grade 1 (n = 22)	Grade 2 (n = 39)	Grade 3 (n = 41)	External tests (n = 20)
Level 1. Reading the data	18.2	33.3	22.0	42.0
Level 2. Reading between the data	81.8	64.1	63.4	58.0
Level 3. Reading beyond the data		2.6	9.8	
Level 4. Reading behind the data			4.9	

read on the vertical scale the frequency of children with that particular preference. Consequently, this question only requires Level 1 (reading the data) activity on the part of the children and involves determining the position of an element along a common scale (Cleveland & McGill, 1984).

The second question posed in Fig. 10.2 (ordering the meals according to frequency) requires that the child first makes a direct reading of the frequency that corresponds to each meal (reading the data) and then compares the frequencies corresponding to the different meals to find out which are the most and least frequent categories; this requires that the child interprets and integrates the data in the graph, an activity that corresponds to the second level (*reading between the data*) in the Curcio (1989) classification.

In Fig. 10.5, we find an example of activity where children can reach the reading levels 3 and 4. Since the activity suggests that the children invent some questions about the graph, their question may just be reduced to one of the previous levels; however, children may ask a question related to a comparison of the two graphs, and in this case, they will work at level 3 (comparing trends). When the child questions the generalization of findings or the data collection, he would be working at the Shaughnessy et al. (1996) level of *reading behind the data.*

In Table 10.3, we classify the activities found in the different grades and the external assessment items according to the reading levels requested from the student and can observe that the most common reading level is level 2, where children do not only make a literal reading of the graphs, but need to compute, order or compare some of the graph data. Level 1 (literal reading) is the second most frequent level and the other two levels only appear on isolated items. These reading levels are reflected in the external assessment that only considers the two first levels with a frequency that approximately reproduces that of textbooks. We compare this variable by editor in Table 10.4 observing few differences between the editors (i.e. Level 2 is the most frequent level among editors with level 1 being the second most frequent level of questioning).

Table 10.4 Percentage of activities according to reading levels by editor

Reading level	Anaya (n = 12)	Edebé (n = 41)	Santillana (n = 24)	SM (n = 25)
Level 1. Reading the data	16.7	26.8	16.7	36.0
Level 2. Reading between the data	83.3	65.9	79.2	52.0
Level 3. Reading beyond the data		7.3	4.2	4.0
Level 4. Reading behind the data				8.0

10.4.2.2 Graph Semiotic Complexity

We classified the activities presented to children according to the graph complexity in the classification by Batanero et al. (2010) and discovered examples of activities in the three upper levels. An activity based on a level 2 graph is displayed in Fig. 10.6, where a list of isolated values of a quantitative variable is represented, but there is no need to group the data or work with the frequency distributions.

Most graphs introduced in the previous figures correspond to Level 3, where the distribution for only one variable is displayed. For example, in Fig. 10.2 the graph corresponds to the distribution of a qualitative variable; in these graphs, the children have to distinguish between the variable (type of meal) and the frequency in each category; therefore, implicitly the idea of distribution is used in the graph.

Level 4 graphs are also included in both the books and the external assessment tests. For example, in Fig. 10.4 two different distributions (time to get up and time on weekdays and on Sunday) are represented in the two pictograms. The interpretation or building of these graphs is more complex as children should first interpret each separate graph and then compare the two distributions.

In Table 10.5, we classify the activities found in the different grades and the external assessment according to graph complexity. Surprisingly, the most common complexity in the graph is level 3, representing a distribution and therefore, children are expected to work (at least implicitly) with the idea of distribution. The frequency of this level decreases in grade 2 and 3, since in these grades the introduction of attached bar graphs lead to representing two or more distributions in the same graph (level 4). Moreover, in grade 3 line graphs are introduced and then the frequency of representation of lists of data, for which line graphs are a natural representation, increases in this grade. Again, the three upper levels of graph complexity are taken into account in the external test with a frequency that parallels that of textbooks.

When examining this variable across different editors (Table 10.6), we observe that level 4 (representing two or more distribution) is not considered in SM. These same editors, however, are the only ones to include a level 1 graph activity.

Table 10.5 Percentage of activities according to graph complexity in the different grades and external tests

Graph complexity	Grade 1 (n = 22)	Grade 2 (n = 39)	Grade 3 (n = 41)	External tests (n = 20)
Level 1. Representing isolated data		2.6		
Level 2. Representing a list of data	27.3	15.4	36.6	50.0
Level 3. Representing a distribution	72.7	53.8	36.6	33.3
Level 4. Representing two or more distributions		28.2	26.8	16.7

Table 10.6 Percentage of activities according to graph complexity by editor

Graph complexity	Anaya (n = 12)	Edebé (n = 41)	Santillana (n = 24)	SM (n = 25)
Level 1. Representing isolated data				4.0
Level 2. Representing a list of data	25.0	19.5	37.5	28.0
Level 3. Representing a distribution	41.7	56.1	29.2	68.0
Level 4. Representing two or more distributions	33.3	24.4	33.3	

10.4.3 Context of the Task

The context of a task refers to the part of the students' world in which the tasks are placed or according to Roth (1996, p. 491) "a real-world phenomenon that can be modelled by mathematical form". Context makes mathematics significant to the child, since it helps the student understand and value the applications of mathematics in everyday life. Moreover, students might connect the context to their experiences and, as a result, they might add some informal strategies to their mathematical knowledge in order to solve the problem. Taking into account this consideration, we classified the activities in the textbooks and external assessment according to the PISA contexts (OECD, 2015). We took into account the following types of contexts:

- Personal—Tasks classified in the personal context category focus on activities of the child, his or her family or the group of children in the classroom. Contexts that may be considered personal include those involving books or food preference, furniture, school work or materials, school elections, games, physical measurement, sports or transport.
- Occupational—Tasks classified in this category are focused on the world of work and may involve such things as commerce, measuring and classifying materials, building, architecture, communications or other job-related activities.
- Societal—The focus on this category is the child's community (whether local, national or global). They may involve voting systems, public transport, shows or museums, sport competitions, government, demographics, advertising or surveys.

Although children are also involved in these activities in their everyday life, the societal context category focuses on the community perspective.

- Scientific—Activities in this category relate to the application of mathematics to the natural world and issues and topics related to science and technology. Particular contexts might include weather, animals, biology, medicine, space science, genetics, measurement and the world of mathematics itself.

In Table 10.7, we note that personal context tasks (which are more frequent in grades 1 and 2) are replaced by societal contexts tasks in grade 3 and external tests. These societal context tasks become the most frequent context with the possible intention of showing children the utility of statistics in the whole society. All the contexts considered in our classification are included in the books, although some differences are evident in Table 10.8 across different editors.

10.4.3.1 Activity Requested

Usually for any one graph, more than one question is posed; each requiring a different activity from the child. We classified these activities according to the following categories that were identified by Arteaga (2011) and Díaz-Levicoy et al. (2016):

- *Building or completing a graph* using the information given in a data list or in a table. The child is given a list or a table with data and is requested to build a graph, where usually the title and scales are already represented. Other times, the graph is partially built. The children should identify the frequency that corresponds to each category or the value for a given list of values and draw a given element of the graph (e.g. a bar, one or more icons) to represent the data.
- *Reading or interpreting the graph.* In these activities, a graph is presented to the child who is asked to read specific elements (e.g. the frequency corresponding

Table 10.7 Percentage of tasks according to context in the different grades and external tests

Context	Grade 1 (n = 22)	Grade 2 (n = 39)	Grade 3 (n = 41)	External tests (n = 20)
Personal	54.5	71.8	29.3	37.5
Societal	4.5	20.5	39.0	54.2
Occupational	9.1	2.6	17.1	
Scientific	31.8	5.1	14.6	8.3

Table 10.8 Percentage of tasks according to context by editor

Context	Anaya (n = 12)	Edebé (n = 41)	Santillana (n = 24)	SM (n = 25)
Personal	58.3	41.5	58.3	56.0
Societal	25.0	29.3	20.8	20.0
Occupational	16.7	7.3	16.7	4.0
Scientific		22.0	4.2	20.0

Table 10.9 Percentage of activities in the different grades and external tests

	Grade 1 (n = 32)	Grade 2 (n = 67)	Grade 3 (n = 78)	External tests (n = 48)
Complete	18.8	10.4	3.8	31.3
Build	18.8	10.4	17.9	2.1
Read and interpret	43.8	35.8	29.5	35.4
Order or compute	12.5	28.4	21.8	31.3
Translate the data to a graph or table		6.0	7.7	
The graph is only used as an example		3.0	9.0	
Invent questions or describe the graph		4.5	9.0	
Collect data	6.3	1.5	1.3	

to a category, the title or the scale), compare several elements in the graphs (e.g. finding the most frequent value) or discuss a sentence referring to the graph. Some examples are presented in Figs. 10.2 and 10.5.

- *Order the data or perform some computation.* In this activity, a small data set is given to the child who is requested to perform some computations (e.g. finding the total or the difference between two variables, like in Fig. 10.6). In this category, we also included activities where the student has to order the data according to the numerical order (Fig. 10.2, second question). In these two activities, the main goal is practising arithmetic competence; however, the student still has to be able to read the graph correctly.
- *Translating* the data represented in a graph to a different graph or to a table, for example, changing the data represented in a bar graph to a bar graph.
- Other activities include *Using the graph as an example* (e.g. to explain how a particular graph is built), *Inventing a question* that can be answered with a particular graph, *Describing the graph* or *Collecting some data*, as part of the process of building the graph.

While the external tests usually request to complete or read the graph or perform computations with the graph data (Table 10.9), we discovered a variety of other activities in the books; in particular, building the complete graph while in the tests a part of the graph is given to the child who only has to complete it. There is not much difference in the activities in the different grades or in the different editor versions (Table 10.10) apart from the more creative activities, such as inventing a question that neither appears in grade 1 nor in Anaya.

Table 10.10 Percentage of activities by editor

	Anaya (n = 21)	Edebé (n = 71)	Santillana (n = 42)	SM (n = 43)
Complete	14.3	1.4	16.7	11.6
Build	9.5	26.8	2.4	11.6
Read and interpret	38.1	29.6	45.2	30.2
Order or compute	33.3	18.3	21.4	25.6
Translate the data to a graph or table	4.8	5.6	4.8	7.0
The graph is only used as an example		7.0	4.8	4.7
Invent questions or describe the graph		8.5	2.4	7.0
Collect data		2.8	2.4	2.3

10.5 Implications for Research and Teaching

In our analysis, we observed a change in comparison to our previous study (Díaz-Levicoy et al., 2016) in that pictograms, which are not explicitly recommended in the curricular guidelines, are now introduced in grade 1, instead of waiting until grade 3. This change may be positive, given that the icons used to represent the data in these graphs help children understand what is represented. Moreover, Cruz (2013) analysed the interpretation of pictograms by 21 children in grade 3 of Primary Education in Lisbon after a teaching process and obtained 95% correct answers to Level 1 questions and 77.3% to Level 2 questions in Curcio's (1989) classification. However, research by Cruz (2013) found that the construction of pictograms was extremely difficult for grade 3 children, suggesting that the building of these graphs be postponed; moreover, we need more research to evaluate the graphical competence of children before recommending the reading of pictograms since grade 1. We therefore recommend the reading of pictograms from grade 3 with the caution of making the area of icons proportional to each category frequency, and ensuring that equally sized icons represent only one case in order to not mislead the children.

A positive change observed is the presence of some level 3 *reading beyond the data* activities from grade 2 and level 4 *reading behind the data* in grade 3 that were not found for these levels in our previous research. In these reading levels, children do not only perform computations or comparisons with the data, but they use their statistical reasoning to make some predictions or inferences about information not directly reflected in the graph (level 3) or question the information presented in the graph (level 4); consequently, these activities will reinforce the children's statistical reasoning. Finally, there is not much difference in the graph complexity or activities presented and the task contexts were not analysed in the previous study.

Our conclusion is that the development of graphical competence in the Spanish current curricular guidelines is reflected and favoured by the textbooks and external assessment with a variety of tasks that take into account the research literature recommendations for the teaching of graphs. These curricular materials introduce a rich variety of type of graphs, activities and context (according to those considered by the OECD, 2015). The reading levels described by Curcio (1989) and Shaughnessy et al. (1996), as well as the graph semiotic complexity, (Batanero et al., 2010) are adequately ordered in progressive difficulty in the different grades, which no doubt will favour children's acquisition of graphical competence.

A warning is that some editors put too much emphasis on computation with the graph data, which is shown in the high percentage of *reading between the data* (level 2) when compared with other levels. In these activities, children should compare different data in the graph or perform computations with the graph data, and therefore, indirectly they are oriented to reinforce the child's arithmetic knowledge. Although in the lower levels of education, this is a reasonable goal, more interpretative activities should be included in order to develop the child's statistical literacy and reasoning. A second concern is that we also found important differences in the textbooks as regards the different variables analysed in our study, and therefore, we highlight the responsibility of teachers when selecting the most adequate book for his or her students.

Acknowledgements Acknowledgement. Projects EDU2016-74848-P (AEI, FEDER).

References

Arteaga, P. (2011). *Evaluación de conocimientos sobre gráficos estadísticos y conocimientos didácticos en futuros profesores* (assessing knowledge on statistical graphs and didactic knowledge in prospective teachers) (Unpublished Ph.D Dissertation). University of Granada, Spain.

Batanero, C., Arteaga, P., & Ruiz, B. (2010). Statistical graphs produced by prospective teachers in comparing two distributions. In V. Durand-Guerrier, S. Soury-Lavergne, & F. Arzarello (Eds.), *Proceedings of the Sixth Congress of the European Society for Research in Mathematics Education.* Lyon: ERME. Retrieved from www.inrp.fr/editions/editions-electroniques/cerme6/.

Bertin, J. (1967). *Semiologie graphique*. Paris: Gauthier-Villars.

Bertin, J. (1981). *Graphics and graphic information-processing*. Berlin: Walter De Gruyter.

Cleveland, W. S., & McGill, R. (1984). Graphical perception: Theory, experimentation and application to the development of graphical methods. *Journal of the American Statistical Association, 79*(387), 531–554.

Consejería de Educación de la Junta de Andalucía. (2008). *Orden de 5 de agosto de 2008, por la que se desarrolla el Currículo correspondiente a la Educación Infantil en Andalucía.* Sevilla: Autor.

Consejería de Educación, Cultura y Deporte de la Junta de Andalucía. (2015). *Orden de 17 de marzo de 2015, por la que se desarrolla el currículo correspondiente a la Educación Primaria en Andalucía.* Sevilla: Autor.

Cruz, A. (2013). *Erros e dificuldades de alunos de 1.º ciclo na representação de dados estatísticos* (1rt cycle students errors and difficulties in representing statistical data) (Master's Thesis). University of Lisbon.

Curcio, F. R. (1989). *Developing graph comprehension*. Reston, VA: NCTM.

Díaz-Levicoy, D., Batanero, C., Arteaga, P., & Gea, M. M. (2016). Gráficos estadísticos en libros de texto de Educación Primaria: un estudio comparativo entre España y Chile/Statistic graphs in primary education textbooks: a comparative study between Spain and Chile. *Bolema, 30*(55), 713–737.

Gal, I. (2002). Adult's statistical literacy: Meaning, components, responsibilities. *International Statistical Review, 70*(1), 1–25.

Krippendorff, K. (2013). *Content analysis: An introduction to its methodology*. London: Sage.

Ministerio de Educación y Cultura, MEC. (2007). *Real Decreto 1630/2006, de 29 de diciembre, por el que se establecen las enseñanzas mínimas del segundo ciclo de Educación infantil (Royal Decreet establishing the basic curriculum for Pre-school)*. Madrid: Autor.

Ministerio de Educación, Cultura y Deporte, MECD. (2014). *Real Decreto 126/2014, de 28 de febrero, por el que se establece el currículo básico de la Educación Primaria (Royal Decreet establishing the basic curriculum for Primary Education)*. Madrid: Author.

National Council of Teachers of Mathematics, NCTM (2000). *Principles and standards for school mathematics*. Reston, VA.

National Governors Association Center for Best Practices & Council of Chief State School Officers (2010). *Common core state standards for mathematics*. Washington, DC: Author.

OECD. (2015). *PISA 2015 Assessment and analytical framework*. Paris: Author.

Roth, W.-M. (1996). Where is the context in contextual word problems? Mathematical practices and products in grade 8 students' answers to story problems. *Cognition and Instruction, 14*(4), 487–527.

Shaughnessy, J. M. (2007). Research on statistics learning and reasoning. In F. K. Lester (Ed.), *Second handbook of research on mathematics teaching and learning* (pp. 957–1009). Charlotte, NC: Information Age Publishing.

Shaughnessy, J. M., Garfield, J., & Greer, B. (1996). Data handling. In A. J. Bishop, K. Clements, C. Keitel, J. Kilpatrick, & C. Laborde (Eds.), *International handbook of mathematics education* (pp. 205–237). Dordrecht, The Netherlands: Kluwer Academic Publishers.

Watson, J. M. (2006). *Statistical literacy at school: Growth and goals*. Mahwah, NJ: Lawrence Erlbaum Associates.

Appendix. Textbooks Analysed

Bernabeu, J., Carvajal, A., Garín, M., Puente, P., Téllez, M. J., Fernández, A, … López, S. (2014a). *Matemáticas 1º de primaria*. Madrid: SM.

Bernabeu, J., Garín, M., Rodríguez, M., Navarro, A., Pérez, M. N., Morales, F., et al. (2014b). *Matemáticas 3º de primaria*. Madrid: SM.

Bernabeu, J., Bellido, A., Garín, M., Martín, G., Herrero, N., Morales, Fm, et al. (2015). *Matemáticas 2º de primaria*. Madrid: SM.

Díaz, C., Ferri, T. M., Hidalgo, O., Marsá, M., & Pérez, E. (2014). *Matemáticas 1º de primaria*. Madrid: Anaya.

Díaz, C., Ferri, T. M., Hidalgo, O., Marsá, M., & Pérez, E. (2015a). *Matemáticas 2º de primaria*. Madrid: Anaya.

Díaz, C., Ferri, T. M., Hidalgo, O., Marsá, M., & Pérez, E. (2015b). *Matemáticas 3º de primaria*. Madrid: Anaya.

Rodríguez, M., & Pérez, C. (2015). *Matemáticas 1º Primaria*. Madrid: Santillana.

García, P., & Rodríguez, M. (2015). *Matemáticas 2º Primaria*. Madrid: Santillana.

García, P., Rodríguez, M., & Pérez, C. (2015). *Matemáticas 3º Primaria*. Madrid: Santillana.

Gómez, V., & Forcadell, E. (2015a). *Matemáticas 1 Primaria*. Barcelona: Edebé.

Gómez, V., & Forcadell, E. (2015b). *Matemáticas 2 Primaria*. Barcelona: Edebé.

Gómez, V., Barberá, P., Carvajal, S., Forcadell, E., & Lorente, N. (2015). *Matemáticas 3 Primaria*. Barcelona: Edebé.

Part IV
Teaching Statistics and Probability: Tasks and Materials

Chapter 11
Initiating Interest in Statistical Problems: The Role of Picture Story Books

Virginia Kinnear

Abstract This chapter presents results of a study that explored children's interest in picture story books that were used to contextualise and initiate statistical problems and statistical problem solving. The chapter presents the results of a small study conducted with 5-year-old children in a public school in Australia and identifies the characteristics of the books the children were interested in. It discusses the role of picture story books in initiating interest in the data context and task context of a statistical problem, and the unique challenges in identifying books for contextualising statistical problems.

11.1 Introduction

How can educators stimulate children's interest and engagement in statistical problem solving? Advocates for engaging young children in statistical learning at school argue that school-based, everyday practices with statistical investigation are essential for critical, flexible reasoning with data (Franklin et al., 2007; Lehrer & Schauble, 2002). Young children possess many conceptual resources and can move towards more sophisticated reasoning with appropriately designed and implemented learning experiences (Perry, Dockett & Harley, 2012). In statistics learning, familiarity with the context of a problem is known to influence data analysis and interpretation (Gal, 2005), and knowledge of the context is known to support and influence determining the relevance of data in problem solving (Pfannkuch & Wild, 2004). In statistical investigations, it is the design of a task and the real-world context it provides that impacts children's statistical learning (Kinnear, 2013). The use of picture story books to contextualise a statistical problem for investigation can therefore play a dual role in providing the context for the task that invites interest and shapes statistical engagement, and also provide the context knowledge children can use to

V. Kinnear (✉)
Deakin University, Victoria, Australia
e-mail: v.kinnear@deakin.edu.au

A. Leavy et al. (eds.), *Statistics in Early Childhood and Primary Education*,
Early Mathematics Learning and Development,
https://doi.org/10.1007/978-981-13-1044-7_11

find a solution to the problem. In this dual role, the choice of picture story book used to stimulate statistical problems becomes an important one for educators to make.

This chapter presents the results of a small study conducted with 5-year-old children in a public school in Australia. One aspect of the study examined the role of picture story books in initiating children's interest in a statistical problem, and in the role that the picture story book played in providing the context for that problem. The study's theoretical perspective, Models and Modelling (Lesh & Doerr, 2003), provided a framework for task design principles used in the design of the statistical problems which were developed as data modelling activities, aimed at providing young children with access core statistical concepts and processes. Three different picture story books were used, and study evaluated their role in initiating children's interest in the statistical context of the problem and in handling the data to solve the statistical problem. This chapter reports on the identified characteristics of the books children were interested in, and how knowledge of these characteristics could be used to inform the selection of picture story books to initiate interest in statistical problem-solving activities.

11.2 Connecting to Statistics: Real-World Problem Solving as Task and Data Context

Statistical problems are by definition real world, as statistics is used to solve, describe, measure and understand real-world problems (Scheaffer, 2006). Real-world problems therefore provide the impetus for and end point of a statistical investigation, supply the setting (as context) for the problem and at the same time, engaging the problem-solver's real-world knowledge of that setting when finding a solution. The interdependence of real-world context and statistics is captured by Langrall, Nisbet and Mooney's (2006) definition of statistical context as "real world phenomena, settings or conditions from which data are drawn or about which data pertain" (p. 1). The definition highlights that as a statistical problem is based in the real-world context, all data needed to solve the problem pertains to, and is drawn from that context. A core concept in statistics, and therefore a core consideration for engaging young children in statistical problem solving, is the context a statistical problem is embedded in, the *data context*. In this study, picture story books provided the data context for the statistical problem children were to solve.

The context of a statistical problem is inextricably linked to solving a problem drawn from it (Pfannkuch, 2011). When data are collected in order to solve a problem, our knowledge of their context is engaged in order to understand and interpret it (Moore, 1990). When young children are finding a solution to a statistical problem, they engage their existing knowledge and experience of the problem setting, the data context, including knowledge of the way data have been created, defined and measured (Pfannkuch, 2011). The data context provides meaning for the data, and so becomes the framing structure for data analysis and reasoning

(Garfield & Ben-Zvi, 2007). The data context of a statistical problem therefore influences statistical sense making and reasoning processes.

The data context chosen for a statistical investigation and how it is presented introduces the role of context in the statistical problem-solving task. *Task context* is therefore "the presentation of data or the way they are encountered" (Langrall, Nisbet, Mooney, & Jansem, 2011, p. 50). The task context influences the way data are approached, engaged, analysed and interpreted and therefore how statistical problems are reasoned and what knowledge is engaged to find a solution (Kinnear, 2013). The multiple dimensions of task context and the data context therefore influence *statistical* analysis and reasoning. Pfannkuch (2011), drawing from the work of Herschkowitz, Schwartz and Dreyfus (2001), made distinctions between "data context" and "learning-experience-contexts" in informal inferential reasoning. Task context, as part of the learning-experience-context included "task sequence and motivating story" (p. 28) and was identified as a key influence in facilitating statistical reasoning. The story and sequence of a story in a task context therefore has a role in children's task perception, and can influence their engagement and reasoning (Langrall et al., 2006). The interrelationship between the data context and the task context in statistical reasoning and problem solving is therefore pivotal to bringing children into statistical contexts, concepts and processes when problem solving. The study therefore investigated the role picture story books played in initiating interest in the data context and the data modelling problem that was drawn from its context.

11.3 Engaging Data Contexts Through Task Design

11.3.1 Picture Story Books: Real-World Contexts as Task and Data Context

Children should encounter data in ways that support their interaction with, not on, data (Makar & Rubin, 2009), therefore the task design for a statistical problem must be mindful of the impact of the task's context in the presentation of the problem and the way children perceive, approach and work with it to find a solution. In mathematics, picture story books have been found to provide a stimulus and motivation for children to investigate problems by offering a meaningful framework for active construction of mathematical knowledge (Elia, van den Heuvel-Panhuizen, & Georgiou, 2010). There are opportunities to make meaningful connections with young children's prior knowledge through the contents of a book, and a book can create a real issue for a child that needs to be addressed. Picture story books can therefore offer a context that supports young children's interest in, and emotional connection to mathematics, and present problems to be investigated or solved (Van den Heuvel-Panhuizen & van den Boogaard, 2008). They can provide a familiar and accessible framework for children, with "cognitive hooks" (Lovitt & Clarke, 1988, p. 439) for exploring the relationship between pieces of information and garnering children's interest in the problem at hand.

Research and pedagogical attention in recent decades has shifted attention to the potential for picture story books to support children's engagement with, and learning in, mathematics. There is still, however, limited empirical evidence for principles that can support educators in selecting picture books for use in mathematics teaching (Flevares & Schiff, 2014), and none for the selection of books for contextualising statistical problems. The limited literature and research that is available suggests that mathematics learning is successful when depicted in picture story books as a familiar part of everyday life and within contexts that are meaningful for children (Casey, Erkut, Ceder, & Young, 2008; Hong, 1996; Moyer, 2000; Whitin & Wilde, 1995). Books need to be of interest to children.

Interest is a principal concern for task design in the Models and Modelling perspective as it is a means of realising a match between the goals of the educator and the child in ways that move a child to engage in the task (Lesh & Doerr, 2003). An aim of Models and Modelling activities is to facilitate a child's interest in a task through its design in a way that places the child "squarely within the activity" (Middleton, Lesh, & Heger, 2003, p. 415). In Models and Modelling, interest begins with initiation into the modelling task, termed the *elicitation stage*, with designed experiences that aim to challenge children with the need to develop a model to solve a problem (Lesh & Doerr, 2003). Whatever is chosen as the initiating stimulus for the problem context is the stimulation for interest in the modelling activity itself. It is in this framework that picture story books have the potential to both provide the context for a statistical problem (as a data modelling problem) and initiate children's interest in the problem they are to solve (the task that requires the development of a model to solve the problem). The potential dual role a picture story book can play in stimulating interest in, and providing contextual knowledge for solving a statistical problem, means that its role differs from that of a book specifically chosen for use in mathematics learning.

11.3.2 Choosing Picture Story Books for Statistical Problems

The differences between statistics and mathematics as disciplines impact on the role of the content of a picture story book, as the concepts, processes and outcomes of statistics and mathematics differ (Moore, 1998). As a result, one of the difficulties in choosing books for initiating and contextualising statistical learning arises from the nature of the discipline itself; the context that stimulates the driving question (Leavy & Hourigan, 2015) and demands a solution also supplies the setting for the problem and at the same time, engages the problem-solver's knowledge of that setting in finding a solution. Picture story books have the capacity to provide a meaningful context for a statistical problem and act as a cognitive lure for young children's statistical learning. Statistical problem solving is, however, a contextualised activity, and until or unless elements of the picture story book are drawn on by the children, the statistical content of the book as *data context* is largely unknown. It is only in finding a solution to the problem that the "statistical content" of the book, that is, the knowledge children choose to employ to problem solve, is visible. Trying to choose picture story books for statistics by examining comparable research on the use of children's literature and mathematical learning is problematic for two reasons. First, with the exception of English (2009b; 2010; 2011, 2013), Kinnear (2013) and Hourigan and Leavy (2015), children's picture story books have not been used to initiate statistical problem solving or data modelling activities in published research. Second, research that evaluates young children's responses to the characteristics for classifying picture story books that initiate interest in and contextualise *statistical* problems has not been undertaken.

There have been a number of classification schemes developed to provide criteria for selecting published children's literature for teaching and learning mathematics (e.g. Hellwig, Monroe, & Jacobs, 2000; Hunsader, 2004; Marston, 2010; Nesmith & Cooper, 2010; Schiro, 1997; Van den Heuvel-Panhuizen & Elia, 2012; Whitin & Whitin, 2004). Although helpful in providing a range of book characteristics, the statistical role of a picture story book is not readily supported by these frameworks. They principally focus on "trade books" written specifically for mathematics teaching, or on identifying known mathematical concepts that are visible, clearly identifiable, or where the potential for mathematics specific concepts to be drawn out or used as a springboard for other mathematical learning is readily apparent. Published texts to support teachers' selection of mathematical fiction books offer few or no recommended texts for handling or analysing data (e.g. Burns & Sheffield, 2004; McKenney & Revves, 2012; Whitin & Whitin, 2004). Some work in storytelling in mathematics has considered how narrative can spark interest or pose, make sense of problem situations for problem solving (e.g. English, 2010; Skoumpourdi & Mpakopoulou, 2011; Zazkis & Liljedahl, 2009). The issue remains, however, that criterion for identifying content for supporting children's learning of statistical concepts and processes is not accommodated by existing classification frameworks for selecting books for mathematics teaching. Picture story books that are used to contextualise statistical problems play a crucial role in statistical problem solving, as they act as a source of contextualised data knowledge for children to use in statistical problem solving, and stimulate interest in engaging with the problem in the first place. A book stimulus needs to encourage children to make sense of a problem-solving situation and move them to recognise the need for a statistical model to be developed to solve the problem (Lesh & Doerr, 2003).

11.4 The Study

In the study, picture story books fulfilled two contextual roles. First, each book provided the *data context* that stimulated the statistical problem and bound the data that were available and able to be used to solve the problem. As a result, the narrative and picture content of the book had the potential to influence the knowledge the children drew from and the reasoning they used when working to find a solution to the problem. Second, as an integral part of the *task context*, each book initiated the children into the statistical problem and had the potential to influence the children's interest in, and connection to it. In the study, three picture story books were used to initiate three separate and consecutively implemented statistical problems (as data modelling problems). The books were chosen to support a recycling theme underpinning the data modelling problems. Prior to engaging with the statistical problem, the children were read the story so that they could focus on enjoying the picture story book as literature (Hunsader, 2004). The children's spontaneous responses to questions and comments about the picture story book on this first reading were captured as data.

This study proceeded on the same assumptions as the studies conducted by Moschovaki and Meadows (2005), Van den Heuvel-Panhuizen and van den Boogaard (2008), and Van den Heuvel-Panhuizen and Elia (2013), using similar methods to capture children's spontaneous responses however it is distinguished in two ways. In the initial reading by the teacher of the picture story book, the whole class was grouped informally and seated on the floor, and the teacher did not ask questions or make remarks about the story or illustrations. As the book was read, the children responded spontaneously with comments. Immediately following the reading, the children were invited to ask questions or to provide comments about the story. Questions or comments were written onto the classroom whiteboard with the name of the child who proposed the question or comment noted next to it. The children were then invited to answer any questions that had been raised. The children's responses were extended beyond their initial spontaneous comments or questions to include questions or comments the children initiated immediately following the book reading. Findings emerged from the thematic analysis of the children's spontaneous responses and questions and comments for each of the three picture books. This chapter presents these findings and discusses the characteristics of the picture story book that were found to be of interest to the children.

11.4.1 Study Participants, Data Collection and Analysis

The study participants were members of one class of 14 children, comprised of five girls and nine boys aged from 5 years to 5 years 3 months (mean age 5 years 2 months) in their first term of their first year of formal schooling in a State government primary school in South Australia. A qualitative design-based research methodology, informed by the Models and Modelling perspective (Lesh & Doerr, 2003), underpinned the study which was conducted over a 10-week school term. The data collection for the findings for this chapter was collected as whole class digital video-taping episodes that were then transcribed by the researcher from the audio recordings that were embedded in the video recording. The transcriptions included descriptions of visual information gleaned from the video. The use of video enabled multiple aspects of the children's interaction to be captured, such as the children's facial expressions and body movements, vocal emphases when speaking, silences, and other non-verbal communications (Lesh & Lehrer, 2000) that would indicate interest. Further data were provided by teacher and researcher meetings and researcher recorded personal reactions and reflections to the data collection process. These multiple data were used to support the ongoing evaluation of the data collection process and future analysis (Neuman, 2003). The data were systematically reviewed and worked through (Cobb, Confrey, diSessa, Lehrer, & Schauble, 2003) enabling "evidenced-based claims and results to be examined in concert with the underlying design theory" (Wang & Hannafin, 2005, p. 11).

11.4.2 Picture Books and the Data Modelling Activities

The three data modelling activities implemented in the study used the following picture story books. A brief synopsis, the data modelling focus and problem for each data modelling activity are described in Table 11.1 in order to frame the picture book reading and purpose. The data relating to the children's use of the picture story books content in their statistical problem solving are not presented in this chapter.

11.5 Children's Spontaneous Responses, Questions and Comments

11.5.1 Baxter Brown's Messy Room (English, 2009a)

The children listened attentively as the picture book was read, and their interest was sustained and consistent. During the story reading, spontaneous reactions were visible as the children asked questions, pointed to the illustrations, laughed as the plot unfolded, and used gestures such as bringing a hand to the mouth during a moment of tension in the story. The picture story book was written in a style reminiscent of familiar children's "lift the flap" books, for example, the "Spot the Dog" series (Hill, 1983, 1994, 2003). In lift the flap books, a question is posed in the book and the picture answer is found when a paper flap is lifted. The children reacted spontaneously as a group to the questions posed about where Baxter Brown might be, sometimes in improbable places as in the following example:

Teacher *One morning, Mr and Mrs Brown noticed that Baxter Brown was missing. Is he sleeping in the washing machine?*
Children (laughing) no!

There was some excitement on his discovery under the rubbish in his room:

Teacher *It was a tail! Was it Baxter Brown's tail? It looks like we have to wade through all this junk to find out, thought Mr and Mrs Brown. They stepped*

Table 11.1 Summary of book synopsis, data modelling focus and problem for each data modelling activity

Book	Plot synopsis	Data modelling problem	Data modelling focus
Baxter Brown's Messy Room (English, 2009a). Written specifically to support data modelling research	Baxter Brown is a white fluffy dog with a room that is so messy from all the rubbish he has collected that he is lost in it. Baxter Brown has to solve his rubbish problem and at the end of the story the question "What do you think Baxter Brown should do?" is asked	To help Baxter Brown clean up his room, the children were asked to sort real objects into pre-determined task categories of recycle, reuse and throw away. Next using pictures, they worked out what objects in his messy room could be recycled, or reused or what could be thrown and represented their solution in a data display	Generating and selecting attributes and organising, displaying and representing data
Michael Recycle (Bethel, 2008). A commercial publication	Michael Recycle is a caped superhero whose mission is to save planet earth. He flies into towns that are messy and teaches the people about recycling so they can clean up where they live. The story is written in rhyme	Michael Recycle needed help to sort a group of real objects he had found. Children drew pictures of the objects would be sorted these into categories	Identifying and displaying pictorial data in a more abstract form
Litterbug Doug (Bethel, 2009). A commercial publication	Litterbug Doug is a lazy, messy character who does not recycle and has rats for friends. His enormous, smelly piles of rubbish are upsetting the town folk, until Michael Recycle arrives to teach him how to recycle and clean himself up. As a result, on the last page of the story, Litterbug Doug becomes the Litter Police, cleaning up the town park	Litterbug Doug, in his new role as the Litter Police, had tidied up the town by collecting rubbish in the park. A data table represented how much rubbish Litterbug Doug had collected in the town's park on three days; Monday, Tuesday and Wednesday. The children were asked to predict what Litterbug Doug collected on the fourth day, Thursday, which was left blank on the data table	Reading, interpreting and extending data represented in an abstract format (a data table) to make predictions

	over bones, apple cores, newspapers, cereal packets, old toys, biscuits, old shoes, empty drink cans, milk cartons and plastic bags. All that junk! Finally, Mr and Mrs Brown reached the long, thick, white tail that was waving above the rubbish. Is that you Baxter Brown?
Ted	(gasps audibly and quickly points) aah! Yes it is!
Teacher	*The tail kept swaying, we have to remove some of this rubbish to see if it is him!*
Blake	(kneels up) it is!
Lee	(looks anxiously at Blake) yes or no?

The spontaneous responses, questions and comments offered by the children showed persistent interest in the problem revealed in the plot and the illustrations which focused on the problem that was revealed. The children were concerned about the dilemma Baxter Brown was in—how it arose and how it might be resolved, with speculative comments such as; "why did he, I know why he didn't clean up, because um, he just didn't think he might get lost in it". Some children hypothesised about why he had not taken action before he got himself into such a mess; "maybe he was going to have a real rest then he forgot that he had all that rubbish". The question of what Baxter Brown should do to solve his rubbish problem was posed at the end of the story when he "What should I do?" and the story narrator asking of the reader "What do you think Baxter Brown should do?" There was an immediate response to the question with multiple suggestions such as; "he should throw it away!", "he has to clean his room up!" and "he should put all the stuff in the bin", as the children enthusiastically suggested solutions to Baxter Brown's problem.

11.5.2 Michael Recycle (Bethel, 2008)

When *Michael Recycle* (2008) was read, generally the children listened quietly; however, a few children began to fiddle with their clothing and turn their head to noise distractions outside of the classroom. In contrast to the initial reading of *Baxter Brown's Messy Room* (2009), none of the children responded by pointing, laughing or using body movements that might indicate interest, such as moving forward or leaning towards the book. Some children began to lie back or roll to their side as the story came to an end. The only spontaneous responses evidenced were as the picture story book was introduced and the title read, with one child stating; "They rhyme. Michael Recycle!" When invited to offer questions or comments about the picture story book, the children's responses revealed interest in the character's altruistic behaviour, evidenced by comment such as: "he helps people", and questions such as; "how come he tries to, um how come he helps people?" indicating an interest in the "goodness" of the character and his motivations. The principal interest the children displayed, however, was in the gender identity of the main character, interest initiated by one child's use of the personal pronoun *she,* which immediately generated animated debate among the children, moving to point to the book, rising up on their knees and leaning forward to listen. Children's reasoning relied on the illustrations in the book as evidence to support their decisions about gender.

11.5.3 Litterbug Doug (Bethel, 2009)

During the reading of *Litterbug Doug* (2009), the children listened quietly and calmly. They spontaneously responded to descriptions of the objects Litterbug Doug had collected, indicating disgust and enjoyment, evidenced through responses such as "yuk" and "ooh!" to the illustrations of rubbish piles, toilets and rats. The character Michael Recycle was immediately recognised when he appeared in the plot and illustrations, and several children pointed excitedly, shuffled forward, and the pitch of their voice rose as they called out "Michael Recycle!"

Teacher	*But then something happened, that none could explain, it wasn't a bird and it wasn't a plane. A green caped crusader stupendously swooped, descending to earth with a great loop the loop.*
Children	(unanimous statements) Michael, Michael Recycle!
Mia	(pointing to the book) there's Michael Recycle!
Ted	Coming to tell him

The children's interest in the arrival of Michael Recycle contrasted with their disinterested response to the reading earlier that week of the Michael Recycle picture story book, which may be explained by recognition of what was now a familiar character and the character's anticipated role in the story.

When invited to offer questions or comments about the picture story book, the children's responses showed that they were interested in the resolution of Litterbug Doug's rubbish problem. They were curious about how Litterbug Doug had got into the problem he had, his failure to "be good", and how he was reformed, asking: "why did he, didn't he clean up his rubbish?"; "how did he get all that rubbish?" and "how did he like, um, how did he be good in the end?" The children's responses and questions also suggested that they considered the problem was his, and he had failed to clean up because he was messy, tired, lazy and needed help. Reasons for these judgements included "because Michael Recycle didn't help him and he was too tired"; "he doesn't clean up his room and he doesn't like cleaning up his house and he likes being messy"; and "because he was lazy and too tired because he was, he needed help". There was significant interest also in how Litterbug Doug was able to "become good" and that "being good" resulted from recycling, being physically clean, taking responsibility and learning how to recycle.

Analysis of the findings revealed characteristics of affective and cognitive interest to the children. The importance of the children's interest is twofold. First, stimulating interest in a problem is an expressed aim of designing modelling activities and begins with initiation into the modelling task (Lesh & Doerr, 2003). Effective use of Models and Modelling design principles of personal meaningfulness, and model construction are pivotal for initiating a modelling activity (Lesh & Doerr, 2003). Second, initiating tasks are designed to encourage children to make sense of the situation based on their personal knowledge and experiences and stimulate the need for a model to be constructed, modified or refined (Lesh & Harel, 2003). The use of picture story books aimed to stimulate both interest in the data context and the task context of the modelling activities (for this study, a statistical problem) and to provide a source of context knowledge for all the children that could be drawn on when finding a solution to the problem.

11.6 Characteristics of Picture Story Books that Stimulate Interest in the Data Context

The unique role of picture story books in contextualising statistical problems is not accommodated by existing classification schemes for picture story books for mathematics teaching; however, the following categories of interest were generated from the findings.

11.6.1 Limiting and Misdirecting Interest

The children's lack of interest in the plot in the picture story book *Michael Recycle* (Bethel, 2008) was unique in the study, and contrasted with the interest generated by the other two picture story books [*Litterbug Doug* (Bethel, 2009) and *Baxter Brown's Messy Room* (English, 2009a)]. The book failed to generate any responses that indicated enjoyment of the plot, which reached a climax that was fully resolved by the characters as the story concluded; Michael Recycle cleaned up the village successfully, and without fuss. The lack of spontaneous responses suggests that the story did not stimulate mental processing (Moschovaki & Meadows, 2005; Van den Heuvel-Panhuizen & van den Boogaard, 2008) or provide information that was interesting or personally meaningful to engage the children's attention (De Young & Monroe, 1996). The story had a predictable story line, and it is possible that this failed to provide an authentic connection for the children (Nesmith & Cooper, 2010).

Interest was visible in the animated debate about whether the main character was male or female, debate that was sustained and fuelled by the children's perceived ambiguity in the book's illustrations. Illustrations have an informational function (Van den Heuvel-Panhuizen & Elia, 2013) and are part of the whole picture book. They can represent story-related components, so they have the potential to cognitively engage and interest children (Elia et al., 2010). The illustrations in *Michael Recycle* may have tapped into core questions young children have about gender identity, and their responses highlight the importance of the relationship between text and illustration. Pictures with a representational function have been found to evoke mathematical thinking and utterances (Van den Heuvel-Panhuizen & Elia,

2013). Ambiguity between text and illustration may be a useful tool for stimulating discussion; however, such ambiguity in the illustrations could serve to misdirect children's attention away from other elements of the story designed to generate interest in the context of a statistical problem.

11.6.2 Capturing Interest: Uncertainty and the Unresolved Problem

In contrast to *Michael Recycle* (Bethel, 2008), the responses to *Baxter Brown's Messy Room* (English, 2009a, b) revealed that the children enjoyed the element of uncertainty in the plot. The mystery of where Baxter Brown *could* be hiding was followed by relief and excitement when he was discovered, visible in physical responses that provided information not expressed in speech (Broaders, Cook, Mitchell, & Goldin-Meadow, 2007). Enjoying uncertainty is in keeping with elements of an interesting and engaging story (De Young & Monroe, 1996). The children responded to the unresolved problem in *Baxter Brown's Messy Room* which was partially but not fully resolved when he was found. The rubbish problem in the plot was presented as one left hanging at the end of the story itself, inviting the children to solve it. The unresolved climax at the end of the story challenged the children to work out how could Baxter Brown's problem be *resolved*, and this served as a springboard to the data modelling problem. The only incidences of spontaneous comments or questions generating predictive solutions to Baxter Brown's problem were found in the children's responses to *Baxter Brown's Messy Room* (English, 2009a).

Stories with elements such as uncertainty such as not knowing how a story will end, achieve interest through the cognitive challenges in prediction or anticipation that this created (De Young & Monroe, 1996). Baxter Brown's unresolved problem is in contrast to the other two picture story books *Michael Recycle* (Bethel, 2008) and *Litterbug Doug* (Bethel, 2009), where the main character's problem was neatly solved by the end of the book. Events in a good narrative story have the power to impact affective responses and stimulate mental acts such as guessing and supposing (Fisher, 2005). The findings indicate that the children were sensitive to and responded to the cognitive challenge brought by uncertainty, and that this feature generated predictions. The findings for Baxter Brown suggest that a story that arouses spontaneous interest in an unresolved problem in the plot may stimulate interest in the statistical problem itself as well as the data context.

11.6.3 Capturing Interest: Personal Connection

The findings further suggest that Baxter Brown was a meaningful story for the children. Ainley (2006) points to the prominence in younger children of a range of affective emotional interest responses such as enjoyment and concern that help form coordinated relationships between interest as affect, motivation and cognition. Interest is also supported when a child identifies with a character, and personalising a storyline to connect to the audience's prior knowledge makes it engaging and enjoyable (De Young & Monroe, 1996). Baxter Brown elicited personal connections with the children by experiencing a problem that is a common one for young children; a love of collecting objects that leads to a need to tidy one's room. *Baxter Brown's Messy Room* (English, 2009a) therefore combined the elements of uncertainty and personal connections for the children in a way that stimulated interest. The picture story book was written in a familiar genre style, so interest may have been generated by the children finding connections to the story character and plot.

11.6.4 Capturing Interest: The Resolved Problem and the Role of the Character

In contrast to *Baxter Brown's Messy Room*, the children did not spontaneously respond to the resolved problems presented in the two picture story books, *Litterbug Doug* (Bethel, 2009), and *Michael Recycle* (Bethel, 2008). These books differed both in the way that the problem in the plot was presented and resolved. In each story, the plot reached a climax that was fully resolved by the characters within the story itself; Michael Recycle cleaned up the village successfully, and Litterbug Doug was reformed and became a litter policeman. The resolution of a dilemma or problem within the story is in keeping with the core elements of a fictional story for engaging young children (De Young & Monroe, 1996). Although a fictional story may resolve the problem by the last page, this may not be a feature that best fits a picture story book that initiates a statistical problem, where interest is requisite to the developing a problem solution. The children's failure to find interest in a resolved problem, however, was alleviated to an extent by their interest in how a problem *was* resolved, particularly when the character affected by the problem was of interest, as the character Litterbug Doug was found to be.

A significant finding from the children's responses to *Litterbug Doug* (Bethel, 2009) was that he was a worrisome character for them, messy, lazy and forgetful, he needed to be taught how to "be good". The children did not respond to Litterbug Doug's reformed role as the litter police but focused on the characteristics that concerned and interested them which was twofold; first, the problematic behaviour that had led him to be living in piles of rubbish with rats for friends and second, how he was reformed in the story. The children directed their attention by asking questions to find out more about Litterbug Doug, and to work out how his problem had been solved. They reasoned that Litterbug Doug's rubbish problem was the result of his own actions. The children had ongoing concern for the "goodness" of Litterbug Doug's character: his failure to be good, his "coming good" and the consequences for both. Characterisation is an element of engaging stories identified by De Young and Monroe (1996), and the findings as to the children concerns for Litterbug Doug indicate that they identified with the character as one they could care about and follow through the story.

A focus on the moral dimension of the character is important and ties to research on the moral concerns of young children (Nucci, 2001). This element of the picture story book appears to have tapped into matters that children care deeply about. Their concern for Litterbug Doug's reform, as a problem resolved within the story, contrasted with the interest in the unresolved problem that was generated by Baxter Brown. Unlike Baxter Brown where problem resolutions were predicted, the resolution of Litterbug Doug's problem was explained within the story. The children's responses indicate that their interest was in how resolution had *occurred.* Interest in the worrisome Litterbug Doug's problem resolution contrasts further with the children's lack of interest in the virtuous Michael Recycle, who as an "already good" super hero, flew in and reformed Litterbug Doug into a litter policeman. Both the picture story books *Litterbug Doug* and *Michael Recycle* had the plot problem resolved in the story. The difference in the children's interest between the two books appears to be one of whether the problem is one that involves a character of interest.

11.7 Stimulating Interest in Statistical Problems

Statistics is an effective tool to investigate everyday problems; however, finding suitable literature to support statistical investigations has already been identified as problematic for teachers and researchers who wish to consciously use children books to support children's statistical learning (Hourigan & Leavy, 2015). From a statistical perspective, the children's interests in this study highlight some possible insights into children's *data context* interests which can be fruitful in initiating interest in a *task context* and the subsequent need to use the data context to support finding a solution to the statistical problem it generated. Research reveals that personal interest and connection to the context of a problem, that is, the task design, activates children's interest in the task (Clarke & Roche, 2009; Paparistodemou & Meletiou-Mavrotheris, 2010). Research highlights that educators need to be able to assess of the quality of books used in teaching mathematics (LeSage, 2013), and that literature can support the development of driving question for statistical investigations (Hourigan & Leavy, 2015). What is needed is further research on the characteristics of picture story books that stimulate interest in the data context and initiate interest in the task context. Questions that remain unanswered are what characteristics of picture story books stimulate children's interest in a data context, and how can these be best utilised by researchers and educators when selecting books that aim to bring children into meaningful statistical engagement with statistical problems. This small study provides a starting point.

References

Ainley, M. (2006). Connecting with learning: Motivation, affect and cognition in interest processes. *Educational Psychology Review, 18*(4), 391–405. https://doi.org/10.1007/s10648-006-9033-0.

Bethel, E. (2008). *Michael Recycle*. Mascot, Australia: Koala Books.

Bethel, E. (2009). *Litterbug Doug*. Mascot, Australia: Koala Books.

Burns, M., & Sheffield, S. (2004). *Math and literature: Grades K–1*. Sausalito, CA: Maths Solutions.

Casey, B., Erkut, S., Ceder, I., & Young, J. M. (2008). Use of story-telling context to improve girls' and boys' geometry skills in kindergarten. *Journal of Applied Developmental Psychology, 29,* 29–48. https://doi.org/10.1016/j.appdev.2007.10.005.

Clarke, D., & Roche, A. (2009). Opportunities and challenges for students provided by tasks built around 'real' contexts. In R. Hunter, B. Bicknell, & T. Burgess (Eds.), *Crossing divides: Proceedings of the 32rd Annual Conference of the Mathematics Education Research Group of Australasia* (Vol. 1). Palmerston North, NZ: MERGA.

Cobb, P., Confrey, J., diSessa, A., Lehrer, R., & Schauble, L. (2003). Design experiments in educational research. *Educational Researcher, 32*(9), 9–13. https://doi.org/10.3102/0013189X032001009.

De Young, R., & Monroe, M. C. (1996). Some fundamentals of engaging stories. *Environmental Education Research, 2*(2), 171–187. https://doi.org/10.1080/1350462960020204.

Elia, I., van den Heuvel-Panhuizen, M., & Georgiou, A. (2010). The role of pictures in picture books on children's cognitive engagement with mathematics. *European Early Childhood Research Journal, 18*(3), 275–297. https://doi.org/10.1080/1350293X.2010.500054.

English, L. D. (2009a). *Baxter Brown's Messy Room.* Retrieved from http://www.ici.qut.edu.au/projects/discoveries/statisticalreasoning.jsp.

English, L. D. (2009b). Promoting interdisciplinarity through mathematical modelling. *ZDM, 49*(1&2), 161–181. https://doi.org/10.1007/s11858-008-0106-z.

English, L. D. (2010). Young children's early modelling with data. *Mathematics Education Research Journal, 16*(3), 59–60.

English, L. D. (2011). Data modelling in the beginning school years. In P. Sullivan, & M. Goos (Eds.), *Proceedings of the 34th Annual Conference of the Mathematics Education Research Group of Australia (MERGA).* Alice Springs, NT: MERGA.

Fisher, R. (2005). *Teaching children to think* (2nd ed.). Cheltenham, UK: Nelson Thomas.

Flevares, L., & Schiff, J. (2014). Learning mathematics in two dimensions: A review and look ahead at teaching and learning early childhood mathematics with children's literature. *Frontiers in Psychology, 5*(459), 1–12. https://doi.org/10.3389/fpsug.2014.00459.

Franklin, C., Kader, G., Mewborn, D. S., Moreno, J., Peck, R., Perry, M., et al. (2007). *Guidelines for assessment and instruction in statistics education (GAISE) report: A pre-K–12 curriculum framework*. Alexandria, VA: American Statistical Association.

Gal, I. (2005). Statistical literacy. In D. Ben-Zvi & J. Garfield (Eds.), *The challenge of developing statistical literacy, reasoning, and thinking* (pp. 47–78). Dordecht, The Netherlands: Kluwer Academic Publishers.

Garfield, J. B., & Ben-Zvi, D. (2007). How students learn statistics revisited: A current review of research on teaching and learning statistics. *International Statistical Review, 75*(3), 372–296. https://doi.org/10.1111/j.1751-5823.2007.00029.x.

Hellwig, S. J., Monroe, E. E., & Jacobs, J. S. (2000). Making informed choices: Selecting children's trade books for mathematics instruction. *Teaching Children Mathematics, 7,* 138–143.

Herschkowitz, R., Schwartz, B., & Dreyfus, T. (2001). Abstraction in context: Epistemic actions. *Journal for Research in Mathematics Education, 32*(2), 195–222.

Hill, E. (1983). *Where's Spot?* Harmondsworth, UK: Puffin Books.

Hill, E. (1994). *Spot bakes a cake.* London, UK: Frederick Warne.

Hill, E. (2003). *Spot goes to the beach.* London, UK: Frederick Warne.

Hong, H. (1996). Effects of mathematics learning though children's literature on math achievement and dispositional outcomes. *Early Childhood research Quarterly, 11*(4), 447–494. https://doi.org/10.1016/S0885-2006(96)90018-6.

Hourigan, M., & Leavy, A. (2015). Practical problems: Using literature to teach statistics. *Teaching Children Mathematics, 22*(5), 283–291.

Hunsader, P. D. (2004). Mathematics trade books: Establishing their value and assessing their quality. *The Reading Teacher, 57,* 618–629.

Kinnear, V. A. (2013). *Young children's statistical reasoning: A tale of two contexts* (Ph.D. dissertation). Queensland University of Technology. Retrieved from https://eprints.qut.edu.au/63496/.

Langrall, C. W., Nisbet, S., & Mooney, E. S. (2006). The interplay between students' statistical knowledge and context knowledge in analysing data. In A. Rossman & B. Chance (Eds.), *Proceedings of the 6th International Conference on Teaching Statistics (ICOTS6, Salvador, Brazil).* Voorburg, The Netherlands: International Statistics Institute.

Langrall, C., Nisbet, S., Mooney, E., & Jansem, S. (2011). The role of context expertise when comparing data. *Mathematical Thinking and Learning, 13*(1&2), 47–67. https://doi.org/10.1080/10986065.2011.538620.

Leavy, A., & Hourigan, M. (2015). Crime scenes and mystery players! Using driving questions to support the development of statistical literacy. *Teaching Statistics, 38*(1), 29–35.

Lehrer, R., & Schauble, L. (2002). Children's work with data. In R. Lehrer & L. Schauble (Eds.), *Investigating real data in the classroom: Expanding children's understanding of math and science* (pp. 1–26). New York: Teachers College Press.

LeSage, A. C. (2013). *Don't count on the quality of children's' counting books.* Retrieved from http://lesage.blogs.uoit.ca/wp-uploads/LeSage_Evaluating-Early-Counting-Lit_ICET-Conference_2013.pdf.

Lesh, R., & Lehrer, R. (2000). Iterative refinement cycles for videotape analyses of conceptual change. In A. E. Kelly & R. Lesh (Eds.), *Handbook of research design in mathematics and science education* (pp. 665–708). Mahwah, NJ: Lawrence Erlbaum Associates.

Lovitt, C., & Clarke, D. (1988). *The mathematics curriculum and teaching program (MCTP): Professional development package activity bank* (Vol. 2). Carlton, Vic: Curriculum Development Centre.

Makar, K., & Rubin, A. (2009). A framework for thinking about informal statistical inference. *Statistical Education Research Journal, 8*(1), 82–105.

Marston, J. (2010). Developing a framework for the selection of picture books to promote early mathematical development. In L. Sparrow, B. Kissane, & C. Hurst (Eds.), *Shaping the future of mathematics education: Proceedings of the 33rd Annual Conference of the Mathematics Education Research Group of Australasia (MERGA)* (pp. 383–390). Freemantle, WA: MERGA.

McKenney, S., & Revves, T. (2012). *Conducting educational design research.* Oxford, UK: Routledge.

Middleton, J. A., Lesh, R., & Heger, M. (2003). Interest, identity and social functioning: Central features of modeling activity. In R. Lesh & H. M. Doerr (Eds.), *Beyond constructivism: Models and modeling perspectives on mathematics teaching, learning and problem solving*. Mahwah, NJ: Lawrence Erlbaum Associates.

Moore, D. (1990). Uncertainty. In L. A. Steen (Ed.), *On the shoulders of giants: New approaches to numeracy* (pp. 95–137). Washington, DC: National Academy Press.

Moore, D. S. (1998). Statistics among the liberal arts. *Journal of the American Statistical Association, 93*(444), 1253–1259. https://doi.org/10.1080/01621459.1998.10473786.

Moschovaki, E., & Meadows, S. (2005). Young children's spontaneous participation during classroom book reading: Differences according to various types of books. *Early Childhood Research and Practice, 7*(1), 1–17.

Moyer, P. S. (2000). Communicating mathematically: Children's literature as a natural connection. *The Reading Teacher, 54*(3), 246–255.

Nesmith, S., & Cooper, S. (2010). Trade books in the mathematics classroom: The impact of many, varied perspectives on determinations of quality. *Journal of Research in Early Childhood Education, 24*(4), 279–297. https://doi.org/10.1080/02568543.2010.510086.

Neuman, W. L. (2003). *Social research methods: Qualitative and quantitative approaches*. Boston: Allyn and Bacon.

Nucci, L. P. (2001). *Education in the moral domain*. New York: Cambridge University Press.

Paparistodemou, E., & Meletiou-Mavrotheris, M. (2010). Engaging young children in informal statistical inference. In C. Reading (Ed.), *Data and context in statistics education: Towards an evidence-based society. Proceedings of the 8th International conference on Teaching Statistics (ICOTS8)*. Voorburg, The Netherlands: International Statistical Institute.

Perry, B., Dockett, S., & Harley, E. (2012). The early years learning framework for Australia and the Australian curriculum: Mathematics—Linking educators' practice through pedagogical inquiry questions. In B. Atweh, M. Goos, R. Jorgensen, & D. Siemon (Eds.), *Engaging the Australian national curriculum: Mathematics—Perspectives from the field* (pp. 153–174). MERGA: Online publication.

Pfannkuch, M. (2011). The role of context in developing informal statistical inferential reasoning: A classroom study. *Mathematical Thinking and Learning, 13*(1&2), 27–46. https://doi.org/10.1080/10986065.2011.538302.

Pfannkuch, M., & Wild, C. (2004). Towards an understanding of statistical thinking. In D. Ben-Zvi & J. Garfield (Eds.), *The challenge of developing statistical literacy, reasoning, and thinking* (pp. 17–46). Dordecht, Netherlands: Kluwer Academic Publishers.

Scheaffer, R. L. (2006). Statistics and mathematics: On making a happy marriage. In G. Burrill (Ed.), *Thinking and reasoning with data and chance: Sixty-eighth yearbook* (pp. 309–321). Reston, VA: National Council of Teachers of Mathematics.

Schiro, M. S. (1997). *Integrating children's literature and mathematics in the classroom: Children as meaning makers, problem solvers, and literacy critics*. New York: Teachers College Press.

Skoumpourdi, C., & Mpakopoulou, I. (2011). The Prints: A picture book for pre-fromal geometry. *Early Childhood Education Journal, 39,* 197–2016. https://doi.org/10.1007/s10643-011-0454-0.

Van den Heuvel-Panhuizen, M., & van den Boogaard, S. (2008). Picture books as an impetus for kindergartners' mathematical thinking. *Mathematical Thinking and Learning, 10,* 341–373. https://doi.org/10.1080/10986060802425539.

Van den Heuvel-Panhuizen, M., & Elia, I. (2012). Developing a framework for the evaluation of picture books that support kindergartners' learning of mathematics. *Research in Mathematics Education, 14*(1), 17–47.

Van den Heuvel-Panhuizen, M., & Elia, I. (2013). The role of picture books in young children's mathematics learning. In L. D. English & J. T. Mulligan (Eds.), *Reconceptualizing early mathematics learning* (pp. 227–252). Dordecht: Springer.

Wang, F., & Hannafin, M. (2005). Design-based research and technology-enhanced learning environments. *Educational Technology Research and Development, 53*(4), 1042–1629. https://doi.org/10.1007/BF02504682.

Whitin, D. J., & Whitin, P. (2004). *New visions for linking literature and mathematics*. Reston, VA: National Council of Teachers of Mathematics.

Whitin, D. J., & Wilde, S. (1995). *It's the story that counts: More children's books for mathematical learning, K–6*. Portsmouth, NH: Heinemann.

Zazkis, R., & Liljedahl, P. (2009). *Teaching Mathematics as Storytelling*. Rotterdam: Sense Publishers.

Chapter 12
Teachers' Reflection on Challenges for Teaching Probability in the Early Years

Efi Paparistodemou and Maria Meletiou-Mavrotheris

Abstract The present research focuses on the early childhood teachers' reflection on designing and implementing probability tasks. Five early childhood teachers participated in this research, which was organized in three stages: design of lesson plan, classroom implementation and reflection. The researchers analysed the design of each lesson, observed teachers implementing their lesson and interviewed them while they reflected on their instruction. Teachers discussed critical incidents that occurred through their teaching, and they reflected on challenges for teaching probability. An initial analysis of the collected data indicates that early childhood teachers appreciated the importance of using tools and real-life scenarios in their classrooms for teaching stochastics. The study also provided some useful insights into the varying levels of attention that teachers paid to different kinds of activities during their lesson implementation, and into the different types of instructional material they used. Moreover, the findings also show that early childhood teachers' attention to different aspects of probability tasks can be developed through a reflective process on their teaching.

12.1 Introduction

Although the pedagogical and mathematical understanding of teachers both in primary and secondary education has been widely explored and appreciated (Mason, 1998), this is not the case in early childhood. It can be argued that the underlying assumption is that mathematical activity, and tasks in these early years are trivial, simple and playful. Tasks are usually based on the use of materials and hands-on experience, something that is perceived as 'fun maths' rather than as 'real maths'

E. Paparistodemou (✉)
Cyprus Pedagogical Institute, Nicosia, Cyprus
e-mail: e.paparistodemou@cytanet.com.cy

M. Meletiou-Mavrotheris
European University Cyprus, Nicosia, Cyprus
e-mail: M.Mavrotheris@euc.ac.cy

A. Leavy et al. (eds.), *Statistics in Early Childhood and Primary Education*, Early Mathematics Learning and Development,
https://doi.org/10.1007/978-981-13-1044-7_12

(Moyer, 2001). This distinction has important connotations for early childhood education since the mathematics taught at this level is often concealed both due to its intuitive nature (Paparistodemou, Noss & Pratt, 2008) and the context in which it is used. Pre-primary teachers need to recognize the mathematical meaning underlying different tasks and develop such pedagogical competences that will allow them to facilitate children's mathematical learning.

The development of students' stochastical literacy has become an overarching goal of statistics education internationally. This broadening of the curriculum to encompass statistical literacy, reasoning and thinking has put considerable demands on teachers (Hannigan, Gill & Leavy, 2013). In particular, they must design lessons with engaging contexts (Chick & Pierce, 2008), focus on conceptual understanding (Watson, 2001) and pose critical questions (Reston, Jala, & Edullantes, 2006). In statistics education, there is a need to foster children's ability to reason inferentially by introducing them to a reasoning process in which multiple statistical concepts are used as arguments to support an inference (Makar & Rubin, 2009; Paparistodemou & Meletiou-Mavrotheris, 2008; Watson, 2001). Makar and Rubin (2009) introduced a framework for conceptualizing inferential reasoning, in which the use of non-deterministic language constitutes an important component. Nonetheless, the literature on teacher education and statistics education shows that teachers are not really inclined to include probabilistic concepts in their instruction (e.g. Paparistodemou, Potari, & Pitta, 2006).

This paper aims to investigate the way early childhood teachers conceive stochastics education in the early years. Specifically, the research question we address is as follows:

- What do teachers attend to when they design, implement and reflect on probability tasks in early childhood education?

12.2 Theoretical Perspective

12.2.1 Learning and Teaching Probability in the Early Years

Statistics education research suggests that enhancing teachers' subject matter knowledge about uncertainty and statistical inference must be given high priority. Building teachers' knowledge of pedagogical structures and tools by itself is not sufficient. Lee and Mojica (2008), for example, found that although a group of middle school teachers involved their students in authentic statistical inquiry that included use of simulation tools, they missed the chance to develop students' understanding of the frequentist approach to probability because of limited subject matter knowledge. Deep understanding of probability is also needed for identifying student errors and implementing effective teaching practices (Maher & Muir, 2014; Paparistodemou, Potari, & Pitta, 2006). Such understanding can be developed through well-designed professional development. For example, Theis and Savard (2010) helped high school teachers design and implement a technology-based instructional intervention. They found that the use of simulation software within the intervention allowed teachers to adopt more inquiry-oriented strategies and to begin to incorporate frequentist probability for teaching probabilistic concepts rather than relying solely on theoretical probability.

Although subject matter knowledge is necessary for effective teaching of uncertainty, it is not sufficient. Leavy (2010) worked with a group of prospective teachers who demonstrated relatively strong subject matter knowledge about informal inference. However, they had trouble using this knowledge to develop pedagogical contexts for advancing children's learning. In particular, they had difficulty choosing sufficiently complex data, creating engaging contexts, handling unexpected student responses and scaffolding children's thought processes. In other studies, gaps in pedagogical content knowledge have been framed as contributing factors to teachers' failure to emphasize important probability concepts when writing lesson plans (Chick & Pierce, 2008) and designing productive learning environments for students (Groth, 2010). Consequently, statistics education researchers have, in recent years, engaged in professional development efforts aimed at facilitating the development of both subject matter knowledge and pedagogical content knowledge of teachers (e.g. Groth, Kent, & Hitch, 2015; Leavy, 2010; Serradó, Meletiou-Mavrotheris, & Paparistodemou, 2014).

Mathematical activity is not only characterized by the content knowledge, but requires a broader view of knowledge like the mathematical know-how, how to teach the mathematical concept or procedure (Boaler, 2003, 2009). Noss and Hoyles (1996) and Hoyles (2002) claim that mathematical activity is designed to foster mathematical meanings through construction, interaction and feedback, while students can also scaffold their own thinking through communicating with tools. However, tools are often considered in early years more as a means to motivate children rather than to challenge them mathematically. One reason could be that the teachers have not constructed the flexible mathematical knowledge required to be able to plan activities that will encourage young children's mathematical activity. Moreover, they may lack the knowledge required to consider children's ideas and intuitions both in their planning and in their actual teaching. Shulman (1986) defines this knowledge as 'pedagogical content knowledge', while Ball and Bass (2000) argue that teaching mathematics entails work with microscopic elements of mathematical knowledge in order to make sense of a child's apparent errors or to appreciate a child's insights.

The importance of introducing probability tasks in the early mathematics classroom is related to the idea of constructing stochastical knowledge based on children's intuitions. Nowadays, probability and statistics have an important role to play in everyone's daily life and particularly in children's lives where most of the games they engage with include the idea of chance. Jones et al. (1999) gave a framework for assessing probabilistic thinking. They concluded that there are four levels for each of the following probabilistic concepts: sample space, probability of an event, probability comparisons and conditional probability. Moreover, research on learning stochastics (Paparistodemou & Noss, 2004) has shown that young children use spatial representations for expressing stochastic ideas. Research (Paparistodemou &

Noss, 2004; Pratt, 2000) also shows that the design of activities targeting children aged 4–8 is critical for them to be able to express their probabilistic ideas. Kilpatrick (2001) defines mathematical proficiency by identifying five aspects: conceptual understanding, procedural fluency, strategic competence, adaptive reasoning and productive disposition. These aspects of mathematical proficiency are equally important to the development of students' reasoning about the stochastic. A necessary area of investigation is the role of the teacher: if we consider the necessity of educating students who are used to thinking stochastically, it is necessary to rethink the role of the teacher in the teaching/learning process. Keeping in mind that teachers have an important role in designing and implementing stochastical tasks, Petocz and Reid (2002) indicate the importance of developing learning environments that can engage students' interest, broaden their understanding of statistics and help them enrich their own lives. They suggest that the development of learning environments must be 'total' and that the learning of stochastics should be less focused on the curriculum itself and certainly less focused on the traditional concern of material to be 'covered' or 'examined'. Rather, the focus should move towards supporting students to develop 'holistically'. Boaler (2003) argues that a rich mathematical activity ought to include elements such as creativity, inquisitiveness, making of connections and a dynamic view of mathematical representations.

The above analysis indicates that stochastical activity is a complex activity, which extends beyond the notion of stochastical knowledge, and requires students to develop holistically, through their involvement in a number of tasks focused on the different components of stochastical reasoning.

12.2.2 Mathematics Teaching and Reflection

We consider pedagogical activity as the means that can support children's stochastical activity. It includes classroom management, taking into account children's contributions both to planning and to actual teaching, curriculum issues or classroom interaction. We conceive pedagogical activity as being integrated in stochastical activity, an approach that has been found effective in mathematics teacher education (Cooney, 1999; Tripp, 2012; Vandercleyen, Boudreau, Carliera, & Delens, 2014) and in the analysis of classroom teaching (Potari & Jaworski, 2002). Several different approaches have been used in teacher education to promote reflection. One approach, which is relevant to our study, is the identification and description of critical incidents by the teachers (Goodwell, 2006; Hanuscin, 2013). When following this approach, the researchers also offered teachers critical incidents they observed for further discussion. The approach is similar to the one described by Scherer and Steinbring (2006) who argued that the joint reflection between teachers and researchers on concrete classroom situations is of major importance for teacher development. Mason (1998) considered that the key notions underlying real teaching are 'the structure of attention and the nature of awareness'. Schön (1987) named this knowledge as knowing-in-action. However, he highlighted that this knowledge is not adequate on its own for teaching and that teachers need to evolve through a progression from knowing-in-action to reflecting-on-action to reflecting-in-action. Reflecting-on-action occurs when the teacher starts looking critically at events after they have occurred while reflecting-in-action appears while teaching or planning for teaching. These stages involved a metacognitive awareness in which knowledge and action are linked. For analysing the complexity of teaching mathematics, Potari and Jaworski (2002) used the teaching triad (management of learning, mathematical challenge, and sensitivity to students) as an analytical device and as a reflective agent for teacher professional development. Management of learning describes the teacher's role in the constitution of the classroom-learning environment, for example in the planning of tasks and activities. Mathematical challenge describes the challenges offered to students to engender mathematical thinking and sensitivity to students describes the teacher's knowledge of students and attention to their needs.

The present research focuses on teachers' planning, teaching and reflection on young children's (4–6 year old) stochastic activities. It concentrates on the way in which teachers realize the activity in terms of the mathematical challenge it offers and the development of children's stochastic ideas (Potari & Jaworski, 2002).

12.3 Methodology

12.3.1 Research Design: Participants and Context of Study

Five early childhood teachers (all females) participated in this research, which was organized in three stages: Stage 1—lesson planning, Stage 2—lesson implementation and Stage 3—reflection. In Stage 1, the teachers were engaged in lesson planning. They selected a topic from the national mathematics curriculum on probability and statistics and developed a lesson plan and accompanying teaching material aligned with the learning objectives specified in the curriculum. The lesson plans were shared with the researchers for comments and suggestions and were revised based upon received feedback. In Stage 2, the teachers implemented the lesson plans in their classroom, with the support of the researchers. Once the classroom implementation was completed, in Stage 3, teachers were interviewed and prepared and submitted a reflection paper, where they shared their observations on students' reactions during the lesson, noting what went well and what difficulties they faced and making suggestions for improvement.

12.3.2 Instruments, Data Collection and Analysis Procedure

Multiple forms of data were collected:

(i) *Observations and artefacts collected during Stage 1*: early childhood teachers' submitted work (students' *entry slips*-student's responses on pre-knowledge tasks, lesson plans, etc.), researchers' observations and field notes;

(ii) *Observations and artefacts collected during Stage 2*: each early childhood teacher intervention that took place lasted for 80 min (two teaching periods). Researchers were present, observing closely and videotaping the lesson, keeping field notes and collecting samples of student work.

(iii) *Individual interviews and reflection reports during Stage 3*: upon completion of Stage 2, the researchers conducted semi-structured interviews with each of the teachers. Qualitative data were also obtained from the reflection papers prepared by the teachers.

For the purpose of analysis, we did not use an analytical framework with predetermined categories to assess how teachers' perceptions evolved. We identified, through careful reviewing of the transcripts, reports, and other data collected during the study, recurring themes or patterns in the data. To increase the reliability of the findings, the activities were analysed and categorized by the researchers. Inter-rater discrepancies were resolved through discussion.

12.4 Results: The Existence of Randomness

Findings concerning the study research question on what early childhood teachers attend to when designing, implementing and reflecting on probability tasks will be discussed with regards to one of the three elements of the teaching triad: mathematical challenge, which we call here 'stochastical challenge'.

The learning objectives included in all five teachers' lesson plans and did refer to challenges for teaching probability like the probabilistic concepts of certain, impossible and probable events. Moreover, the lesson plans included activities that involved children in authentic statistical inquiry. Nonetheless, similarly to researchers such as Lee and Mojica (2008), we also found that teachers missed the opportunity to develop their students' understanding of the randomness due to limited subject matter knowledge. The activities included in their lesson plans indicated that they tended not to recognize the concept of randomness, as it appeared difficult for them to see the stochastical idea underlying tasks (the context).

In presenting the main insights from our study, we focus on three of the five participating teachers (Alice, Miranda, Sally),[1] as they are the more typical cases in the study referring to the existence of randomness.

12.4.1 Stage 1: Lesson Planning

In Alice's case, the aim of the lesson plan she designed was for 'children to make predictions based on impossible, possible and certain events'. She decided to use the context of children making hats for a party:

[1] All the names are pseudonyms.

Alice	You know, children need an everyday scenario… I decided to ask them to make hats with dots. The dots will be blue and/or red… they [the children] will select a spinner, make a prediction as to whether it would be possible for the hat to have blue dots or not, mark it on their board and then make their hat…
Researcher	So, will they use their spinners?
Alice	If they like to…yes. I know that some students will know beforehand how to make their hat …But… maybe it is not necessary…

The other activities included in Alice's lesson plan also indicated lack of attention to the concept of randomness. Alice was focusing only on children making decisions on what might have happened and what not. We had a discussion with Alice to find out why she did not 'allow' children to spin the spinners. She referred to management issues, but after we pointed out the importance of children experiencing randomness, Alice decided to ask children to spin the spinner ten times, record the colour coming up each time and then to make their hat based on the recorded outcomes. Figure 12.1 shows how Alice organized her classroom based on the above task. We can see that children had different hats of blue and red dots based on their recorded outcomes.

In Miranda's case, there was confusion between *predicting* and *guessing*. She decided to give some pictures to the children and to ask them to make a guess as to whether these pictures were taken from a forest, a sea or a field. Similarly to Alice, Miranda's aim was also for children to use the term possible. However, students were not directed in this activity towards the stochastical connotation of the term, since the element of randomness was totally missing from the lesson. Thus, although children might have wondered about where each picture was taken, they did not really approach the idea of probability. In the following conversation, the researcher asked Miranda to explain how she perceived the concept of probability:

Researcher	What does the concept of probability mean to you?
Miranda	You know that something is certain or not or *something is possible.*
Researcher	What does it mean when we say that something is possible?

Fig. 12.1 Material used in Alice's task

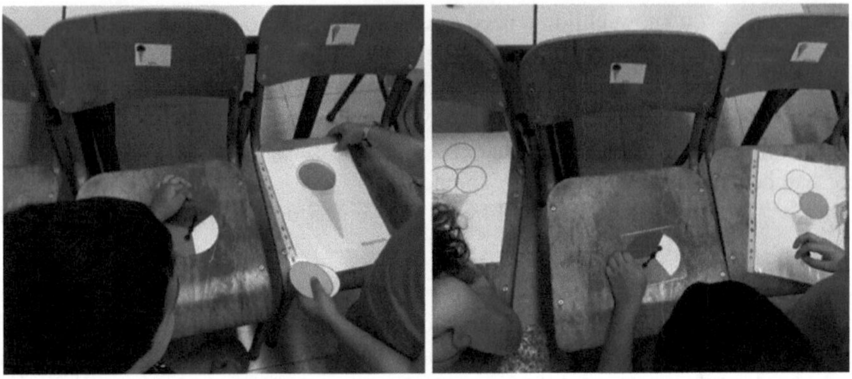

Fig. 12.2 Children playing with a spinner in Sally's task

Miranda You are not sure about something…but someone else might know…
Researcher So you guess what the other person knows?
Miranda You mean that in this activity I know and children do not know. But, is this probability?
Researcher Is randomness somewhere there?
Miranda So…maybe this task is not about the concept of probability?

We can argue here that the idea of probability was confused with 'what something may be' instead of 'what may happen'. Miranda appeared not to realize that probability is related to the likelihood of an *event* occurring, and she instead related it with the likelihood of something being a fact. We identified some 'apprehension' in Miranda's words. We could argue that the teacher has based her definition on her own intuitions and has not been able to identify the stochastical meaning of probability.

12.4.2 Stage 2: Lesson Implementation

In Sally's case, we recognized an awareness of the concept of randomness. Sally was very specific in her two aims for the lesson, which were the following: 'Children should be in a position to select the appropriate sample spaces for certain, impossible and fair events', and 'Children should experience the concept of randomness'.

At the beginning of Sally's lesson, children selected their favourite flavour of ice cream. They had this selection stacked on their chair and were given a 'fair' spinner with three flavours. They worked in pairs, with one child spinning the spinner and the other one comparing the flavour that had come up in the spinner with his/her favourite flavour (Fig. 12.2).

While children were working in pairs, Sally had the following conversation with them:

Sally So, did you get your favourite flavour?

Child 1 No, I didn't.
Sally Why was that?
Child 1 I only turned the spinner once.
Child 2 Can we turn it many times?
Sally Do you think you will get your favourite flavour like this?
Child 2 I might.
Sally Turn it to find out!

It was good to see Sally letting the children experience randomness. Another activity she planned and implemented was to show children different spinners, which had different sample spaces (e.g. whole brown (as a chocolate flavour), whole red (as a strawberry flavour), whole white (vanilla flavour), ¼ brown-3/8 red-3/8 white). Children were asked questions such as: '*Which spinner should you use in order to get your favourite flavour? Why? If I spin this spinner will I get strawberry? Why or why not?*'

The interesting point here is that Sally used the 'why' question and her activity was addressing the aims of the lesson. She seemed to be very aware of the epistemological characteristics of the stochastical activity in her lesson plan. In the planning of the activity, she stated that children would *experience* the concept of randomness and she tried in various ways to achieve this, although at some instances during her lesson, she presented the spinners to the children and just asked them what colour they would get without having them spin any of the spinners. Sally often asked the question why, but her questions were sometimes phrased in such a way that they provoked specific answers. For example, 'For getting strawberry should we use this spinner?' Posing such type of questions resulted in children providing yes or no answers rather than using words like certain, impossible, possible.

12.4.3 Stage 3: Reflection

Sally's case is interesting from the point of view that although we recognized from her lesson plan an awareness of stochastic activities, we can see from her teaching that she put a lot of effort in understanding the probability concept and the presence of randomness in her activities. On this point, in her interview she stated:

Sally This activity was aimed at children seeing different sample spaces and making decisions with regard to which colour would win. It is not an easy task. Well, I thought that it was easy but this idea has something different from other concepts. You don't know from the beginning what will be the answer, so you have to be 'alert'. It was a lesson where I had to make a discussion with them….It wasn't like the one I did with numbers…I had to know well my lesson plan and what I aimed to achieve, and then be able to change the plan while teaching…
Researcher What did you want children to learn?

Sally Certain, Possible, Different sample spaces…Well, the activity went well. It was not difficult for them… and I think children built a good understanding of the idea of randomness….

Researcher What was the difficult for you?

Sally You know, I spent a lot of time to design the lesson…finding the scenario, making the materials…but the 'key' was for me to understand…you know…to feel confident that I know…this subject [probability] seems easy but it is not so easy to teach… you need to *know.*

Researcher What was it that you needed to know?

Sally There were many times that in order to continue teaching I had to grasp from their predictions, continue with *having the experience* and then find a way lets say…giving them [to the children] a chance to 'control' this 'uncontrolled idea' [randomness]… For example, Dani was sure that he would get brown in a fair spinner…that was not the case…At that point I had to be able to give him an experience of randomness and also help him build his new prediction…

Sally seemed able to recognize some critical incidents in her lessons and to reflect on them. She paid particular attention to the provision of opportunities for children to *experience randomness*, something that was critical in building on children's intuitions.

Similarly, Alice reflected on her lesson:

Alice I spent hours trying to find the scenario, making the materials and I felt that everything was ready…In this lesson that was not the case… I had also to be able during my teaching to find ways to help children adjust their predictions, making questions and giving the right spinner…I also want to mention that using technology to help children visualize what would happen in many turns was critical for helping them understand that the spinners were fair, even when they did not get what they expected

In her teaching intervention, Alice used at the end of the lesson an applet with a fair spinner, in order for the class to get an approximately equal number of occurrences for each colour. It was very encouraging that she recognized, through her students' predictions, the need for using technology to help children understand fairness.

12.5 Discussion

These findings suggest that combining mathematical know-how (Boaler, 2003) with mathematical content knowledge is very important for early childhood educators to be able to teach stochastic ideas like randomness. Teachers discussed critical incidents they observed (Goodwell, 2006; Hanuscin, 2013). Critical incidents were crucial to joint reflection between teachers and researchers on concrete classroom situations (Scherer and Steinbring, 2006). Teachers in our study recognized that they needed to know not only what randomness is and how to deal with this concept, but also how to make it visible in their lessons. They pointed out that successful introduction of the idea of randomness in the early mathematics classroom requires knowledge of how to build upon children's prior intuitions regarding random phenomena, as well as how to react during the lesson to students' predictions regarding stochastic events and to the actual outcomes of such events.

All of our early childhood teachers appreciated the importance of using materials and finding everyday scenarios for teaching probability. In the follow-up interviews and reflection papers, they all exhibited understanding of the fact that these tools acted as a means for not only motivating children, but also for challenging them in reasoning about probability. However, quite often the stochastical challenges they posed in their classroom were rather trivial in both the planning and the classroom implementation stages. Teachers did set learning objectives that were related to probabilistic concepts, but these objectives appeared to have certain limitations in at least some of the cases. At some instances, the learning objectives were too general, emphasized procedures rather than conceptual understanding, and/or were disconnected from the designed tasks. For example, in some of the stochastic lesson plans, one of the aims was for children to use and understand the words 'certain', 'probable' and 'impossible'. However, although teachers and children were using verbally these words in various scenarios, the scenarios lacked the idea of randomness. Furthermore, teachers often used closed-ended questions rather than engaging the class in investigative activities. Most of the teachers' attention during the planning stages was focused on what Ball and Bass (2000) spoke about the curriculum knowledge as including knowledge of educational goals, where specific concepts appear. This type of knowledge may help teachers appropriately sequence the introduction of statistical ideas in a curriculum (Godino, Ortiz, Roa, & Wilhelmi, 2011). However, there is considerable variability in how teachers interpret curriculum materials. The findings show that it was through actually teaching the lesson and after reflection that the knowledge components/challenges became more evident for these early-stage teachers. Similar findings were obtained in a previous study conducted by Paparistodemou et al. (2006), which had investigated prospective teachers' awareness of young children's stochastic activities. The fact that similar tendencies were observed in both pre-service and in-service teachers might lead us to conclude that awareness is not only a matter of teaching experience, but also of the extent to which teachers engage with the process of noticing and understanding their teaching.

In the case of Sally and Alice, we saw some degree of awareness of important concepts like randomness. At some instances, these two teachers indicated some understanding of randomness and of how it relates to other probabilistic concepts, but this was not always the case. We could say that their awareness remains at the level of action (mathematical), and it does not indicate a greater degree of awareness that of the awareness in discipline (Mason, 1998).

Summing up, this study shows the kind of experiences that a small group of early childhood teachers' incorporated into the design, implementation and evaluation of teaching related to stochastics. It can be argued that these teachers did build some relations between theory and teaching practice, but that their transition and reflection to more specific stochastical and pedagogical issues appeared to be a difficult endeavour. This calls upon special attention and reflection on behalf of mathematics teacher educators to tackle this problem.

12.6 Limitations of the Study

A number of issues warrant a cautious interpretation of the results. Turning to the participants of the study, the early childhood teachers were all women who participated in a voluntary capacity in the study. The researchers acted as their advisors throughout all of the activities. This involvement was an integral part of the design process and the implementation of tasks.

12.7 Implications for Teacher Education and Future Research

In conclusion, it appears that the 'mathematical know-how' procedure (Boaler, 2003) is a difficult endeavour in designing lessons with engaging contexts (Chick & Pierce, 2008). The early childhood teachers in this study had rich ideas on the context, but they needed extra effort to understand the stochastical idea hidden in the tasks. The findings of the study show that this procedure calls upon a balance between the stochastical challenge, the management of learning and sensitivity to students' behaviour. We suggest a way of developing early childhood teachers' implementation of designing, implementing and reflection process by identifying critical incidents, analysing aspects of these incidents and finally by reconsidering their planning and suggesting changes in the light of these experiences. Moreover, it is really important for early childhood teachers to realize the importance of accepting and experiencing the concept of randomness in their tasks. The successful introduction of critical incidents as a tool for reflection described by Potari et al (2011) supports further the potentiality of this approach to mathematics teacher education.

Further research is needed to investigate approaches that teacher educators can use to support teachers in directing balanced attention to the different dimensions that constitute the teaching of stochastics. Our work shows that the relationship between theory and practice becomes an element of both teacher education and researcher development. Further research has a crucial role to play in supporting early childhood teachers to 'transform' theoretical ideas in school experience.

References

Ball, D. L., & Bass, H. (2000). Interweaving content and pedagogy in teaching and learning to teach: Knowing and using mathematics. In J. Boaler (Ed.), *Multiple perspectives on the teaching and learning of mathematics* (pp. 83–104). Westport, CT: Ablex.

Boaler, J. (2003). Exploring the nature of mathematical activity: Using theory, research and 'working hypotheses' to broaden conceptions of mathematics knowing. *Educational Studies in Mathematics, 51*(1–2), 3–21.

Boaler, J. (2009). *The elephant in the classroom. Helping children learn and love maths.* London: Souvenir Press.

Chick, H. L., & Pierce, R. U. (2008). Teaching statistics at the primary school level: Beliefs, affordances, and pedagogical content knowledge. In C. Batanero, G. Burrill, C. Reading, & A. Rossman (Eds.), *Joint ICMI/IASE study: Teaching statistics in school mathematics. Challenges for teaching and teacher education. Proceedings of the ICMI Study 18 and 2008 IASE Round Table Conference.* Monterrey: ICMI and IASE. Retrieved from http://www.ugr.es/~icmi/iase_study.

Godino, J. D., Ortiz, J. J., Roa, R., & Wilhelmi, M. R. (2011). Models for statistical pedagogical knowledge. In C. Batanero, G. Burrill, & C. Reading (Eds.), *Teaching statistics in school mathematics: Challenges for teaching and teacher education* (pp. 271–282). Dordrecht, The Netherlands: Springer.

Goodwell, J. (2006). Using critical incident reflections: A self-study as a mathematics teacher educator. *Journal of Mathematics Teachers Education, 9*(3), 221–248.

Cooney, T. J. (1999). Conceptualizing teachers' ways of knowing. *Educational Studies in Mathematics, 38*(1–3), 163–187.

Groth, R. E. (2010). Teachers' construction of learning environments for conditional probability and independence. *International Electronic Journal of Mathematics Education, 5*(1), 32–55. Retrieved from http://mathedujournal.com/dosyalar/IJEM_v5n1_2.pdf.

Groth, R. E., Kent, K., & Hitch, E. (2015). Journey to centers in the core. *Mathematics Teaching in the Middle School, 21*(5), 295–302.

Hannigan, A., Gill, O., & Leavy, A. M. (2013). An investigation of prospective secondary mathematics teachers' conceptual knowledge of and attitudes towards statistics. *Journal of Mathematics Teacher Education, 16*(6), 427–449. https://doi.org/10.1007/s10857-013-9246-3.

Hanuscin, D. L. (2013). Critical incidents in the development of pedagogical content knowledge for teaching the nature of science: A prospective elementary teacher's journey. *Journal of Science Teacher Education, 24*(6), 933–956.

Hoyles, C. (2002). From describing to designing mathematical activity: The next step in developing a social approach to research in mathematics education? *Educational Studies in Mathematics, 46*(1), 273–286.

Jones, G., Langrall, C., Thornton, C., & Mogill, A. (1999). Using students' probabilistic thinking in instruction. *Journal of Research in Mathematics Education, 30,* 487–519.

Kilpatrick, J. (2001). Understanding mathematical literacy: The contribution of research. *Educational Studies in Mathematics, 47*(1), 101–116.

Leavy, A. M. (2010). The challenge of preparing preservice teachers to teach informal inferential reasoning. *Statistics Education Research Journal*, *9*(1), 46–67. Retrieved from http://iase-web.org/documents/SERJ/SERJ9(1)_Leavy.pdf.

Lee, H. S. & Mojica, G. S. (2008). Examining how teachers' practices support statistical investigations. In C. Batanero, G. Burrill, C. Reading, & A. Rossman (Eds.), *Joint ICMI/IASE study: Teaching statistics in school mathematics. Challenges for teaching and teacher education. Proceedings of the ICMI Study 18 and 2008 IASE Round Table Conference.* Monterrey: ICMI and IASE. Retrieved from http://www.ugr.es/~icmi/iase_study.

Maher, N., & Muir, T. (2014). “I don’t really understand probability at all”: Final year pre-service teachers’ understanding of probability. In J. Anderson, M. Cavanagh, & A. Prescott (Eds.), *Curriculum in focus: Research guided practice. Proceedings of the 37th annual conference of the mathematics education research group of Australasia* (pp. 437–444). Sydney: MERGA.

Makar, K., & Rubin, A. (2009). A framework for thinking about informal statistical inference. *Statistics Education Research Journal, 8*(1), 82–105.

Mason, J. (1998). Enabling teachers to be real teachers: Necessary levels of awareness and structure of attention. *Journal of Mathematics Teachers Education, 1*(3), 243–267.

Moyer, P. S. (2001). Are we having fun yet? How teachers use manipulatives to teach mathematics. *Educational Studies in Mathematics, 47*(2), 175–197.

Noss, R., & Hoyles, C. (1996). *Windows on mathematical meanings: Learning cultures and computers*. Dordrecht: Kluwer.

Paparistodemou, E., & Meletiou-Mavrotheris, M. (2008). Enhancing reasoning about statistical inference in 8 year-old students. *Statistics Education Research Journal, 7*(2), 83–106.

Paparistodemou, E., & Noss, R. (2004). Designing for local and global meanings of randomness’ continuum. In *Proceedings of the twentieth eighth annual conference of the international group for the psychology of mathematics education* (Vol. 3, pp. 497–504), Bergen, Norway.

Paparistodemou, E., Noss, R., & Pratt, D. (2008). The interplay between fairness and randomness in a spatial computer game. *International Journal of Computers for Mathematical Learning, 13*(2), 89–110.

Paparistodemou, E., Potari, D., & Pitta, D. (2006). Prospective teachers’ awareness of young children’s stochastic activities. In A. Rossman & B. Chance (Eds.), *Proceedings of the seventh international conference on teaching statistics*, Salvador, Brazil. International Statistical Institute and International Association for Statistical Education. Retrieved from https://iase-web.org/documents/papers/icots7/2A1_PAPA.pdf.

Petocz, P., & Reid, A. (2002). How students experience learning statistics and teaching. In B. Phillips (Ed.), *Proceedings of the 6th international conference of teaching statistics (ICOTS 6)*, Cape Town, South Africa. International Statistical Institute and International Association for Statistical Education. Retrieved from http://citeseerx.ist.psu.edu/viewdoc/download?doi=10.1.1.215.7406&rep=rep1&type=pdf.

Potari, D., & Jaworski, B. (2002). Tackling complexity in mathematics teaching development: Using the teaching triad as a tool for reflection and analysis. *Journal of Mathematics Teacher Education, 5*(4), 351–380.

Potari, D., Psycharis, G., Kouletsi, E. & Diamantis, M. (2011). Prospective mathematics teaching noticing of classroom practice through critical events. *Proceedings of CERME- 7*, Rzeszów, Poland.

Pratt, D. (2000). Making sense of the total of two dice. *Journal for Research in Mathematics Education, 31*(5), 602–625.

Reston, E., Jala, L., & Edullantes, T. (2006). Probing college statistics teachers’ instructional goals and classroom practices within a statistical literacy framework. In A. Rossman & B. Chance (Eds.), *Working cooperatively in statistics education: Proceedings of the seventh international conference on teaching statistics*, Salvador, Brazil. Voorburg, The Netherlands: International Association for Statistical Education and International Statistical Institute.

Scherer, P., & Steinbring, H. (2006). Noticing children’s learning processes—Teachers jointly reflect on their own classroom interaction for improving mathematics teaching. *Journal of Mathematics Teachers Education, 9*(2), 157–185.

Schön, D. A. (1987). *Educating the reflective practitioner: Toward a new design for teaching and learning in the professions*. San Francisco, CA: Jossey-Bass.

Serradó, A., Meletiou-Mavrotheris, M., & Paparistodemou, E. (2014). Early statistics: A course for developing teachers’ statistics technological and pedagogical content. *Statistique et Enseignement, 5*(1), 5–29.

Shulman, D. (1986). Those who understand: Knowledge growth in teaching. *Educational Researcher, 15*(2), 4–14.

Theis L., & Savard A. (2010). Linking probability to real-world situations: How do teachers make use of the mathematical potential of simulation programs? In C. Reading (Ed.), *Data and context in statistics education: Towards an evidence-based society. Proceedings of the 8th international conference on teaching statistics*, Ljubljana, Slovenia. Voorburg, The Netherlands: International Statistical Institute. Retrieved from https://iase-web.org/documents/papers/icots8/ICOTS8_C126_THEIS.pdf.

Tripp, D. (2012). *Critical incidents in teaching: Developing professional judgment*. New York: Routledge.

Vandercleyen, F., Boudreau, P., Carliera, G., & Delens, C. (2014). Pre-service teachers in PE involved in an organizational critical incident: Emotions, appraisal and coping strategies. *Physical Education and Sport Pedagogy, 19*(2), 164–178.

Watson, J. M. (2001). Profiling teachers' competences and confidence to teach particular mathematics topics: The case of chance and data. *Journal of Mathematics Teacher Education, 4*(4), 305–337.

Chapter 13
Design, Implementation, and Evaluation of an Instructional Sequence to Lead Primary School Students to Comparing Groups in Statistical Projects

Daniel Frischemeier

Abstract Concerned citizens need statistical skills to be able to participate in public decision-making processes. The first steps in the development of these skills can be made as early as primary school. Here, students can have initial experience in the data analysis cycle including posing statistical questions, preparing for data collection, collecting data, and analyzing data to answer the statistical question(s). Some challenges are enabling students to explore large datasets and leading them to more sophisticated activities such as group comparisons. In the first part of this chapter, we will describe the design and implementation of a teaching unit on early statistical reasoning for German primary school students in Grade 4 (age 9–10). Different representation levels (enactive, iconic, symbolic) are used to enhance students' understanding of statistical displays. TinkerPlots was used to help students analyze large and real datasets and to enable them to compare groups. In the second part of this chapter, we will present results from an accompanying qualitative research study investigating how Grade 4 students compared groups after the teaching unit described in the first part of this chapter.

13.1 Introduction

Statistical reasoning is important for everyday life because many decisions in politics, economics, and society are based on statistics. So it is critical that we educate statistically literate students. The cornerstone for this can already be set in primary school. The recommendations for "data, frequency and chance" in Grades 1–4 for primary school in Germany include posing statistical questions, collecting data, getting to know how to represent data, and creating and interpreting representations of data (Hasemann & Mirwald, 2012). These recommendations imply that students need to conduct a data analysis cycle such as PPDAC (Wild & Pfannkuch, 1999), where students are asked to plan data collection for a given statistical problem (P),

D. Frischemeier (✉)
University of Paderborn, Paderborn, Germany
e-mail: dafr@math.upb.de

A. Leavy et al. (eds.), *Statistics in Early Childhood and Primary Education*,
Early Mathematics Learning and Development,
https://doi.org/10.1007/978-981-13-1044-7_13

prepare for data collection (P), collect data (D), analyze and interpret data (A), derive conclusions (C), and finally to answer the problem posed at the beginning. In this chapter, we call this application of the PPDAC cycle a statistical project and we concentrate on how to enhance early statistical reasoning with a focus on the "analyzing and interpreting data" component of the PPDAC cycle. In order to support the analysis and interpretation of data in statistical projects, it is necessary to develop competence with statistical displays, to handle and explore data in large and real datasets, and to be able to carry out fundamental statistical activities such as group comparisons. The use of adequate digital tools is especially important for exploring large datasets from real-world contexts. This chapter addresses the following three questions:

- In what manner is it possible to introduce early statistical reasoning elements (in regard to analyzing large datasets) in German primary school?
- In which ways are primary school students (Grade 4) able to compare non-equal-sized groups before and after experiencing our teaching unit?
- In which way does the performance of primary school students (Grade 4) for comparing groups improve after experiencing the teaching unit?

The first part of this chapter will present activities which are developed and designed via the design-based research approach (Cobb, Confrey, diSessa, Lehrer, & Schauble, 2003). These activities are intended to enhance early statistical reasoning about analyzing and interpreting data in statistical projects. The second part of this chapter presents results of an empirical study about comparing groups and how Grade 4 students compared groups before and after experiencing the teaching unit.

13.2 Conducting Statistical Projects in Primary School

Statistical projects, as we conceive of them, include the following six phases: (1) posing a statistical problem, (2) generating statistical questions, (3) preparing for data collection, (4) collecting data, (5) analyzing and interpreting data, and (6) presenting conclusions to solve the statistical problem posed in (1). With our focus on phase (5) there are three fundamental components in the "analyzing and interpreting data" phase: First, students need a good understanding of statistical displays; second, they need to be able to explore real and motivating questions in statistical projects and explore large datasets; third, they need to be able to compare groups. We refer to the three components as *understanding statistical displays*, *exploring large datasets,* and *comparing groups* below.

13.2.1 Understanding Statistical Displays

One central goal related to developing understanding of statistical displays in the sense of "reading between" and "reading beyond" the data (Friel, Curcio, & Bright 2001) is to attain a global view of distributions (taking into account global features of a distribution like center, spread, or skewness). When dealing with typical statistical displays like bar graphs or stacked dot plots, learners often focus on local features of the distribution (e.g., a single bar in bar graphs or single points in stacked dot plots) and therefore have problems identifying global features like center, spread, or skewness. Leading primary school students to a global view of distributions of numerical variables is a challenge, because students often concentrate on local features like single points or extreme values (the maximum or the minimum) of a distribution (Bakker & Gravemeijer, 2004). For developing a global view on, for example, stacked dot plots, so-called modal clumps (see Konold et al., 2002 and Bakker, 2004) might serve as fruitful precursors for the concepts of center and spread of a distribution. To develop a deeper and competent understanding of conventional statistical displays (like bar charts) and to lead students to a global view on distributions, it might be helpful to approach statistical displays on different representation levels (enactive, iconic, symbolic). By deeper and competent understanding, we mean that students are enabled to make a bar graph with given categorical data and a stacked dot plot with given numerical data and that they are able to answer questions with their statistical displays (bar graph, stacked dot plot) on the first two levels of Friel et al. (2001, p. 130): reading the data ("What is displayed in the graph?", "Which categories do exist?" or "How many pupils are in category x?") and reading between the data ("Which category is the most frequent/seldom one?", "Are more pupils in category x than in category y? If yes, how much more?" or "Which title would fit to the display?").

In this respect as a first step (enactive), students can do "animated" statistics (in the sense of embodied cognition, see Lakoff & Núñez, 2000) in a way that the students are the statistical units themselves (see Biehler & Frischemeier, 2013 and Biehler & Frischemeier, 2015). That means the students generate categories (like boys and girls) by separating themselves, structuring and ordering themselves in these categories, and finally creating a "human" display such as a bar chart. The transition from animated statistics to more abstract levels can be realized with data cards (Harradine & Konold, 2006). Here, students also can use intuitive data operations like "stack," "separate," and "order" for their data exploration process to construct statistical displays with data cards, the so-called data card bar graphs (see Fig. 13.1).

Furthermore initial multivariate data analysis activities can also be carried out with data cards. Finally (as part of the transition from the iconic to symbolic) after the data cards are taken away, the final stage (symbolic level) is a conventional bar graph (see Fig. 13.2).

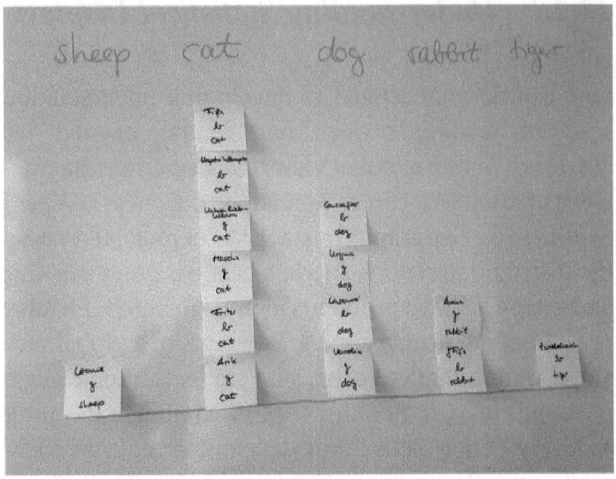

Fig. 13.1 Data cards separated by categories and stacked

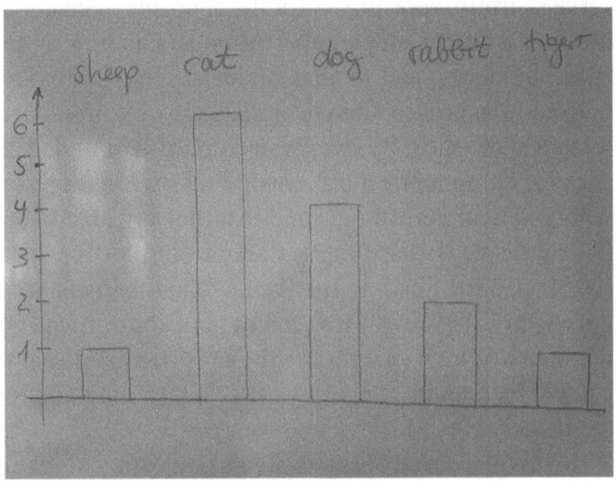

Fig. 13.2 Final stage of data card bar graph

13.2.2 Exploring Large Datasets

For analyzing real, large datasets, the use of educational software is desirable. TinkerPlots (Konold & Miller, 2011) is a educational data analysis software (Biehler, Ben-Zvi, Bakker, & Makar, 2013) for primary and secondary schools that can help students carry out statistical projects in primary school with large datasets. TinkerPlots builds on the work with data cards, stores the data in data card stacks, and offers a plot in which the operations (separate, stack, order) can be performed. TinkerPlots can support operations with data cards, can lead to better comprehension of statistical displays, and can facilitate the exploration of large datasets.

13.2.3 Comparing Groups

Group comparisons are fundamental activities in statistics (Konold & Higgins, 2003) because they incorporate ideas like data, representation, and variation (Burrill & Biehler, 2011). Research (e.g., Watson & Moritz, 1999) shows that at an early stage, students in primary school are able to establish adequate strategies to compare two groups. Also Makar and Rubin (2009) see activities like group comparisons as opportunities to engage young students in making first informal inferences. As previously mentioned, distributions of numerical variables, for example, in the form of stacked dot plots, can be very challenging for students to analyze, because students often tend to focus on single local features (values) and do not have a global view of the distribution. Pre-formal ideas like modal clumps (Konold et al., 2002; Bakker, 2004) can help students identify the center of a distribution and compare distributions by focusing on the shift of modal clumps. The hat plots (Watson, Fitzallen, Wilson, & Creed, 2008), which are also available in TinkerPlots, build on the modal clump concept and bring students to a more formal concept of the middle 50% of the distribution and therefore can help young students identify and compare other important characteristics of distributions of numerical variables such as spread (of the middle 50%).

Frischemeier (2017) found six categories of methods for comparing groups in the work of German pre-service mathematics teachers: center, spread, skewness, shift, p-based, and q-based. In this scheme, center comparisons aim to compare the centers of distributions (mean, median). Comparisons via spread can be done by comparing the ranges or interquartile ranges of distributions. Skewness comparisons take into account the comparisons of skewness of distributions (left-skewed, right-skewed, and symmetric). Furthermore, it is possible to compare the shift of two distributions by comparing the modal clumps of the distributions. Comparisons of two distributions of numerical variables are called p-based, if a cut-point x can be given (e.g., 10 h) and the proportion of cases which are equal or larger than this value x (e.g., 10 h) is compared in both groups (see Biehler, 2001, p. 110). In addition, comparisons of two distributions of numerical variables are called q-based, if distributions are compared in regard to their quartiles or more general in regard to their quantiles. Comparisons in regard to center, spread, skewness, shift, and p-based can already be taught in primary school with slight modifications: Instead of concentrating on formal concepts of center like the mean or formal concepts of spread like the interquartile range, one could use modal clumps as precursors for center and spread.

In the following section, we will outline how we implemented these three components *understanding statistical displays*, *exploring large datasets,* and *comparing groups* in the design and implementation of our teaching unit for Grade 4 in primary school in Germany.

13.3 Design and Implementation of Activities to Enhance Early Statistical Reasoning and to Lead Grade 4 Students to Comparing Groups

The teaching unit we present in this chapter was designed, implemented, evaluated, and revised as part of two bachelor theses (Breker, 2016; Schäfers, 2017) in cooperation with the author of this chapter. The lessons were taught by the bachelor students (Breker, 2016; Schäfers, 2017) themselves. Both bachelor students were studying teaching mathematics for primary school and have been in their fifth semester when conducting their teaching experiment. In evaluating the teaching unit, the bachelor students collected and analyzed written notes and TinkerPlots data files of all students. Furthermore, they took field notes of all of the lessons in the teaching unit for a retrospective analysis.

With the aims of helping Grade 4 students to develop understanding of statistical displays, explore large datasets with software, and to compare groups, we designed the teaching unit using the design-based research approach (Cobb et al., 2003). In regard to general design issues of our teaching unit, we implemented elements of the statistical reasoning learning environment (Garfield & Ben-Zvi, 2008). For instance, we focused on central statistical ideas (group comparisons), used real and motivating datasets (class and school data), used classroom activities (cooperative learning environments), and also integrated the use of appropriate technological tools (TinkerPlots).

The teaching unit was framed in terms of conducting a statistical project with the six phases (listed in Table 13.1). We chose the overarching topic "Our school in numbers." Students investigated the leisure time activities (e.g., sports activities) and preferences (e.g., favorite pet, favorite meal) of their schoolmates.

Table 13.1 Phases of the teaching unit

No.	Phase	Content
1	Statistical problem	Statistical problem was discussed and posed, e.g., "Our school in numbers—we want to get to know more about our school"
2	Generating statistical questions	Statistical questions for the inquiry were generated, e.g., "What does the distribution of the variable *favorite pet* look like for our class/school?"
3	Preparing for data collection	A questionnaire was created
4	Data collection	Data were collected in all classes at school
5	Analyzing and interpreting data	Students were: • introduced to data analysis on different representation levels • introduced to data analysis with TinkerPlots • introduced to group comparisons
6	Presenting conclusions and results	The results of the inquiry were presented in the form of posters

Table 13.2 Several steps of phase (5) of the teaching unit

Step	Activity	Dataset used (class/school)	Type of variable used
1	Getting started with animated statistics	Class	Categorical
2	Analyzing categorical data with data cards	Class	Categorical
3	Introduction to categorical data analysis with TinkerPlots	Class	Categorical
4	Analyzing categorical data from our school with TinkerPlots	School	Categorical
5	Introduction to numerical data analysis	Class	Numerical
6	Analyzing numerical data from our class with TinkerPlots	Class	Numerical
7	Analyzing numerical data from our school with TinkerPlots	School	Numerical
8	Comparing non-equal-sized groups with TinkerPlots	School	Categorical, numerical (group comparison)

In this chapter, we will focus on the design and implementation of our unit on "analyzing and interpreting data" (phase (5)). In Table 13.2, we show the different steps, which dataset (small = class; large = school), and which type of variable (categorical; numerical) we used to develop students' understanding of statistical displays, to allow them explore large datasets, and to help them make group comparisons.

The eight steps shown (Table 13.2) are intended to lead Grade 4 students stepwise to group comparisons. Using these steps, we aim for three major learning goals to be accomplished in phase (5) of our teaching unit; we want students to: (i) develop a competent understanding of statistical displays like bar graphs or stacked dot plots; (ii) use adequate software to explore larger datasets; and (iii) compare groups using modal clumps and, later at a more formal stage, hat plots.

In the following section, we will describe our implementation of the teaching unit in a Grade 4 class from a rural area primary school in North Rhine-Westfalia (Central Germany). The teaching unit consists of 13 lessons, and each lesson lasts 45 min. The teaching unit is tested for the second time (Breker, 2016; Schäfers, 2017). In this chapter, we will report on the teaching unit of Breker (2016) only because the analysis of the implementation of the teaching unit of Schäfers (2017) is still ongoing. Fourteen students participated in the teaching unit of Breker (2016), and twelve students participated in the teaching unit of Schäfers (2017). All students had little previous knowledge relating to data analysis: They had collected data and documented data in tallies in Grade 3. Furthermore, they were introduced to reading pie graphs in Grade 3.

Step1: Getting Started with Animated Statistics

To introduce Grade 4 students to initial data operations, such as separate and stack, we worked on an enactive level and did animated statistics. In regard to the question

"What does the distribution of the variable favorite pet look like for our class?" the students separated themselves into groups (dog, cat, rabbit, etc.) to build categories. Then, they stacked (in the sense of standing one after another in a line) themselves in the categories. Hence, a human bar graph was created.

Step 2: Analyzing Categorical Data with Data Cards

As a next step, we brought the students to a more abstract level (data cards) to bring them closer to the symbolic level. To collect data to answer the question "What does the distribution of the variable favorite pet look like in our class?", each student received a yellow sticky note and wrote his/her name, gender (b: boy; g: girl), and favorite pet on it. An example of Anna's (pseudonym) data card (in this case with the variables name, gender, and height) can be seen in Fig. 13.3.

The students put the sticky notes with their characteristics on the board (Fig. 13.4, left) and used data operations like "stack" and "separate" (they already have internalized from animated statistics) to construct statistical displays with data cards. For instance, at first (similar to their animated statistics), the data cards were separated by favorite pet into the categories sheep, cat, dog, rabbit, and tiger (see Fig. 13.4, right).

Within the categories, the cards were stacked so that a distribution of the variable "favorite pet" became visible (Fig. 13.5, left) as a data card bar graph. In this bar graph, the students could still make a 1:1 assignment (data card = student) and identify their own data cards in the different categories/bars. To arrive at a more abstract, symbolic level, the teacher framed the data card bars with a pen and then removed the data cards to gain a conventional bar graph with a scale on a symbolic level (Fig. 13.5, right).

At this point, the students learned how to read and interpret the bar graphs. The teacher discussed how to read the data and posed questions like "What is displayed in the graph?", "Which categories exist?" or "How many pupils are in category x?" In addition, the teacher led the students to reading between the data by posing questions such as "Which category is the most frequent/seldom one?", "Are more pupils in category x than in category y? If yes, how many more?" or "Which title would fit the display?"

Fig. 13.3 Anna's data card

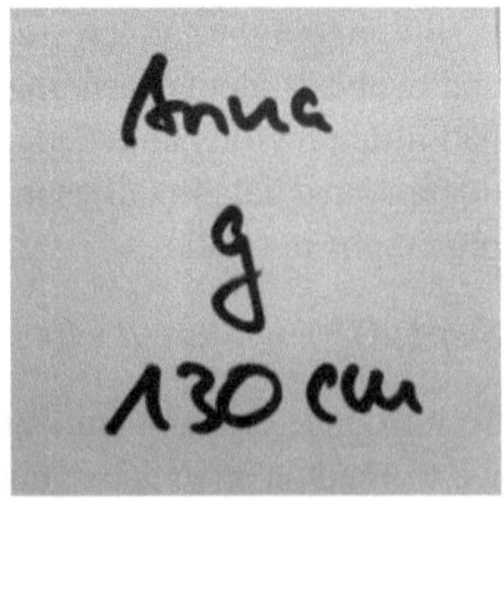

Fig. 13.4 Data cards unordered on board (left), and data cards separated by categories, but still not stacked (right)

Step 3: Introduction to Categorical Data Analysis with TinkerPlots

As the third step, when the students have internalized the data card operations (in the teaching unit by Breker (2016), this process took three lessons) and they learned about reading and interpreting bar graphs, TinkerPlots was introduced alongside the well-known operations (stack, separate, order) with data cards. The teacher demonstrated the fundamental features (that the data are stored in data card stacks in TinkerPlots) and the operations (stack, separate, order) and showed the students how to create bar graphs in TinkerPlots analogous to the data card operations done earlier on the board. The teacher first used square icons with the names of the students labeled inside (see Fig. 13.6, left), so that students could still assign each data card to themselves, and then, the icons were fused rectangular to create a conventional bar graph in TinkerPlots (Fig. 13.6, right).

Step 4: Analyzing Categorical Data from Our School with TinkerPlots

In step 4, students explored larger datasets in TinkerPlots. The same operations they had internalized during exploration with smaller datasets were carried out with

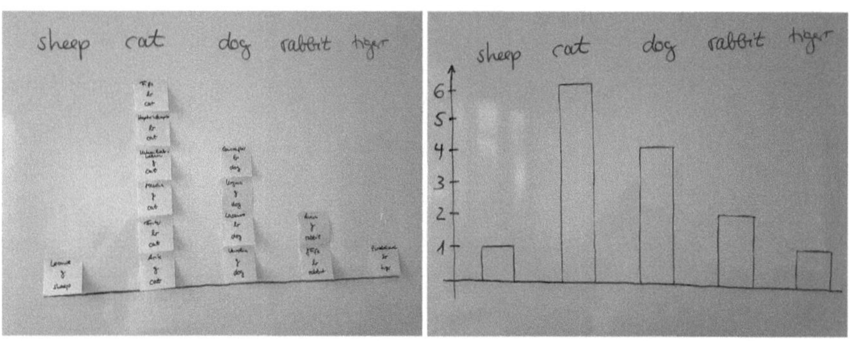

Fig. 13.5 Data cards separated by categories and stacked (left), and final stage of bar graph (right)

the larger school dataset in TinkerPlots. Here, TinkerPlots was used to carry out the operations with data cards, and to support the exploration of large datasets. In Fig. 13.7 (left), we can see a precursor to a bar graph in TinkerPlots with the cases (dots) visible in each bin. As a further developed graph, we can see the distribution of the variable *favorite pet* as a conventional bar graph in TinkerPlots (see Fig. 13.7, right).

Step 5 & 6: Introduction to Numerical Data Analysis and Analyzing Numerical Data from Our Class with TinkerPlots

After dealing with distributions of categorical variables, the students were introduced to the distribution of numerical variables. To accomplish this, students were first asked to collect data on the heights of their classmates. Then, each student received a magnetic dot and was encouraged to place it on the prepared scale on the board. Together with the teacher, the students created a stacked dot plot of the variable height with dots on the board (see Fig. 13.8, left). The teacher then imported the data to TinkerPlots and demonstrated how to create a stacked dot plot of the variable height via the operations of separate and stack in TinkerPlots (Fig. 13.8, right).

Step 7: Analyzing Numerical Data from Our School with TinkerPlots

As shown with the distributions of categorical data (steps 3 & 4), TinkerPlots allows students to handle large amounts of numerical data and to create stacked dot plots. In Fig. 13.9, we see the distribution of the variable *height* as a stacked dot plot. The teacher first drew attention to single local values like the maximum and minimum of the distribution and then emphasized that a global holistic perspective on the distribution is important. For this reason, the teacher demonstrated how to identify a modal clump in the distribution with the drawing tool (see Fig. 13.9). The modal clump was seen as characteristic of the distribution of heights and can be seen as precursor for examining global features (like center and spread) of the distribution.

At this stage, we observed the initial difficulties for students in our stepwise development. Many students had problems understanding how to identify the modal clump in the stacked dot plot distribution. Also, many had problems describing and interpreting the distribution with the modal clump and finding a headline for the statistical display shown in Fig. 13.9.

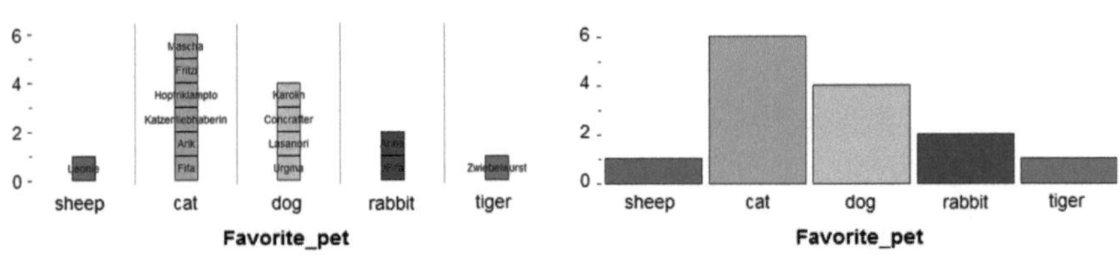

Fig. 13.6 Bar graph with data cards in TinkerPlots (left); final stage of bar graph in TinkerPlots (right)

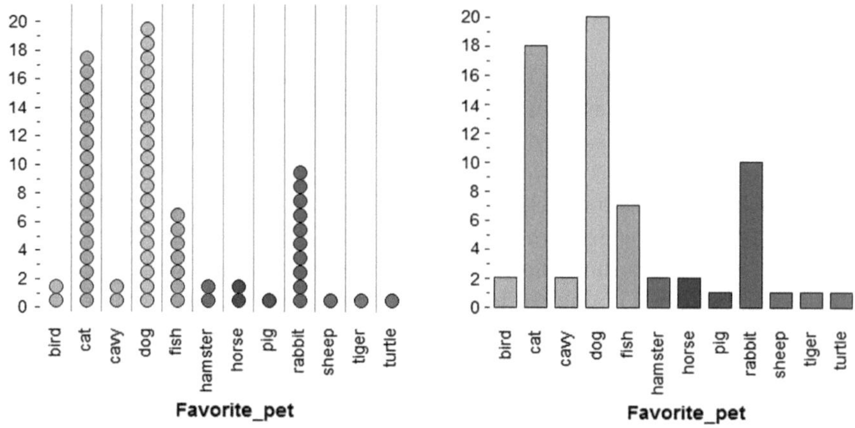

Fig. 13.7 Precursor to a TinkerPlots bar graph (stacked dot plot graph) of distribution of variable favorite pet (left), and final version of bar graph of distribution of variable favorite pet in TinkerPlots (right)

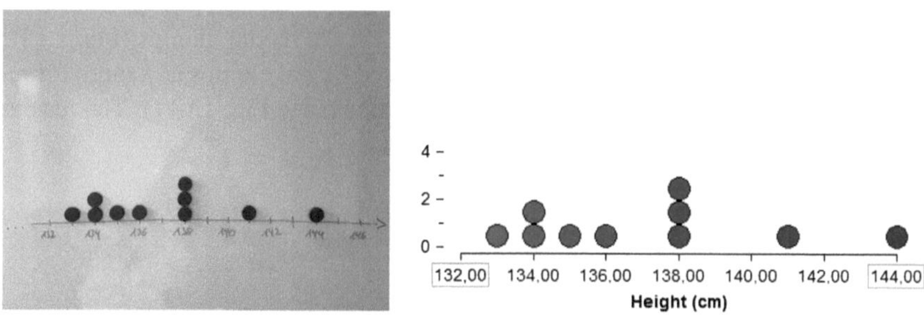

Fig. 13.8 Stacked dot plot of the distribution of the variable height on the board (left), and stacked dot plot of the distribution of the variable height in TinkerPlots (right)

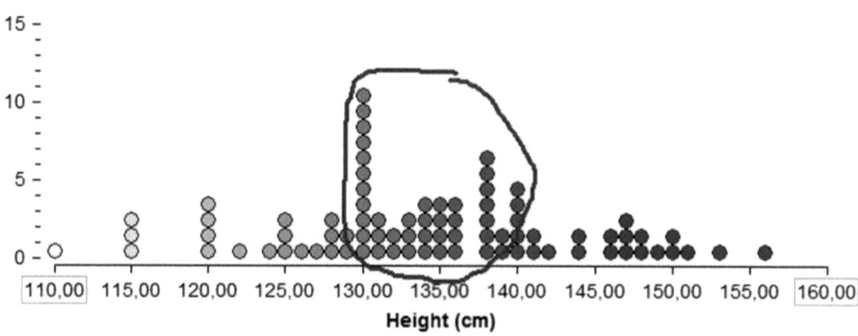

Fig. 13.9 Distribution of the variable *height* in TinkerPlots as a stacked dot plot with identified modal clump

Step 8: Comparing Groups with TinkerPlots

Finally, in step 8, students were introduced to group comparisons and were expected to apply the knowledge gained from the previous steps (describing stacked dot plots, identifying modal clumps, etc.). As an introduction to group comparisons, students were shown (see Fig. 13.10) two distributions of the variable *PackWeight* from the dataset backpacks (which is included in TinkerPlots). Together with the teacher, the students realized that there were 17 Grade 1 and 21 Grade 3 students in the dataset. The students were then shown how to identify the modal clumps in each distribution (grade one/grade three) using the TinkerPlots drawing tool. Afterward, the location of both modal clumps was compared to conclude that in the backpacks dataset Grade 3 students tended to have heavier backpacks than Grade 1 students.

As in step 7, students had problems identifying the modal clumps in the distributions. To practice the identification and the comparison of modal clumps, further group comparisons with distributions given as stacked dot plots were conducted. Finally, the students were shown how to calculate the median and how to use hat plots in TinkerPlots to compare two groups as a more formal comparison procedure than modal clumps (see Figs. 13.11 and 13.12).

Here, the teacher built on the group comparison via modal clumps (Fig. 13.10) and used TinkerPlots to put hat plots on the marked modal clumps as a more formal way to identify the middle 50% of both distributions (see Fig. 13.11). The teacher

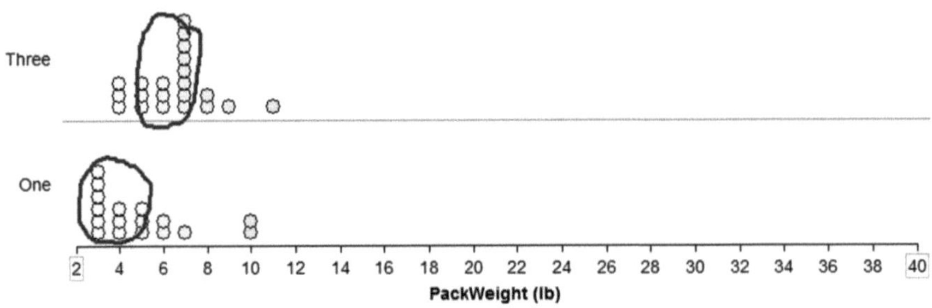

Fig. 13.10 Distributions of the variable PackWeight (Grade 1 and Grade 3) with modal clumps

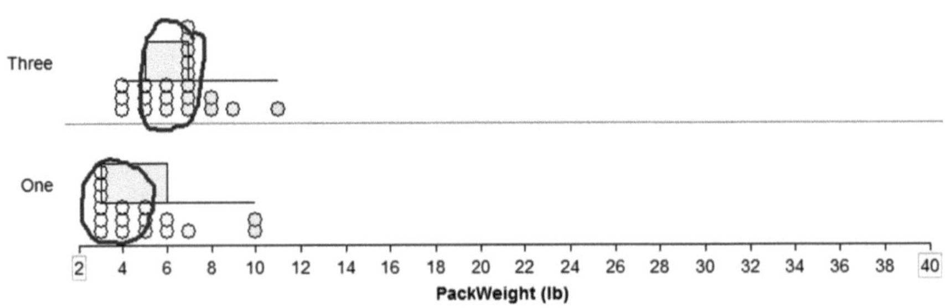

Fig. 13.11 Distributions of the variable PackWeight (Grade 1 and Grade 3) with modal clumps and hat plots

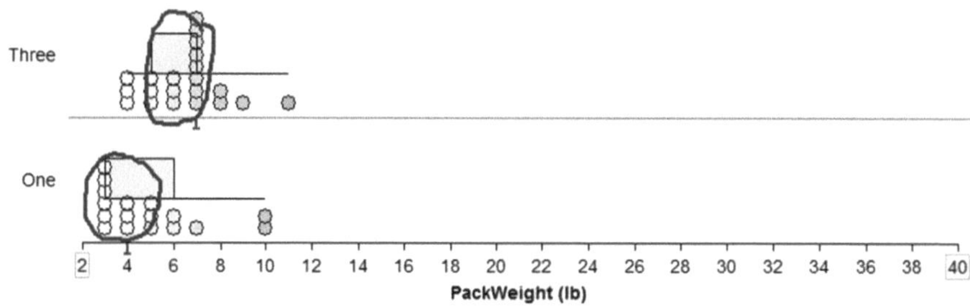

Fig. 13.12 Distributions of the variable PackWeight (Grade 1 and Grade 3) with modal clumps, hat plots, and medians

then pointed out that the crown of the upper hat plot was shifted more to the right than the crown of the lower hat plot and therefore stated that Grade 3 students tended to have heavier backpacks than Grade 1 students. For the median, the teacher used modal clumps to estimate the location of the median to have a rough approximation of the center of the distribution (see Fig. 13.12).

13.4 Study of Primary School Students' Statistical Reasoning When Comparing Groups After Attending to Our Activities

In this section, we concentrate on an empirical study conducted in the frame of the teaching unit we described. One main goal of the teaching unit was to lead Grade 4 students to sound statistical reasoning for comparing groups via modal clumps. So we investigated the ways in which Grade 4 students performed group comparisons when comparing two non-equal-sized groups before and after participating in our teaching unit.

Research Questions, Design & Method of the Study

The following research questions guided our work: *In which ways are primary school students (Grade 4) able to compare non-equal-sized groups before and after experiencing our teaching unit? In which way does their performance for comparing groups improve after experiencing the teaching unit?*

To investigate these research questions, we presented Grade 4 students a group comparison task before and after the teaching unit to observe how they improved and in which ways their group comparison strategies changed. We distributed an exercise sheet with two distributions of the variable "Playing_computer_games_in_hours_per_week" in the form of stacked dot plots. In the task, students were asked: "Do 4th graders tend to play more computer (in hours per week) than 3rd graders? Explain!" (see Fig. 13.13).

In a survey students of grade 3 and grade 4 were asked how many hours they spend on playing computer games per week. The result of the survey can be seen in the diagram below.

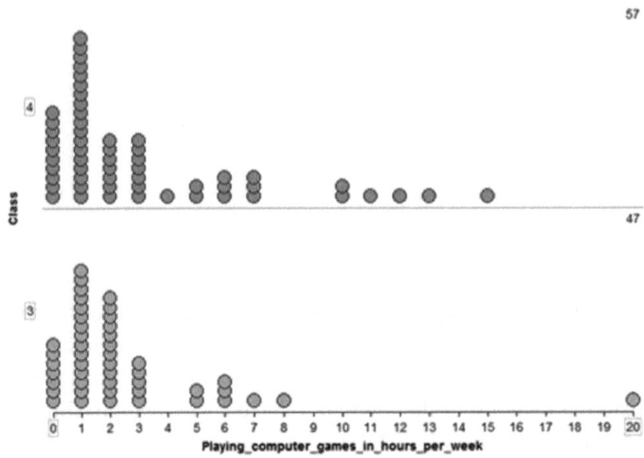

Do 4th graders tend to play more computer (in hours per week) than 3rd graders? Explain!

Fig. 13.13 Group comparison task from pre- and posttest

The correct statement to solve the task is that fourth graders tend to play more computer games (in hours per week) than third graders. There are different ways our students could solve this task based on their experiences from the teaching unit. One possibility would be to identify modal clumps in both distributions and compare the locations of the modal clumps. In this case, we have to mention one limitation of the task, because the modal clumps in each distribution are very close to one another. So the students can only identify little differences between the two distributions in regard to modal clumps. Furthermore, the students could use p-based comparisons to identify differences between both distributions. For example, the students could choose 10 h as cut-point and then compare the frequencies of cases playing more than 10 h/week in each group in regard to the total amount of cases in each group. Another possibility would be to concentrate on extreme values (maximum values of the distributions). This approach could lead to misleading conclusions about which group tends to play more computer games (in hours per week), because the maximum of the distribution for third graders is larger than the maximum of the distribution for fourth graders. The students could also calculate the mean (but this was not discussed in our teaching unit) and find that fourth graders spend on average approximately 0.6 h more on playing computer games than third graders. Another (but incorrect) possibility could be to calculate and compare the total numbers of hours in both groups, but in this case (comparing non-equal-sized groups) such a method is not adequate.

Data Gathering and Analysis

Eleven Grade 4 students (age 9–10 years) participated in our study. They were asked to do the group comparison task prior to participating in the teaching unit and work on the same task again after participating in the unit. Students recorded their conclusions on an exercise sheet (see Fig. 13.13). Thus, our data constitute eleven exercise sheets from the pre-task and eleven exercise sheets from the post-task. We used a qualitative content analysis approach (Mayring, 2015) to analyze our data. One fundamental step in qualitative content analysis is to generate categories. In our case, we had to identify group comparison elements that our students used to compare both groups (for an overview of group comparisons elements, see Table 13.3). This can be done deductively, inductively, or mixed (Kuckartz, 2012). At first, we derived categories from a deductive perspective. We used the categorization system of Frischemeier (2017) as groundwork and adapted it to our purposes: We left out q-based comparisons, since they are not applicable to primary school because they involve comparing quartiles. From an inductive point of view, we broadened the comparison perspective on shift. We added shift to two portions of our categorization system: shift of multiple data points (shift Ps) and shift of modal clumps (shift MC). We have also broadened our perspective on p-based comparisons and distinguish between p-based comparisons taking into account additive reasoning (absolute frequencies) and multiplicative reasoning (relative frequencies). We also added the category "extreme values," for learners concentrating on extreme single values (like maximum or minimum) and also the category "local, but no extreme value." Finally, we added "Total score" as a non-adequate group comparison element for comparing two unequal-sized groups.

In regard to the quality of group comparison elements when comparing unequal-sized groups, we rated the group comparison elements p-based_relative frequencies, skewness, spread, shift Ps, shift MC, and center as sustainable group comparison elements ("+"), because these elements take into account global features of the distributions. Focusing on local features of distributions like extreme values (minimum, maximum) was rated with medium quality ("0"), and comparing unequal-sized groups by adding up the values in each group and comparing the total scores or conducting p-based comparisons using absolute frequencies was rated as low quality ("–"). When students did not mention a group comparison element or when they used a local (but not relaying on extreme value) explanation to compare the two groups, the rating was also low ("–"). In Table 13.3 there is an overview of several group comparison elements with definitions and examples.

Table 13.3 Overview of group comparison elements with definitions and examples

Group comparison Element (Rating)	Definition	Example
None (−)	No element for comparison is used	–
local, but no extreme value (−)	The element used for comparison focuses on local data points (but not on extreme values)	"The 4th graders play more computer because there are more 4th graders playing one hour than 3rd graders." (Johannes)
Total score (−)	Learners calculate the total score of each distribution and compare both sums	"The 3rd graders play 107 h on the computer, the 4th graders play 125 h of computer games" (Peter)
p-based_abs_frequency (−)	Learners choose a cut-point in both distributions and count and compare the number of data cases larger/smaller than the cut-point in both distributions	"They both spend the same amount on playing computer games, because in both groups there is the same amount of children playing computer games from 0–3 h" (Josefine)
Extreme values (0)	Extreme values like maximum/minimum or outliers of a distribution of a numerical variable are used to explain the difference between the two groups	"The 3rd graders play more, because there is one child at 20 h" (Leon)
p-based_rel_frequency (+)	Learners choose a cut-point in both distributions and count and compare the relative frequency of data cases larger/smaller than the cut-point in both distributions	No example in data
Skewness (+)	The difference between the two groups is explained with differences in skewness (right-skewed, etc.)	No example in data
Spread (+)	The difference between the two groups is explained with differences in spread (range or interquartile range)	No example in data
Shift Ps (+)	Learners compare the location of more than one point in both distributions	"The 4th graders play more because the points are more right" (Marc)
Shift MC (+)	Learners identify modal clumps in both distributions and compare the shift between both modal clumps	"The 4th graders play more computer because the clump is more right" (Fiona)
Center (+)	The difference between the two groups is explained with differences in center (mean/median)	No example in data

Table 13.4 Overview of group comparison performance (and change: before/after) of all students

No.	Name	Correct statement (before)	Group comparison element (before)	Correct statement (after)	Group comparison element (after)
1	Peter	Yes	None (−)	Yes	Total score (−)
2	David	Yes	None (−)	Yes	None (−)
3	Lorenz	Yes	None (−)	Yes	local but no extreme value (−)
4	Lars	No	Extreme value (0)	No	Shift MC (+)
5	Marc	No	Extreme value (0)	Yes	Shift Ps (+)
6	Noel	No	local but no extreme value (−)	Yes	Shift Ps (+)
7	Sarah	No	p-based_abs_frequency (−)	Yes	Shift Ps (+)
8	Fiona	Yes	None (−)	Yes	Shift MC (+)
9	Christian	Yes	None (−)	No	None (−)
10	Johannes	No	Maximum (0)	No	Idiosyncratic (−)
11	Josefine	No	Maximum (0)	No	p-based_abs_frequency (−)

We analyzed the exercise sheets for the pre- and post-tasks for all eleven students. Firstly, we have analyzed whether the students made the statement that the fourth-grade students tended to play more computer games per week than the third-grade students. The second step of analysis was to document which group comparison element was used to make the comparison statement.

13.5 Results

The results of our coding process can be seen in Table 13.4. Column three indicates whether in the pre-task a statement like "the 4th grade students tended to play more computer games per week than the 3rd grade students" was given, and column four identifies which comparison element was used to compare both groups in the pretest. Column five indicates whether in the post-task the desired statement was given, and column six shows which comparison element was used in the posttest task.

Five (Peter, David, Lorenz, Fiona, and Christian) of eleven students were able to state that fourth graders tend to spend more hours on playing computer games than third graders—even before participating in the teaching unit. After experiencing the teaching unit, seven out of eleven students were able to make this statement. This is, at first glance, not a large improvement. But the improvement becomes clearer

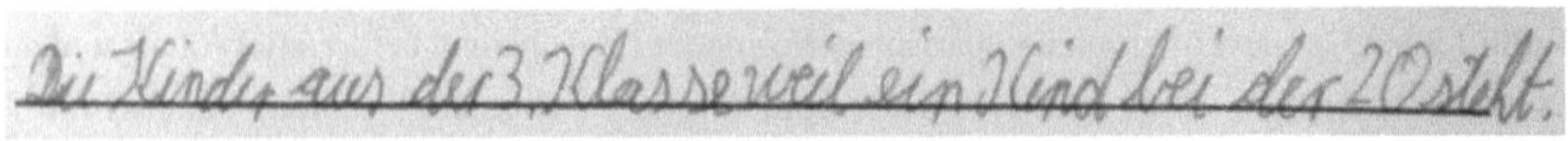

Fig. 13.14 Written note of Marc (pretest)

when examining the quality of the group comparison elements used by students to solve the task. As part of the pretest, none of the five students (Peter, David, Lorenz, Fiona, and Christian) who made the statement ("4th graders tend to spend more hours on playing computer games than 3rd graders") used an adequate group comparison element to state the difference between both groups. In the posttest, four (Marc, Noel, Sarah, and Fiona) of the seven students (Peter, David, Lorenz, Marc, Noel, Sarah, and Fiona) who provided the statement ("4th graders tend to spend more hours on playing computer games than 3rd graders") used adequate group comparison elements to state the differences between both groups. They used the shift of multiple points (Marc, Noel, Sarah) or the shift of modal clumps of the distributions (Fiona) to compare both groups.

Examining the work of Marc on the group comparison tasks reveals how the quality of the group comparison element he used changed following participation in the teaching unit. In the pretest, Marc compared both groups and wrote on the exercise sheet that "the children in class 3 (tend to spend more time on playing computer games), because there is one child at 20 (maximum of the distribution of the variable in class 3)" incorrectly, as can be seen in Fig. 13.14. Thus, he took into account local features of the distribution; in this case, the maximum (20 h/week) concluded that the pupils of class 3 tend to spend more time playing computer because the maximum of the distribution of the variable "Playing_computer_games_in_hours_per_week" is larger in class 3 than in class 4.

In the posttest, Marc showed a more sophisticated comparison view ("The children of class 4 play more, because the points are more at the higher numbers") as we can see in Fig. 13.15. Here, Marc compared both groups correctly and identified a shift (shift Ps) of multiple points to the right in the distribution of the variable in class 4 compared to the distribution of the variable in class 3.

In all, we can say that many of our Grade 4 students were able to make the statement ("4th graders tend to spend more hours on playing computer games than 3rd graders") even before being confronted with group comparison tasks, but the

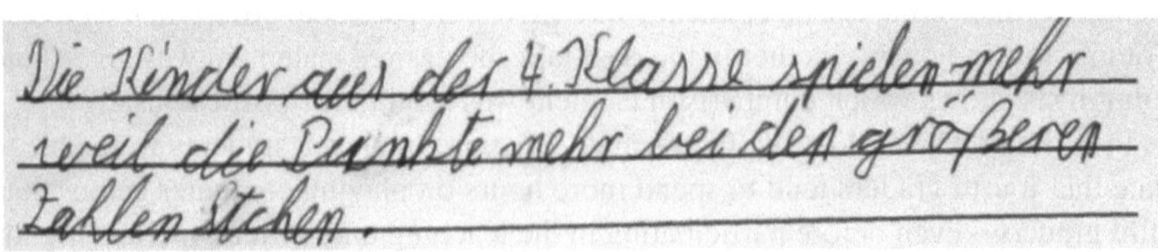

Fig. 13.15 Written note of Marc (posttest)

group comparison elements used by the students lacked quality before the teaching unit. This quality seemed to improve after engagement in the teaching unit.

13.6 Conclusion and Discussion

In the introduction to this chapter, we outlined which competencies are demanded from German primary school students in the field of data analysis (Hasemann & Mirwald, 2012). From a retrospective view and taking into account the results of our study, we can say that the understanding of statistical displays and also the comparison of distributions can be supported already in primary school by approaching statistical displays via different representation levels (enactive, iconic, symbolic) and using educational software like TinkerPlots to support young children to explore data in larger datasets. In addition, the research presented in this chapter shows us the potential to engage young students' sophisticated statistical reasoning with some pedagogical support at an early stage and gives us design ideas for instructional sequences to lead young children to group comparisons. One major implication is that foundational work with concrete activities and materials should be done before and alongside using TinkerPlots. Specifically, it seemed to help that the students at first identified themselves as statistical units (animated statistics, enactive level), then put their characteristics on data cards (enactive and iconic level), and finally were enabled to use the data operations stack, separate, order, and fuse (rectangular) in TinkerPlots to explore larger datasets.

With regard to the theme of this book, the instructional sequence we have presented in Sect. 13.3 of this chapter and the use of TinkerPlots shows how the statistical reasoning of students can be already enhanced at primary school. In addition, group comparisons can serve as fruitful activities taking into account fundamental statistical ideas and therefore confronting young children with these fundamental statistical ideas at an early stage. Our study (see Sect. 13.4 of this chapter) shows that some students can compare groups already at primary school level. Makar and Rubin (2009), as mentioned in the introduction of this chapter, see the comparison of two distributions as one way to engage young students in making informal inferences. This fact and the observations and results of our study may lead us to the conclusion that it is important to confront young children with activities like group comparisons very early. The instructional sequence presented in this chapter can be used and adapted for other teaching projects not only in Germany but also internationally to develop young students' statistical reasoning, especially in regard to comparing groups. One recommendation for practice and pedagogical suggestions arising from our experiences of the teaching units [see Breker (2016) and Schäfers (2017)] is to introduce students to TinkerPlots in these three steps and on these three representational levels, so that the operations with data cards can be transferred to data operations in TinkerPlots and so that students are led to a competent understanding of statistical displays.

As we have already learned in Konold et al. (2002), in Bakker (2004), and in Bakker and Gravemeijer (2004), modal clumps offer fruitful precursors to identifying global characteristics of distributions like center or spread and also can help students compare two distributions. In our instructional sequence, we have first concentrated on summarizing numerical data (stacked dot plots) using modal clumps and then we have used modal clumps as tools to compare distributions. Our main focus was on the second component: the use of modal clumps for comparing groups. To lead students to compare groups using modal clumps, we have designed an instructional sequence which we have described in Sect. 13.3 above. But working with modal clumps also causes difficulties for Grade 4 students. As experienced in our teaching projects, students struggled to identify modal clumps in distributions of numerical data in the form of stacked dot plots. A further difficulty we experienced is that Grade 4 students have problems describing and interpreting the modal clump of the distribution of a numerical variable. One further pedagogical suggestion is that the identification of modal clumps has to be discussed intensively in the classroom for different examples (small dataset, medium dataset, and large dataset) and different types of distributions (skewed, symmetric). Looking forward, in our point of view hat plots may offer an adequate next step after modal clumps, but from the experiences of our teaching units we recommend introducing modal clumps first, to develop the notation and a sense of the meaning of a hat plot. After the modal clump (drawn by hand) has been internalized as a pre-formal stage of identifying the center or the spread of the distribution, hat plots can help to set the stage for group comparisons in TinkerPlots and later one could "set the stage for explorations of alternative ways to define modal clumps (e.g., 1/3–2/3 of the range, 40th–60th percentile)" (Konold et al., 2002, p. 6). The framework (see Table 13.3) listing different elements for comparing groups, their definitions, and examples can be used and refined for further research projects focusing on students' statistical reasoning when comparing groups.

In our research pointed out above, we have only concentrated on the written notes of our students, but we have not interviewed our students or collected video data from the working processes of our students. With a laboratory interview study, we intend to get insight into the cognitive processes of primary students when comparing groups and to develop a sense of the difficulties primary school students face when comparing groups. So we plan to conduct an interview study where primary school students are confronted with a group comparison task in TinkerPlots and are asked to think aloud when comparing both groups with TinkerPlots. With this study, we want to investigate how students identify modal clumps in their distributions, whether they see modal clumps as representatives for the whole distributions, and which other group comparison elements they might use to identify differences between two groups.

Acknowledgements I am very grateful to Rebecca Breker and Christina Schäfers who have been collaborators on the design and the realization of the instructional unit on group comparisons. Furthermore, I thank the editors and the reviewers for their very helpful feedback on previous versions of this chapter.

References

Bakker, A. (2004). Design research in statistics education—On symbolizing and computer tools. *Dissertation*. University of Utrecht.

Bakker, A., & Gravemeijer, K. (2004). Learning to reason about distributions. In D. Ben-Zvi & J. Garfield (Eds.), *The challenge of developing statistical literacy, reasoning and thinking* (pp. 147–168). Dordrecht, The Netherlands: Kluwer Academic Publishers.

Biehler, R. (2001). Statistische Kompetenz von Schülerinnen und Schülern - Konzepte und Ergebnisse empirischer Studien am Beispiel des Vergleichens empirischer Verteilungen. In M. Borovcnik, J. Engel, & D. Wickmann (Eds.), *Anregungen zum Stochastikunterricht* (pp. 97–114). Hildesheim: Franz Becker.

Biehler, R., Ben-Zvi, D., Bakker, A., & Makar, K. (2013). Technology for enhancing statistical reasoning at the school level. In M. A. Clements, A. J. Bishop, C. Keitel-Kreidt, J. Kilpatrick, & F. K.-S. Leung (Eds.), *Third international handbook of mathematics education* (pp. 643–689). New York: Springer Science + Business Media.

Biehler, R., & Frischemeier, D. (2013). Spielerisches Erlernen von Datenanalyse - Von Datenkarten und lebendiger Statistik zur Software TinkerPlots - Ein Workshop im Rahmen einer Lehrerfortbildung für die Primarstufe. *Stochastik in der Schule, 33*(3), 1–8.

Biehler, R., & Frischemeier, D. (2015). Förderung von Datenkompetenz in der Primarstufe. *Lernen und Lernstörungen, 4*(2), 131–137.

Breker, R. (2016). *Design, Durchführung und Evaluation einer Unterrichtseinheit zur Entwicklung der Kompetenz „Verteilungen zu vergleichen" in einer 4. Klasse unter Verwendung der Software TinkerPlots und neuer Medien.* (Bachelor of Education), University of Paderborn.

Burrill, G., & Biehler, R. (2011). Fundamental statistical ideas in the school curriculum and in training teachers. In C. Batanero, G. Burrill, & C. Reading (Eds.), *Teaching statistics in school mathematics-challenges for teaching and teacher education* (pp. 57–69). Netherlands: Springer.

Cobb, P., Confrey, J., diSessa, A., Lehrer, R., & Schauble, L. (2003). Design experiments in educational research. *Educational Researcher, 32*(1), 9–13.

Friel, S. N., Curcio, F. R., & Bright, G. W. (2001). Making sense of graphs: Critical factors influencing comprehension and instructional implications. *Journal for Research in Mathematics Education, 32*(2), 124–158.

Frischemeier, D. (2017). *Statistisch denken und forschen lernen mit der Software TinkerPlots*. Wiesbaden: Springer Fachmedien Wiesbaden.

Garfield, J., & Ben-Zvi, D. (2008). *Developing students' statistical reasoning. Connecting research and teaching practice*. The Netherlands: Springer.

Harradine, A., & Konold, C. (2006). *How representational medium affects the data displays students make.* Paper presented at the Seventh International Conference on Teaching Statistics. Salvador, Brazil.

Hasemann, K., & Mirwald, E. (2012). Daten, Häufigkeit und Wahrscheinlichkeit. In G. Walther, M. van den Heuvel-Panhuizen, D. Granzer, & O. Köller (Eds.), *Bildungsstandards für die Grundschule: Mathematik konkret* (pp. 141–161). Berlin: Cornelsen Scriptor.

Konold, C., & Higgins, T. L. (2003). Reasoning about data. In J. Kilpatrick, W. G. Martin, & D. Schifter (Eds.), *A research companion to principles and standards for school mathematics* (pp. 193–215). Reston, VA: National Council of Teachers of Mathematics.

Konold, C., & Miller, C. (2011). *TinkerPlots 2.0*. Emeryville, CA: Key Curriculum Press.

Konold, C., Robinson, A., Khalil, K., Pollatsek, A., Well, A., Wing, R., & Mayr, S. (2002). *Students' use of modal clumps to summarize data.* Paper presented at the Sixth International Conference on Teaching Statistics. Cape Town, South Africa.

Kuckartz, U. (2012). *Qualitative Inhaltsanalyse. Methoden, Praxis, Computerunterstüzung*. Weinheim, Basel: Beltz Juventa.

Lakoff, G., & Núñez, R. (2000). *Where mathematics come from: How the embodied mind brings mathematics into being*. New York: Basic books.

Makar, K., & Rubin, A. (2009). A framework for thinking about informal statistical inference. *Statistics Education Research Journal, 8*(1), 82–105.

Mayring, P. (2015). Qualitative content analysis: theoretical background and procedures. In A. Bikner-Ahsbahs, C. Knipping, & N. Presmeg (Eds.), *Approaches to qualitative research in mathematics education* (pp. 365–380). Dordrecht: Springer.

Schäfers, C. (2017). *Durchführung und qualitative Evaluation einer redesignten Unterrichtsreihe zur Entwicklung der Komptenz "Verteilungen zu vergleichen" in einer Jahrgangsstufe 4 unter Verwendung der Software TinkerPlots.* (Bachelor of Education), University of Paderborn.

Watson, J., Fitzallen, N., Wilson, K., & Creed, J. (2008). The representational value of HATS. *Mathematics Teaching in Middle School, 14*(1), 4–10.

Watson, J. M., & Moritz, J. B. (1999). The beginning of statistical inference: Comparing two data sets. *Educational Studies in Mathematics, 37*(2), 145–168.

Wild, C. J., & Pfannkuch, M. (1999). Statistical thinking in empirical enquiry. *International Statistical Review, 67*(3), 223–248. https://doi.org/10.1111/j.1751-5823.1999.tb00442.x.

Chapter 14
Data Representations in Early Statistics: Data Sense, Meta-Representational Competence and Transnumeration

Soledad Estrella

Abstract Citizens today are immersed in a very complex and technological world, which requires them to analyse and discuss alternatives, as well as to argue and make decisions. Educating students as citizens requires familiarising them with statistics and giving them educational opportunities so they can make decisions based on data. In order for students to be able to understand key statistics concepts better and to begin to develop statistical thinking, early development of the capabilities of exploring and learning from data is beneficial. This chapter shows that by strengthening teachers' reflections in lesson study groups, teachers innovate in the classroom and, for example in grade K (5 years old) and grade 2 (7 years old), they are then able to get pupils involved in resolving exploratory data analysis situations. The chapter goes on to present diverse data representations produced by pupils, details some components—statistical, numerical and geometric—of the different representations and identifies some transnumeration techniques used by students to understand the behaviour of the data. Our findings which are based on the consideration that external representations are cognitive tools that give meaning to discovering, communicating and reasoning with data account for the early understanding of fundamental statistical concepts, the richness of the graph and table meta-representations created by the pupils, and the data sense making they develop.

14.1 Introduction

Citizens today are immersed in data, and as a result, promoting statistical literacy has become a fundamental responsibility. Many school curricula have introduced statistics and probability courses to help students become competent in everyday decision-making regarding data. This literacy involves being able to read and interpret data in tables, graphs and summaries, and being able to use such tools to make arguments that include evidence of their validity (Ben-Zvi & Garfield, 2004). In order for a citizen

S. Estrella (✉)
Pontificia Universidad Católica de Valparaíso, Valparaíso, Chile
e-mail: soledad.estrella@pucv.cl

A. Leavy et al. (eds.), *Statistics in Early Childhood and Primary Education*,
Early Mathematics Learning and Development,
https://doi.org/10.1007/978-981-13-1044-7_14

to participate in democratic society in an informed manner, the early and progressive development of abilities with graphs is necessary. This includes the reading, writing and interpretation of graphical representations. Statistical literacy gives a person the ability to critically interpret and evaluate information, surveys and statistics studies appearing in media coverage, as well as to appreciate the value of statistics in everyday life, civic life and professional life as consumers of data (del Pino & Estrella, 2012).

In this chapter, we look at the challenging process of representing (modelling) for pupils in the first years of school. This involves exploration of a set of raw data before they then go on to build their own representations to reveal and provide evidence of the behaviour of the data, its patterns and relationships.

14.2 Background

14.2.1 Statistical Literacy

The ideas of Cobb and Moore (1997) have been repeated many times in statistics education, particularly the notions that "data are not just numbers, they are numbers with a context" (p. 801) and "in data analysis, context provides meaning" (p. 803). We believe that data are the centre of statistics and they must, therefore, occupy the same position in school statistics teaching and learning. In order to achieve meaningful statistics instruction, the students must learn about and develop data sense, i.e. a certain numerical sense within a context that provides meaning.

Working with problems and data in context is an opportunity to build authentic literacy and to begin to cement statistics thinking, as expressed by Pfannkuch and Wild (2000). The introduction of exploratory data analysis (EDA) (Tukey, 1977) changed the teaching paradigm for statistics learning, because, as opposed to simply answering set questions, it proposed more flexible and exploratory analysis in search of what can be found in the data. EDA applies methods and ideas needed to organise, represent and describe data, using visual representations such as tables, diagrams, graphs and others, as well as numerical summaries.

It takes time to develop statistical ideas linked to this literacy, and it is therefore advisable to begin in the first years of school (English, 2010, 2013; Franklin & Garfield, 2006; Shaughnessy, 2006). Several research studies into statistics education have looked at graphical representations (e.g. Aoyama, 2007; Estrella, Olfos, Morales, & Vidal-Szabó, 2017; Friel, Curcio, & Bright, 2001; Pérez-Echeverría, Martí, & Pozo, 2010), proposing exploratory data analysis as an effective focus for gaining literacy in this area (e.g. Ben-Zvi & Arcavi, 2001; Ben-Zvi & Garfield, 2004; Burgess, 2011; Estrella & Olfos, 2013; National Council of Teacher of Mathematics [NCTM], 2000, 2009; Shaughnessy, 2006). Regarding graphical representations, Pfannkuch and Wild (2004) believe that even simpler tools such as statistical graphs can be considered statistical models, since they are a statistical way to represent and think about reality.

14.2.2 Representations in Statistics Education

Chilean primary school curricula suggest that students should be able to argue and discuss solutions they present to certain problems, supporting their reasoning with the use of different ways to communicate their ideas, including representations (Chilean Ministry of Education [MINEDUC], 2012). The mathematics curriculum states that students should be able to use a wide variety of representations and apply them fluidly. However, this requirement limits students' learning to simply using and reproducing pictorial representations, such as diagrams, figures and graphs, to communicate (MINEDUC, 2012). As with many primary school curricula, it fails to consider the production of original and unique representations by the students themselves.

Our perspective aims to revitalise the spirit of EDA to place focus on original representations that emerge from decisions made by the students in search of better comprehension of how data behave, which is not necessarily achieved with representations from standardised school reproductions established by the curriculum. Statistics learning includes opportunities to select methods to graph and analyse data, and in statistics, the choice of how to analyse the data is equally (if not more) important than the accounts and calculations used to carry out the procedure (Garfield & Franklin, 2011, p. 136). Very few studies explore students' interactions with different representations (e.g. Estrella et al., 2017; Martí, 2009; Pérez-Echevarría, & Scheurer, 2009), although the interpretation and construction of representations can lead to better comprehension of statistical concepts (Duval, 1995; Tippett, 2016). This lack of attention on how to work with representation in schools can limit students' learning regarding representation (diSessa, 2004).

There is very little systematic research that looks at the role of the signs that comprise a representation built by students (e.g. Earnest, 2015; English, 2012; Estrella et al., 2017; Martí, 2009). Research into students with a high level of understanding of the number line and of the structure of a rectangular grid shows that they possess the capacity to acquire graphical abilities quicker than others (Mulligan, Mitchelmore, English, & Crevensten, 2013). Therefore, it is necessary to document, describe and classify the models and representations students use, as well as the explanations they give for their statistical ideas as solution alternatives for a given problem that requires data management. Hence, a contribution to the area of statistics education would be to provide evidence of the components of data representation and thus demonstrate the improved understanding of the behaviour of data shown by some students in the representations they build.

14.2.3 Reading and Building Our Own Representations

The process of deciding what to do with a set of data in order to understand its behaviour is critical. Reading previously constructed data representations has been studied (e.g. Curcio, 1987; Shaugnessy, Garfield, & Greer, 1996), but there are few studies on the creation of data representations (either typical or invented) and the complexity of building them.

The following section describes some concepts that support our teaching proposal and its aim to develop statistical thinking: meta-representational competence (MRC), some components of representation, transnumeration, statistical thinking and data sense.

14.3 Conceptual Framework

14.3.1 Meta-Representational Competence

MRC describes the complete range of abilities that a subject possesses to be able to build and use external representations (diSessa, 2004; diSessa & Sherin, 2000). It includes the ability to select, produce and use representations constructively, and to criticise and modify, understand and explain and even design new representations (see Table 14.1).

MRC recognises the native ability of students to create their own representations, an ability that is gradually developed through cultural practices inside and outside the classroom. MRC also identifies two main categories of native ability in students: first, a wide range of resources to design representations, including perceptual attributes, such as length, size, numerosity, colour and second, judgement of the representation. The term representational competence covers a wide range of activities involved in representing, while the prefix "meta" is added in order to avoid limiting the term to typical representations or those taught in school as methods of reproduction.

Table 14.1 Aspects of meta-representational competence

Aspects of MCR	Focus
Invention	The ideas and abilities of the students that allow them to invent or design new representations
Criticism	The knowledge of the students that allows them to judge and compare the quality of representations in terms of a good representation
Functionality	The reasoning employed by students in order to understand the purpose and use of different types of representations
Learning	The learning and reflection that reveal new awareness by students of their own understanding of representations and of gaps in their knowledge

Note Aspects summarised from diSessa and Sherin (2000, p. 388)

14.3.2 Some Components of Representations

In the construction of data representations, coordination among statistical, numerical and geometric components associated with data organisation is activated. These are assumed to have been built on mathematics abilities in the early development of the students. The ability to represent requires coordination between geometric and numerical quantities (Duval, 2014; Earnest, 2015; Estrella et al., 2017).

As a component of statistical structure, the concept "variable" crosses all areas of statistics and is central to data representations and their behaviour. A variable is understood as any measurable characteristic of a set of individuals that can take different values and can be categorised.

As a component of numerical structure, "frequency" is the cardinal number corresponding to each category of a variable. When part of a representation, it is aligned with other frequencies, but separated from the qualitative variable. Therefore, in determining absolute frequency it is necessary to obtain the cardinal number for the whole set. Through understanding the principle of cardinality, students can relate sets of different sizes depending on their quantity.

Estrella, Olfos, Vidal-Szabô, Morales and Estrella (2018) define some components of geometric structure in a graphical data representation, including *base-line, linearity-graph* and *unit-of-equal-size*. The first establishes the base on which the data representation is placed and then built. The second allows comparison of heights or ordered distribution over a physical space, and the last gives the visual equivalency between each data point. Specifically, the *base-line* component is the base given by a line (either explicit or implicit) on which the data organisation begins in order for the representation to then be built, and *linearity-graph* is the linearity characteristic of data organisation into columns or rows, in which *unit-of-equal-size* is simultaneously respected along with the conservation of the space between each unit of data represented. These components are essential for comparing and visualising a relationship among the data.

14.3.3 Transnumeration

This is a type of statistical thinking that is carried out when beginning a process of transforming data into a representation or changing a representation or coordination of representations, with the intention of gaining better understanding (Pfannkuch & Wild, 2004). This dynamic process involves interpreting the information received from the data representations, returning to the context to make assertions, answering questions or asking new ones. The transnumeration process comprises transnumeration techniques, such as creating a new variable, changing the type of variable, organising the data differently or representing it visually, sorting data, forming groups, graphing a graph or table, calculating central tendency or measure of the spread, calculating frequencies, selecting and analysing a subset of the data (Chick, 2003).

14.3.4 Statistical Thinking

This is a way of thinking that involves the inductive reasoning common to statistical processes (e.g. identification of data patterns). This type of thinking includes generalisation through the relationships between covariant quantities (e.g. correlation), the representation of these relationships in different ways using natural language, symbolic expressions, tables and graphs, and fluid reasoning between representations to interpret and predict the behaviour of functions (Blanton, Levi, Crites, & Dougherty, 2011).

14.3.5 Data Sense

This conceptual framework ends with the notion of data sense, which we have built and presented based on the idea of numerical sense reviewed by Berch (2005). Therefore, the proposed elements of data sense come from key ideas and techniques common in statistical processes (Burrill & Biehler, 2011). They include the ability to do the following:

1. Approximate or estimate *based on data behaviour*.
2. Make data comparisons of numerical magnitude in different *data representations*.
3. Use numbers and quantitative methods to communicate, process and interpret information from the *data* and from *contextual knowledge*.
4. Recognise the *need for data*, searching for links between the new information and prior *conceptual knowledge*.
5. Understand *numbers in context* as reference points for *measuring variability* in the uncertain real world.
6. Move seamlessly between real-world quantities and the world of *statistical data*.
7. Represent *data units* in multiple ways depending on the context and the aim of the *data representation*, moving between different representations to gain a better understanding of the behaviour of the data and to make predictions beyond the data.
8. Think or speak in a sensible manner about the *behaviour of the data* in a *statistical* problem, without making any precise calculations.
9. Recognise research as a *statistical process* that includes a problem, plan, data collection, data analysis and drawing of conclusions, using the *data as evidence*.
10. Be aware that a research process is taking place with a *real experience with data* and the *understanding of statistical concepts*, using everyday decision-making.

14.4 Methodology

14.4.1 Participants and Context

This study details the experiences of five pupils who attend the same Chilean school, and whose performance is average. Three are preschool pupils, 5 years old, (from a class with 27 pupils) and two are primary pupils, 7 years old, (from a class of 38 pupils). The pupils experienced an open-ended data organisation lesson without prior instruction. Both lessons were designed by teachers at their school (one group of four preschool teachers and another group of second grade 4) during eight sessions carried out at the school and mediated by four researchers. The researchers asked the teachers to participate in a course on school-level statistics through lesson study (Estrella, Mena-Lorca, & Olfos, 2018; Isoda, Arcavi, & Mena, 2007; Isoda & Olfos, 2009) and to promote the connection between theory and practice in statistics education.

14.4.2 Statistics Lesson Design by Teachers in Lesson Study

Figure 14.1 shows the lesson study cycle as formulated and used by the groups of teachers. They worked together to formulate their considerations regarding learning statistics and made professional decisions in order to design their lesson plan.

We then analysed the representations built by the pupils with pen and paper when performing an open data organisation task with categorical variables (grade K: "the class's favourite sports activities"; and grade 2: "water consumption in the home"), including some video-recorded transcriptions of pupil interviews. After describing

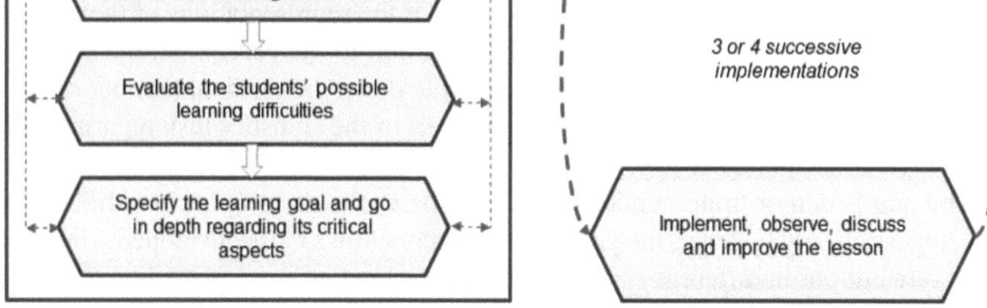

Fig. 14.1 Lesson study cycle implemented for statistics

and analysing the cases, students' MRC is studied based on the representations, transnumeration techniques used in the statistical process, the data sense and the components demonstrated in the representations.

The cases presented here were selected to answer the central statistics question of each lesson, respectively, due to the richness of the productions made by the pupils while completing the lesson, and their degree of participation when expressing their ideas during the lesson and in the interview.

14.5 Statistics Lessons in Preschool and School

14.5.1 Preschool Lesson (Three Preschool Pupils)

A group of teachers of grade K developed a lesson with a theme that would be interesting to the children. The pupils were asked about their favourite sports activity with the idea that the most popular answer would then be organised for the children during one of their break times. This group met for 120 min a week for two months. They planned a lesson and implemented it on three occasions, improving it each time in terms of the teaching for their students' statistics learning. The teachers were involved in a research cycle (Wild & Pfannkuch, 1999) known as Problem, Plan, Data, Analysis and Conclusions (PPDAC). They experienced the entire statistical process and took it to the classroom in order for their pupils to then conduct the same cycle in a statistics learning setting using the EDA paradigm (Ben-Zvi, 2016), applying real data from the school context. The central question of the statistics lesson was: What is our class's favourite sports activity?

Grade K Cases: Juan, José and Maria

The data representation performed by the pupils of grade K contained two critical moments.

Moment 1: The pupils gave evidence of categorisation of the variables (the class's favourite sports activity) and therefore, of implicit comprehension of the concept of variable, by showing in their data representation six activities classified by type (jumping, running, skating, bike-riding, playing basketball and playing football).

Moment 2: After observing the presentation of the representations of their classmates, three pupils showed their data representation to the class, and one of them, on his own initiative, asked for a pencil and wrote down the cardinal number of each categorised activity, thus answering the questions of the statistics task regarding the class's favourite sports activity.

The pupils demonstrated understanding of the statistics component variable and its categorisation. However, the geometric component was less developed, since the data were cut out in different sizes and shapes, and not all were placed on the same base-line (see third category in Fig. 14.2). For this grade, the subsequent emergence of the numerical component was notable. In this case, it was the frequency of the

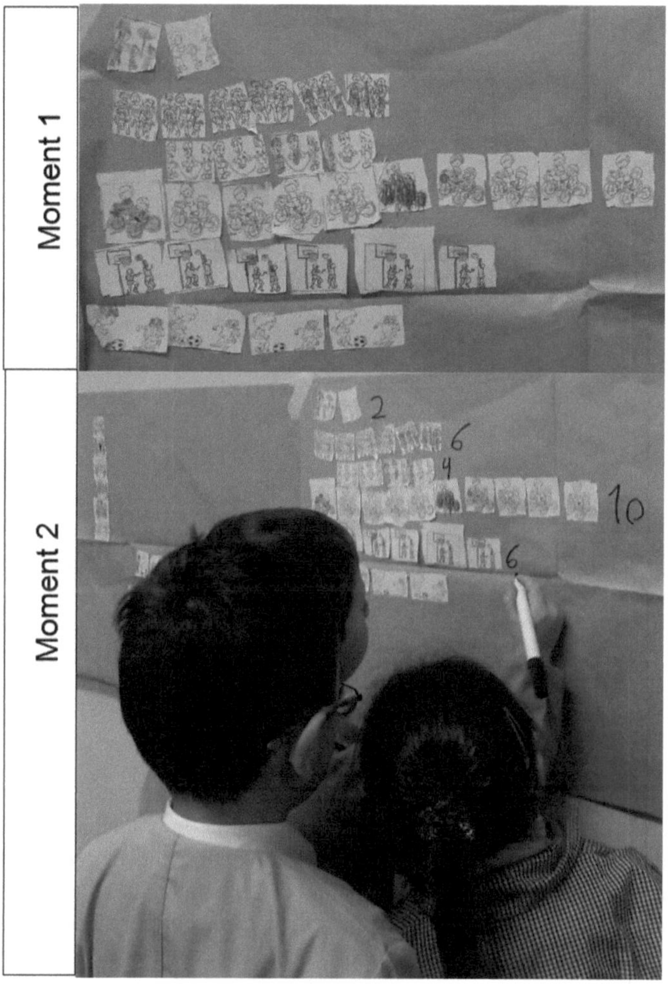

Fig. 14.2 Preschool pupils' data representation, and the moment they calculated and recorded frequency

variable, and in the answer, the pupil not only stated the cardinal number "10", but also the context, "the favourite activity is bike-riding".

At several times, the pupils applied transnumeration techniques (see grey boxes in Fig. 14.3), by ordering the data and grouping them to represent them, creating a new variable (frequency) and analysing a subset of the data.

The preschool pupils participated in a lesson that allowed them to build the categories of the variable, absolute frequency, functionally, and to use subitizing and counting strategies (see Figs. 14.2 and 14.3). This learning experience allowed them to develop their data sense, as the pupils were able to express their ideas on the behaviour of the data without making precise calculations (Moment 1) and to demon-

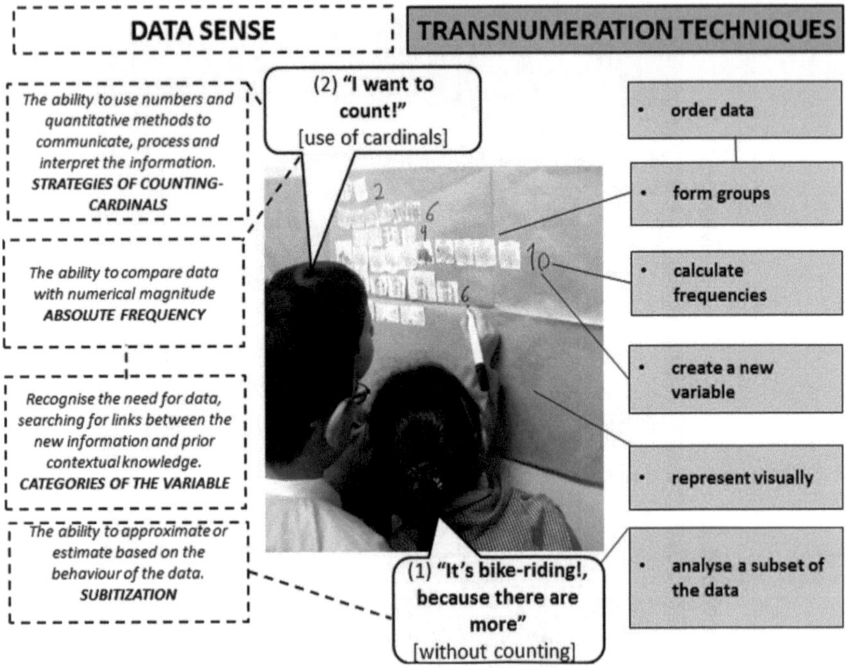

Fig. 14.3 Suggested components of data sense and transnumeration techniques in the task of representing data performed by three preschool children

strate their ability to use numbers and cardination to communicate and process the data in order to obtain information from their contextual knowledge (Moment 2).

With regard to MRC, this group of pupils demonstrated the aspects of invention and learning, as they created and designed their own representations. The learning aspect was shown at the moment when the pupils became aware of their own understanding of the data representation they had built and of the usefulness of absolute frequency to answer the central question.

14.5.2 Second Grade Lesson (Two Pupils)

The group of four grade 2 teachers designed and studied an open-ended problem lesson related to water consumption in the home. These teachers met for 120 min a week for two months. They discussed and planned the lesson, implementing it over three sessions, as with the grade K lesson.

The statistical situation proposed by the four teachers focused on the question: How can we organise data to help reduce water consumption in the home? The teachers chose real data on the daily water consumption of a Chilean family. The

Fig. 14.4 Diagram of data lists in the process of being built

data were given to the students in a worksheet with 36 icons, each representing 1 L of water consumption for a typical family (13 toilets, 10 showers, 8 bathroom sinks, 4 kitchen sinks and 1 garden hose). The students produced different representations of the 36 given data points to answer the central question.

The teachers were involved in the entire statistical process of the PPDAC, which they took to the classroom with the objective that the students would perform the same cycle in an EDA setting and reason with data in a familiar context, such as caring for water use.

The Second Grade Cases: Julia and Her Diagram of Data Lists with Frequency

In the organisation and representation produced by Julia (see Fig. 14.4), it can be seen that the construction process she employed began by her choosing and counting of the icons on water consumption (data) given on the worksheet. She then wrote a word related to the icon and drew a square around it. She then wrote "- 13" (i.e. a dash and the number 13) on the square with the word and drew 13 icons and then another border around them. Based on this, we can interpret that Julia counted and classified the icons to identify and write the category of the variable and the cardinal number. She repeated this procedure for the remaining data in an unusual order (in the west) from right to left.

In Julia's representation, see Fig. 14.4, it can be seen that she organised the data by separating the quantitative from the qualitative. She built vertical lists with repeating icons characterised by: conservation of unity (icons of similar sizes); linearity-graph use in the bars (vertical lists of icons); and base-line (on which the sign of the categorical variable is written with its cardinal number and the bars).

Julia was interviewed and asked about some of her actions and the characteristics of her representation (see Fig. 14.5). She creatively invented a diagram of the data with frequencies and applied her judgement to the representation she built, thus demonstrating the four aspects of MRC: invention, criticism, functionality and learning (Table 14.1). The decisions Julia took in building her data representation allowed her to compare the data lists and reach correct conclusions based on the relationship

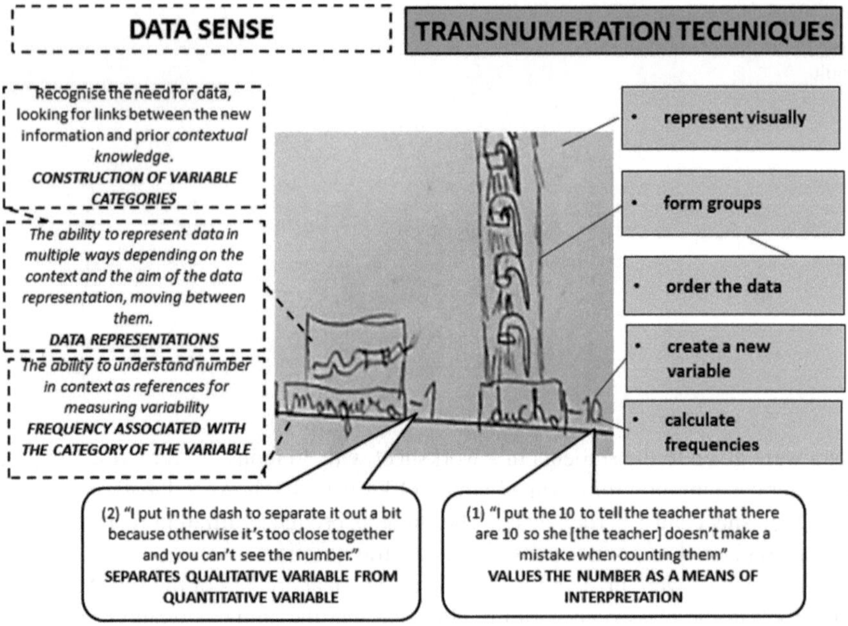

Fig. 14.5 Suggested components of data sense and transnumeration techniques in the task of representing data performed by Julia, grade 2

Fig. 14.6 Icon table of absolute frequencies created by Manuel

of the variable category to its frequency, stating that the use of the highest form of water consumption should be restricted, i.e. the variable category with the highest frequency.

Manuel and His Absolute Frequency Table

Figure 14.6 shows a different representation produced by Manuel, which is a horizontal table comprised of a rectangle with two rows and five columns. The upper cells show the icons that represent the variable categories, and the lower cells show cardinal numbers for each category, i.e. the absolute frequencies.

In the process of building his data representation, Manuel outlined an upper segment and from this drew the cells, filling each with a drawn icon (the categorical variable) and below this he wrote the corresponding cardinal number. This procedure

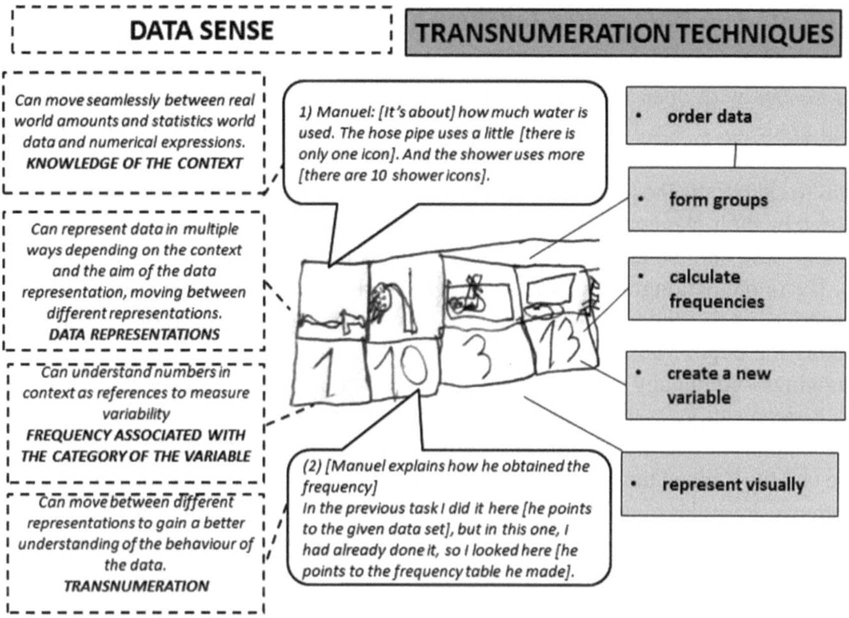

Fig. 14.7 Suggested components of data sense and transnumeration techniques in the task of representing data performed by Manuel, grade 2

was repeated for each category, drawing a representative icon and thus demonstrating a certain linearity-graph aspect of the cells (verticality of the iconic lists).

Manuel classified the data belonging to the same category of a set of data presented in iconic form and then searched for a relationship that would allow him to answer the question. By observing the repeated data icons, he saw that there was a representative of them for each category of the variable and he expressed this by drawing a single icon. Each iconic data symbol represented a category of the variable, and each category was placed on the header upper of the horizontal table of absolute frequencies.

The context of water use awareness gave meaning to the variable "water consumption" and to the categories that emerged, which were visualised as classes (categories of the variable) and ordered horizontally in the table of frequencies (see Fig. 14.6).

In the interview, Manuel indicated understanding of the tabular data representation as a tool that allowed him to simultaneously verify the cardinal value and the frequency without having to count again. By inventing this data representation, Manuel moved away from the individual data point and presented an (aggregated) global summary, and he recognised the situation context and the functionality of his table of frequencies corresponding to this, thus providing indications of strong MRC, i.e. all the aspects (invention, criticism, functionality and learning) described in Table 14.1. The number sense within the context is articulated with *data sense* that emerges from the representation used in his discourse to communicate (see Fig. 14.7).

14.5.3 *The Role of the Teacher in Statistics Lessons*

In lessons with open problems as proposed by both groups of teachers (preschool and grade 2), it was the pupils who spoke, thought, questioned and built answers; they argued, communicated and discussed with their peers. The role of the teacher was to encourage the pupils to listen to each other, to take an interest in the questions posed by their classmates and to make an effort to understand the presentations they constructed and the underlying statistical concepts they used.

By implementing the PPDAC research cycle in an EDA setting, the teachers experienced the entire statistical process and were therefore aware of the challenge facing the pupils. There were some groups of pupils who initially used the data as drawings without any meaning and who could not understand the complexity of the data organisation, as they had not comprehended the contextual meaning of the data (specifically, 21 of the 27 pre-schoolers, and 8 of the 38 2nd-graders). However, in the end they joined the majority of the groups who presented and were able to detect patterns, irregular questions and trends. In order to induce statistical thinking, the teachers conversed with the groups of pupils, promoting criticism, exploration and visualisation of the behaviour of the data, as they had planned, and thus bringing the pupils' emerging statistics ideas into play.

14.6 Conclusions

The first years of school provide an ideal setting for promoting statistics, not only due to the importance of the subject in several different domains of modern society, but also because it reinforces understanding of various mathematical concepts (e.g. numbers, measurement, counting, cardinals, partitions, classification, operations, even distribution, sorting, etc.), while at the same time integrating students into a context in order to awaken the development of their statistical thinking.

Through the products of grade K and grade 2 pupils, we have found evidence of essential components in data representations and of increased understanding of data behaviour acquired by the pupils when freely building their own representations. They used graphs and tables, moving from individual data points to aggregated data. This gave them the opportunity to attain deeper understanding of the characteristics of a data set and its relationships, through individual data manipulation.

The teachers proposed the objective of developing statistics reasoning and not simply learning specific graphical representations, aiming to develop a certain data sense, encouraging the pupils to see the data representations as a whole instead of individually (moving from individual data points to aggregated data).

The experience of preparing lessons as a group allowed the teachers to experience the entire statistical process and then repeat the biggest part of this process with their pupils. This research process included specification of the problem and the central question of the research, data collection, data analysis, data representation, interpretation and discussion of the results, and communication of the conclusions. The pupils then ordered, classified and organised the data, observing that the data vary and detecting the behaviour of most of the data within the context.

In the specific cases described above, there is notable richness in the graph and table meta-representations created by the pupils. Only a few of the pupils, while talking with their classmates, made links between the cardinal numbers and the context (frequency of the associated variable) or between the iconised data and the context (unit-of-equal-size); either this, or they separated the qualitative (categories of the variable) from the quantitative aspects and used the cardinal linearly (frequency). We believe that this shows that the pupils can develop data sense, as the cardinal number gives meaning within context. They demonstrated an interest in the variable, observing its behaviour through the representations they built, and they made sure the icons were of a similar size, spacing them out homogenously, thus allowing comparison between the categories of the variable that emerged from the initial classification. These cases demonstrated the aspects of MRC, transnumeration techniques in action, and the data sense the pupils were developing.

14.7 Projections and Opportunities

Recently, in statistics education systematic research has begun into statistical reasoning in the first years of pre-school, primary school and secondary school. Several questions have arisen in the field of early statistics, providing new opportunities for research and action from a focus on teachers and teacher trainers, such as the following:

Which new demands for statistics teaching are teacher trainers, teachers and first grade pupils facing?
How can early statistic thinking be developed progressively in children?
What type of teaching promotes early conceptual understanding and attitude comprehension in statistics?
Does interdisciplinary lesson study promote effective lessons for thinking statistically?
How does the written curriculum allow early move of EDA to informal statistical inference (ISI)?
How can data sense that provides statistical literacy be promoted from early infancy?
These questions require further research.

Acknowledgements Funding from CONICYT Fondecyt Project N° 11140472 and support from PIA-CONICYT Basal Funds for Centers of Excellence Project FB0003 are gratefully acknowledged.

References

Aoyama, K. (2007). Investigating a hierarchy of students' interpretations of graphs. *International Electronic Journal of Mathematics Education, 2*(3), 298–318.

Ben-Zvi, D. (2016). Three paradigms in developing students' statistical reasoning. In S. Estrella, et al. (Eds.), *Actas de las Jornadas Nacionales de Educación Matemática* (pp. 13–22). Valparaíso, Chile: SOCHIEM, IMA-PUCV.

Ben-Zvi, D., & Arcavi, A. (2001). Junior high school students' construction of global views of data and data representations. *Educational Studies in Mathematics, 45*(1), 35–65.

Ben-Zvi, D., & Garfield, J. (2004). Statistical literacy, reasoning, and thinking: Goals, definitions, and challenges. In D. Ben-Zvi & J. Garfield (Eds.), *The challenge of developing statistical literacy, reasoning and thinking* (pp. 3–15). Dordrecht, The Netherlands: Kluwer Academic Publishers (Springer).

Berch, D. B. (2005). Making sense of number sense: Implications for children with mathematical disabilities. *Journal of Learning Disabilities, 38*(4), 333–339.

Blanton, M., Levi, L., Crites, T., & Dougherty, B. (2011). *Developing essential understanding of algebraic thinking for teaching mathematics in grades 3–5*. In R. M. Zbiek (Series Ed.), Essential Understanding Series. Reston, VA: NCTM.

Burgess, T. A. (2011). Teacher knowledge of and for statistical investigations. In C. Batanero, G. Burrill, & C. Reading (Eds.), *Teaching statistics in school mathematics challenges for teaching and teacher education* (pp. 259–270). Dordrecht, The Netherlands: Springer.

Burrill, G., & Biehler, R. (2011). Fundamental statistical ideas in the school curriculum and in training teachers. In C. Batanero, G. Burrill, & C. Reading (Eds.), *Teaching statistics in school mathematics-Challenges for teaching and teacher education: A Joint ICMI/IASE Study* (pp. 57–69). Dordrecht, The Netherlands: Springer.

Chick, H. (2003). Tools for transnumeration: Early stages in the art of data representation. In *Mathematics Education Research: Innovation, Networking, Opportunity. Proceedings of the 26th Annual Conference of the Mathematics Education Research* (pp. 207–214). Group of Australasia, Geelong. Sydney, NSW: MERGA.

Cobb, G., & Moore, D. (1997). Mathematics, statistics, and teaching. *The American Mathematical Monthly, 104*(9), 801–823.

Curcio, F. R. (1987). Comprehension of mathematical relationships expressed in graphs. *Journal for Research in Mathematics Education, 18*(5), 382–393.

del Pino, G., & Estrella, S. (2012). Educación estadística: Relaciones con la matemática. *Pensamiento Educativo. Revista de Investigación Educacional Latinoamericana, 49*(1), 53–64.

diSessa, A. (2004). Metarepresentation: Native competence and targets for instruction. *Cognition and Instruction, 22*(3), 293–331.

diSessa, A., & Sherin, B. (2000). Metarepresentation: An introduction. *Journal of Mathematical Behavior, 19,* 385–398.

Duval, R. (1995). *Sémiosis et pensée humaine*. Bern: Lang.

Duval, R. (2014). Commentary: Linking epistemology and semio-cognitive modeling in visualization. *ZDM Mathematics Education, 46*(1), 159–170.

Earnest, D. (2015). From number lines to graphs in the coordinate plane: Investigating problem solving across mathematical representations. *Cognition and Instruction, 33*(1), 46–87.

English, L. D. (2010). Young children's early modelling with data. *Mathematics Education Research Journal, 22*(2), 24–47.

English, L. (2012). Data modeling with first-grade students. *Educational Studies in Mathematics, 81*(1), 15–30.

English, L. (2013). Reconceptualizing statistical learning in the early years. In L. English & J. T. Mulligan (Eds.), *Reconceptualizing early mathematics learning* (pp. 67–82). Dordrecht, The Netherlands: Springer.

Estrella, S., Mena-Lorca, A., & Olfos, R. (2018). Lesson study in Chile: A very promising but still uncertain path. In M. Quaresma, C. Winsløw, S. Clivaz, J. da Ponte, A. Ní Shúilleabháin,

A. Takahashi, & T. Fujii (Eds.), *Mathematics lesson study around the world: Theoretical and methodological issues*. Cham: Springer.

Estrella, S., & Olfos, R. (2013). Estudio de Clases para el mejoramiento de la enseñanza de la estadística en Chile. In A. Salcedo (Ed.), *Educación Estadística en América Latina: Tendencias y Perspectivas* (pp. 167–192). Venezuela: Programa de Cooperación Interfacultades, Universidad Central de Venezuela.

Estrella, S., Olfos, R., Morales, S., & Vidal-Szabó, P. (2017). Argumentaciones de estudiantes de primaria sobre representaciones externas de datos: componentes lógicas, numéricas y geométricas. *RELIME, Revista Latinoamericana de Investigación en Matemática Educativa, 20*(3), 345–370. https://doi.org/10.12802/relime.17.2034.

Estrella, S., Olfos, R., Vidal-Szabó, P., Morales, S., & Estrella, P. (2018). Competencia meta-representacional en los primeros grados: representaciones externas de datos y sus componentes. *Revista Enseñanza de las Ciencias, 36*(2), 143–163.

Franklin, C. A., & Garfield, J. (2006). The GAISE project: Developing statistics education guidelines for grades pre-K–12 and college courses. In G. F. Burrill & P. C. Elliott (Eds.), *Thinking and reasoning with data and chance* (68th Yearbook, pp. 345–376). Reston, VA: National Council of Teachers of Mathematics.

Friel, S. N., Curcio, F. R., & Bright, G. W. (2001). Making sense of graphs: Critical factors influencing comprehension and instructional implications. *Journal for Research in Mathematics Education, 32*(2), 124–158.

Garfield, J., & Franklin, C. (2011). Assessment of learning, for learning, and as learning in statistics education. In C. Batanero, G. Burrilll, C. Reading, & A. Rossman (Eds.), *Teaching statistics in school mathematics-challenges for teaching and teacher education* (pp. 133–145). New York: Springer.

Isoda, M., Arcavi, A., & Mena, A. (2007). *El estudio de clases japonés en matemáticas. Su importancia para el mejoramiento de los aprendizajes en el escenario global*. Valparaíso: Ediciones Universitarias de Valparaíso.

Isoda, M., & Olfos, R. (2009). *El enfoque de resolución de problemas en la enseñanza de la matemática a partir del estudio de clases*. Valparaíso: Ediciones Universitarias de Valparaíso.

Martí, E. (2009). Tables as cognitive tools in primary education. In C. Andersen, et al. (Eds.), *Representational systems and practices as learning tools* (pp. 133–148). Rotterdam, The Netherlands: Sense Publishing.

Ministerio de Educación de Chile [MINEDUC]. (2012). *Bases Curriculares de la Educación Básica, Matemática*. Santiago de Chile: Author.

Mulligan, J., Mitchelmore, M., English, L., & Crevensten, N. (2013). Reconceptualizing early mathematics learning: The fundamental role of pattern and structure. In L. D. English & J. T. Mulligan (Eds.), *Reconceptualizing early mathematics learning* (pp. 47–66). New York: Springer.

National Council of Teachers of Mathematics [NCTM]. (2000). *Principles and standards for school mathematics*. Reston, VA: Author.

National Council of Teachers of Mathematics [NCTM]. (2009). *Navigating through data analysis and probability in prekindergarten-grade 2* (Vol. 1). Reston, VA: Author.

Pérez-Echeverría, M. P., Martí, E., & Pozo, J. I. (2010). Los sistemas externos de representación como herramientas de la mente. *Cultura y Educación, 22*(2), 133–147.

Pérez-Echeverría, M. P., & Scheuer, N. (2009). External representations as learning tools: An introduction. In C. Andersen, N. Scheuer, M. P. Pérez Echeverría, E. Teubal (Eds.), *Representational systems and practices as learning tools* (pp. 1–18). Rotterdam, The Netherlands: Sense Publishing.

Pfannkuch, M., & Wild, C. J. (2000). Statistical thinking and statistical practice: Themes gleaned from professional statisticians. *Statistical Science*, 132–152.

Pfannkuch, M., & Wild, C. (2004). Towards an understanding of statistical thinking. In D. Ben-Zvi & J. Garfield (Eds.), *The challenge of developing statistical literacy, reasoning and thinking* (pp. 17–46). Dordrecht, The Netherlands: Springer.

Shaughnessy, J. M. (2006). Research on students' understanding of some big concepts in statistics. In G. Burrill & P. Elliott (Eds.), *Thinking and reasoning with data and chance* (pp. 77–98). Reston, VA: National Council of Teachers of Mathematics.
Shaughnessy, J. M., Garfield, J., & Greer, B. (1996). Data handling. In A. J. Bishop, K. Clements, C. Keitel, J. Kilpatrick, & C. Laborde (Eds.), *International handbook of mathematics education* (pp. 205–237). Dordrecht, The Netherlands: Kluwer Academic Publishers.
Tippett, C. (2016). What recent research on diagrams suggests about learning with rather than learning from visual representations. *Science International Journal of Science Education, 38*(5), 725–746.
Tukey, J. W. (1977). *Exploratory data analysis*. Reading, MA: Addison-Wesley Publishing Co.
Wild, C., & Pfannkuch, M. (1999). Statistical thinking in empirical enquiry. *International Statistical Review, 67*(3), 223–265.

Chapter 15
Supporting Young Children to Develop Combinatorial Reasoning

Lucía Zapata-Cardona

Abstract The goal of this chapter is to discuss young children's approaches to dealing with combinatorial tasks and to present some teachers' strategies to support children's combinatorial reasoning. The discussions are based on clinical interviews with young children (ages 6–8) who were asked to solve a combinatorial task centered on the process of combinatorial counting. Children were interviewed in a private setting and were given some manipulative to help them visualize, explore, model, and solve the combinatorial task. The results revealed by the clinical interviews were contrasted with those disclosed by the literature on children's combinatorial development. Such a contrast suggests that some strategies could be used to support children's combinatorial reasoning. One of the important contributions of this chapter is that it reveals the close relation between young children's combinatorial reasoning and multiplicative reasoning. Consequently, teachers' strategies to support young children's combinatorial reasoning need to be grounded on the development of multiplicative reasoning and to support exploration of combinatorial counting processes. The chapter closes by presenting and discussing some strategies for teachers to support young children in their combinatorial reasoning.

15.1 Statement of Problem

Nowadays, probability is a topic integrated in most elementary school curricula in different countries. Probability includes, among other topics, combinatorial counting which is considered a very fundamental topic in the development of mathematical ideas, and which is based on additive and multiplicative reasoning[1] (Shin & Steffe, 2009). Although combinatorics seems to be a high-level topic for elementary school

[1] Additive reasoning is related to children's first organized attempt to understand and operate with adults' number system and it is mainly based on addition and subtraction while multiplicative reasoning recognizes and uses grouping to manage the number system.

L. Zapata-Cardona (✉)
Universidad de Antioquia, Medellín, Colombia
e-mail: luzapata@ayura.udea.edu.co

A. Leavy et al. (eds.), *Statistics in Early Childhood and Primary Education*,
Early Mathematics Learning and Development,
https://doi.org/10.1007/978-981-13-1044-7_15

curricula, literature has shown that the evolution of combinatorial counting is essential in the establishment of the ideas of chance and probability. Piaget and Inhelder, for example, stated that "the child constructs his notions of probability by his ability to subordinate the disjunctions effected within mixed sets to all the possible combinations, using a multiplicative and not simply an additive mode" (p. 161). Some authors consider that the development of the combinatorial counting is important because it is the basis for more complex subjects (Ura, Stein-Barana, & Munhoz, 2011); others defend that the nature of formal reasoning is based on the combinatorial capability of the learner (Fischbein & Grossman, 1997).

Research literature on teaching and learning statistics has shown that students of all ages struggle with different types of combinatorial counting problems (Lockwood, 2011; Batanero, Navarro-Pelayo, & Godino, 1997) mainly because there is not an upfront way to solve them using procedural reasoning and because they require deep mathematical thinking. In spite of the fact that combinatorics is in the curriculum, there are few resources available for teachers to help them support young children in polishing their combinatorial reasoning. Discussing the path young children go through while exploring, approaching, modeling, and solving combinatorial counting situations could be an important source of reflection for teachers' practice as well as a valuable resource for researchers and statistics educators.

A crucial issue today is that the resources available to support teachers in teaching combinatorics to young children are separated from the daily life of students. Informal knowledge of students is infrequently taken into account when building new knowledge. Consequently, school mathematics is disconnected from the way young children solve problems and do mathematics in their daily lives (Bosch, 2012). Research results have shown, for example, that in the mathematical curriculum of certain educational systems, it is common to find statistics instruction as a set of procedures and algorithms that need to be memorized and applied without any contact with practical situations of the world (Zapata-Cardona & González Gómez, 2017). This form of instruction assumes and accepts statistics as a disarticulated science with no relation to the world experienced by the student. To transform this ingrained practice, daily situations have to be the basis of teaching in elementary education.

Teachers need to understand how young children think and do combinatorial counting tasks. When teachers understand children's ways of thinking and offer them opportunities to construct their own knowledge, it is easy to see how much students are able to learn. In a similar way, learning is meaningful when tasks are connected to students' lives.

Piaget and Inhelder's (1975) work studying the origin of the idea of chance in children has been very influential for those interested in children's combinatorial reasoning. However, in spite of its recognized impact and usefulness, it has raised strong criticism because of the characteristics of the tasks used in the clinical interviews with children. According to some critics, Piaget's research used material that was not very familiar for the children interviewed and this might have affected children's performance in the tasks (English, 1991). Piaget (as cited in Batanero, 2013) indicated that combinatorial reasoning is fully developed during the stage of formal operations (ages 11–15). However, there are some other researchers that stated that some teaching strategies could challenge young children to develop their combinatorial reasoning before coming to the stage of formal operations (Cañadas & Figueiras, 2010; Itzcovich, Ressia de Moreno, Novembre, & Becerril, 2009).

The intent of this chapter is to address the tensions found in the literature relating to the unfamiliarity of the tasks used in research with young children, the lack of detail in the strategies young children use to solve combinatorial counting task, and the limited resources teachers have to support and challenge young children's combinatorial reasoning. The chapter presents a description of the combinatorial counting strategies young children activate when they solve a familiar task. The purpose is not to judge their performance but to illustrate the kind of questions and strategies that researchers and teachers could use to challenge young children's combinatorial reasoning beyond their actual state. This chapter presents the reflections after interviewing three young children when solving a combinatorial counting task related to the multiplication principle.

15.2 Theoretical Framework

Combinatorial reasoning can be defined as the activation of resources (mental or physical) to complete a combinatorial task. Combinatorial counting is essential for the study of discrete mathematics and is the basis for other branches of mathematics. It is fundamental in the study of biology, economics, transportation, agriculture, and others related areas. For some authors, combinatorial reasoning is an important prerequisite for the dynamic and creative power of logical reasoning (Fernández Millán, 2013). Frequently, the study of combinatorics appears in the secondary education curriculum. However, there is an important aspect of combinatorics that can be introduced and successfully worked on with young children in the elementary school mathematics curriculum if it is carried out in conjunction with the strategies to develop multiplicative reasoning.

Some authors (Roa, Batanero, & Godino, 2001) have suggested that the difficulties that even advanced students of mathematics have solving combinatorial problems are related to the way combinatorics is taught. Every so often, teaching is focused on the formula, definition, and combinatorial operation. These authors suggest that in teaching combinatorics, the teacher should privilege problem-solving, systematic enumeration, and tree diagrams. They also indicate that combinatorial reasoning is developed in the stage of formal operations and recognize the strong influence of the environment and personal capacities of the individual. Despite localizing combinatorics at the end of the development period, the multiplication principle can be promoted early on in schooling if teachers support their teaching by using different typologies of problems to develop the multiplicative schemes. In this regard, other authors (Pessoa & Borba, 2012) have shown that combinatorial reasoning is not exclusively a characteristic of the stage of formal operations—as mentioned in Piagetian studies. Pessoa and Borba (2012) provide empirical evidence that children as young as preschool are able to solve, by manipulation of figures, combinatorial problems of different types: arrangement (from a larger set some elements are chosen whose ordering generates different possibilities), permutation (all elements of the set are used, only the order of the presentation varies), combination (these are similar to arrangements in terms of choice of elements, with the difference that the order of the elements does not generate distinct possibilities), and Cartesian product (the total number of all ordered k-tuples from multiple sets).

Scholarly literature in education and psychology reports several studies interested in the way children—from prekindergarten up to high school levels—explore, deal, and come up with solutions for combinatorial situations. The most influential study is that of Piaget and Inhelder (1975) who studied the genesis of the notion of chance by means of clinical interviews with children of a wide range of ages. Piaget and Inhelder's work has been the inspiration for a number of subsequent studies. One of these studies was carried out by English (1991) who explored 50 children's (4–9 years old) strategies to solve combinatorial problems by using seven different forms of the same problem: "find all the possible outfits for toy bears." English found that children could go from trial and error to very sophisticated and efficient algorithmic actions with the potential to generate all possible combinations. One of the important findings of this study was that children in the concrete operational stage, under proper learning conditions, are able to independently develop a method for the multiplication principle prior to formal instruction.

Another study explored the combinatorial abilities of 720 secondary school students (14–15-year-old pupils) finding that children make several mistakes in the combinatorial procedures (Batanero et al., 1997). The authors presented a list of 14 different types of errors participants made during the study. However, one of the most important results arises from comparing the performance of students who received direct instruction with those who did not, and the frequency of errors was reduced in the instruction group. One more study carried out by Cañadas and Figueiras (2010) investigated how students (11–12 years old) solve combinatorial problems using manipulative and how they make a generalization. In the cited study, one of the important results was the different interpretations students gave to multiplication. Students studied multiplication as Cartesian products going beyond the tradition of primary school mathematics instruction that promotes multiplication mainly as repeated addition. Another study carried out by Fuentes and Roa (2014) required 54 compulsory secondary students (12 and 13 years old) to solve a task making all the possible outfits from some shirts, pants, and hats. Fuentes and Roa found that participants were successful 78% of the time and used different strategies like multiplication (59.2% of the time), addition, seriation, and tree diagrams. In another study, Lockwood (2011) examined how students transfer some knowledge developed in solving some combinatorial problems to other types of problems. The study concluded that to help students with the reported difficulties in the literature for combinatorial counting problems, teachers need to pay closer attention to the connections students naturally make.

Studies exploring children's combinatorial abilities have been abundant and they have taken different approaches. Some of them have focused on children's mistakes and some on children's strategies. But what is important to highlight is that some of them have shown that children as young as prekindergarten and elementary school have available the operational structures necessary for dealing successfully with combinatorial counting tasks. This chapter takes into account the reflections and results that the available literature has suggested but the intention is to offer a little more depth and detail in the strategies that young children use when solving combinatorial counting tasks. It also attempts to reveal the type of child–adult interactions that support young children in moving a little beyond their actual capacities.

15.2.1 Multiplicative Reasoning

Usually, a multiplicative structure is constructed prior to operating. Such structure allows the child to shorten a counting activity and later on to internalize conducted actions and operations and use them a priori for the construction of more abstracted combinatorial reasoning. This is also called recursive multiplicative reasoning.

Multiplication in primary school is encouraged under its common intuitive meanings: (1) repeated addition (Cañadas & Figueiras, 2010; Steffe, 1994), (2) ratio and proportion—if a package of cookies has four cookies, how many cookies are there in five packages?—(Cañadas & Figueiras, 2010; Itzcovich et al., 2009), (3) rectangular arrangements—you need to set up a carpet on a surface of 3 by 5 m, how much carpet do you need?—(Itzcovich et al., 2009), and (4) Cartesian product—how many different outfits are you able to create from four shirts and three pants—(Cañadas & Figueiras, 2010; Itzcovich et al., 2009). Nonetheless, there is a disparity in the way multiplicative tasks are stimulated in curriculum materials. Usually, tasks that bring to mind repeated addition, ratio and proportion (formation of groups) or rectangular arrangements are stimulated but those that evoke Cartesian products are left out, having devastating implications for the development of combinatorial reasoning in young children.

15.2.2 Combinatorics and Combinatorial Counting Problems

Combinatorics is defined as "a principle of calculation involving the selection and arrangement of objects in a finite set" (English, 2005, p. 121). It includes areas like combinatorial counting, computations, and probability. In combinatorial counting problems, children are asked to count the number of ways that certain patterns can be formed. However, these are different from simple counting problems such as "how many color pencils do you have?" Counting is understood by Steffe (1983) as "the production of a sequence of number words, such that each number word is accompanied by the production of a unit item" (p. 111). Combinatorial counting problems involve more than children's basic counting schemes and require several connected actions. Initially, children need to deal with units of the indeterminate quantity to be counted in the problem situation; then, they need to properly combine elements from the different sets to create the new counting units; finally, they need to check the counting activity to decide when to stop counting (Shin & Steffe, 2009). In a simple counting situation, the child is asked to count single elements like "how many coins do you have in your pocket." In a combinatorial counting situation, the counting unit is a combination of single units that the child properly creates. Combinatorial counting problems facilitate the development of enumeration processes, conjectures, generalizations, and systematic thinking. Combinatorial activities also help with the development of important concepts such as relations, equivalence classes, mapping, and functions (Batanero et al., 1997; English, 2005).

Within combinatorial counting, problems are those that involve the multiplication principle. This principle declares that if one event can occur in n ways and another event in m ways, then the two events together can occur in $n \times m$ ways. It is important that young children understand and properly use this elemental principle since it is the basis for more complex subjects like combinatorics, probability, and statistics (Ura et al., 2011).

15.3 Methodology

The goal of this chapter is to make evident children's strategies to solve enumerative combinatorial counting situations so that the reflection on such strategies can orient teachers' actions in the classroom when teaching combinatorics to young children. This chapter also pays close attention to how certain questions, actions, and suggestions indicated by the researchers challenge young children's strategies and make young children go beyond their initial strategies. To address this goal, three young children (ages 6–8) were the participants who inspired the reflections discussed here. It will be too ambitious to call this experience a formal study since the participants do not represent all primary school students. The sample was a convenience sample of three girls. They were interviewed in a home setting while they solved the following combinatorial task:

> You have a doll and have four shirts and three pants. If you were to dress the doll in these clothes in how many different ways, could you combine those tops with those bottoms?

The task was presented in verbal form and some manipulative (silhouettes of tops and pants) were given to help visualize, explore, model, and solve the combinatorial task (like the ones shown in Fig. 15.1). The researcher did not reinforce correct choices and avoided referring to the quality of the participants' decisions (as it is recommended by Falk, Yudilevich-Assouline, & Elstein, 2012).

The participants had not had formal academic training in combinatorics during their schooling, which was an advantage in that it made it easier to induce children to express their informal ideas during the interviews. Each child's performance was videotaped, with the camera positioned to capture eye, head, and hand movements, and the use of manipulative.

Fig. 15.1 Silhouettes of tops and pants given to children to model the task

There are different reasons that support the use of attractive manipulative materials. First, images are important in helping children to communicate scientific ideas and support conceptualization. Second, manipulative stimulate children's minds and help them to explore different solutions without giving them the exact way to solve it. Third, children are able to develop concepts related to multiplication and combinations based on their own concrete experience (Ura et al., 2011).

In this study, attractive manipulative materials and an attractive task were used to explore young children's counting combinatorial strategies. By using attractive manipulative materials, teachers and researchers can increase the willingness for children to explore and attempt a solution using their informal knowledge. When children are exposed to tasks that are attractive to them, they increase the possibilities of exhibiting sophisticated solutions (English, 1993; Falk et al., 2012; Ura et al., 2011).

The analysis of data occurred at multiple levels. The researcher reviewed the videos several times to construct a content log. Special attention was paid to young children's strategies and how they reacted to challenging questions from the researcher. The interviews were transcribed verbatim, translated from Spanish into English, and reviewed to refine the understanding and descriptions of key aspects of the children's combinatorial reasoning.

15.4 Results

In this section, some segments of the interviews with the three young children are presented. The order in which the segments are displayed is related to the level of sophistication the young children displayed in the interview. The rudimentary strategies are presented first and then the more elaborate ones. The goal is to look beyond young children' strategies to focus on the potential support they could get from adults (teachers or researchers) to refine their combinatorial reasoning.

The first child is Valery, a seven-year-old girl who was in second grade of elementary school.

Researcher:	Let us suppose you have a doll with different clothes: four shirts and three pants. If you were to dress the doll in these clothes in how many different ways, could you combine those tops with those bottoms?
Valery:	Three ways
Researcher:	How did you do it?
Valery:	I have three outfits. I have three pants that I can dress the doll with and every day I put one on
Researcher:	Show me the three outfits you say
Valery:	[*She pairs up one top with one bottom*] this way
Researcher:	One way. Show me another way
Valery:	This way and this way [*She pairs up two more tops with two bottoms, but one top is left aside as it is shown in Fig.* 15.2]
Researcher:	And with that one [*the top left aside*], what are you going to do with it?
Valery:	If I have another doll out there, I can put it on [*the top left aside*] to it
Researcher:	I see. Thank you so much

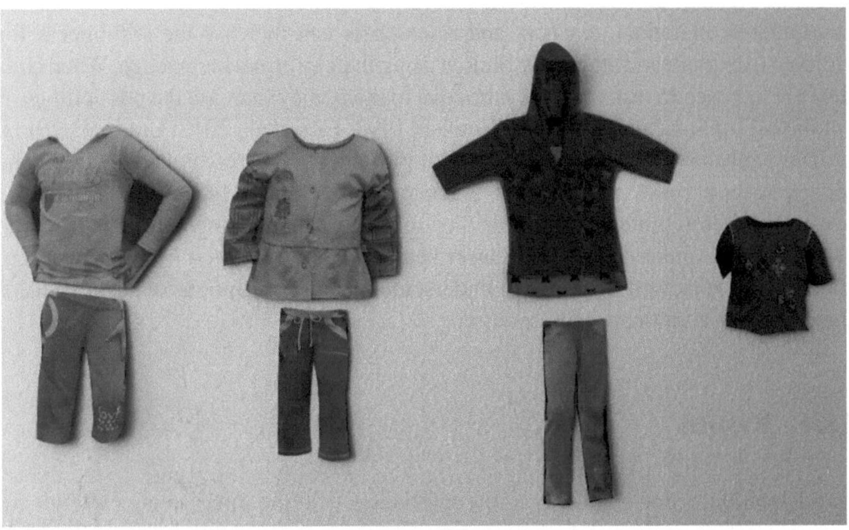

Fig. 15.2 Combinations done by Valery

Valery's counting strategy was very straightforward. She paired up bottoms with tops, and once she ran out of pants, she stopped counting. She left one top aside without using it and when was asked what to do with it she recycled it to use it with another doll. The child used a very simple counting scheme and did not even intend to combine the elements of the sets to create the new counting units. The researcher did not ask further questions in this situation. This is a common strategy used by young children in solving combinatorial counting tasks.

The next child is Eileen. She was a six-year-old girl who was in first grade of elementary school.

Researcher:	You have a doll, four tops and three bottoms. How many ways do you have to dress your doll?
Eileen:	[*She pairs up a top and a bottom*] this one
Researcher:	Do you have any other way to dress the doll?
Eileen:	And these ones [*She pairs up two tops with two pants leaving one top apart*]
Researcher:	What are you going to do with that top? [*Pointing to the top left aside*]
Eileen:	I am going to put it here [*she puts the silhouette of the top on her own chest*]
Researchers:	To whom?
Eileen:	To the doll [*she exchanges the top on her chest with one of the tops that was already paired up with one of the bottoms. She ends up with a different top on her hand*]
Researcher:	So, what are you going to do with this one? [*The one on her hands*]
Eileen:	[*She exchanges the top again with another top already paired up with a bottom. She does that several times completing seven different ways and ends up with one top on her hands*]
Researcher:	So, what are you going to do with this one [*The one on her hands*]?
Eileen:	I will throw it away

In Eileen's interview, she paired up each bottom with each top and she stopped the combinatorial counting when she did not have any more bottoms for the tops. Initially, she formed three different ways and only after being asked what she was going to do with the remaining top, she came up with four more ways. In total, she created seven different ways by using random (unstructured) strategies. She did not use any systematic way to list the combinations or to keep track of the possibilities. Eileen was able to create four more ways because of the researcher intervention. The researcher pushed her to think about what to do with the remaining top and she was able to react to the query by coming up with an action. Eileen's action did not allow her to find all the different ways but at least allowed her to further extend her initial strategy. Comparing Eileen's with Valery's performance, it is evident that Eileen was able to go beyond her initial strategy. Even though both children were asked the same question about what to do with the single remaining top, the question presented the necessary motivation for Eileen to explore more ways to combine clothes. In this sense, the same question challenged only one child.

The third child is Sandy, an eight-year-old girl who was in second grade of elementary school.

Researcher:	You have a doll, three bottoms and four tops. If you were going to combine bottoms and tops in how many different ways, could you dress the doll?
Sandy:	I can dress my doll with this dress [*top*] and with the orange one [*bottom*]. This pink one [*bottom*] with the red one [*top*], and the yellow one [*bottom*] with this one [*top*]
Researcher:	In how many ways could you dress the doll?
Sandy:	Three ways
Researcher:	[*Pointing out to the top that was left without bottom*] And this one, what is going to happen with this one?
Sandy:	That one does not have a pant
Researcher:	So, would you put it on to the doll?
Sandy:	No
Researcher:	And what do you think could happen if we do this? [*Pairing up one of the pants with the shirt that has been left alone*]. One day you dress the doll with this pant and this shirt, and the next day you dress the doll with this other shirt?
Sandy:	Or you could also do this. This one with this one [*she moves the pants around and leaves the tops fixed*] and this yellow one [*pant*] can be also worn with this one
Researcher:	How many outfits do you have then?
Sandy:	Four
Researcher:	Show me the four outfits
Sandy:	I have this one [*she makes some exchanges with the pants*] and also this one. This one with this one
Researcher:	Then, it seems you have found more than four ways
Sandy:	Five then
Researcher:	Show me the five ways
Sandy:	[*she puts together five outfits*] This one with this one, this one with this one, this one too
Researcher:	Do you have more ways?
Sandy:	I have one more.
Researcher:	Which one?
Sandy:	[*she moves two bottoms again getting two new ways*]
Researcher:	Now you have seven. Do you think you have more ways?
Sandy:	Yes [*she moves two pants getting two more ways but one of them is already repeated*], eight and nine
Researcher:	Do you have more forms?
Sandy:	And ten [*she puts together another repeated outfit*]
Researcher:	Do you think that you have repeated some outfits?
Sandy:	This one and this one [*she points out two outfits, one of them was not repeated*]
Researcher:	Do you have more outfits?
Sandy:	No
Researcher:	Then, how many forms in total do you have to dress your doll?
Sandy:	Ten
Researcher:	Thank you so much

In Sandy's interview, she initially paired up bottoms with tops paying attention primarily to the proper combination of colors. Once she ran out of bottoms, she stopped making combinations. She did not consider dressing the doll with the fourth top (the one left out). Sandy paired up one top with one bottom and stopped when she exhausted the elements of the smaller set. The transformation of Sandy's strategy, at the end of the interview, was due to the researcher's stimulating question "And what do you think could happen if we do this? [*Pairing up one of the pants with the shirt that has been left alone*]. One day you dress the doll with this pant and this shirt, and the next day you dress the doll with this other shirt?" After this question, Sandy started randomly (unstructured) matching shirts with pants without being systematic in her approach. In doing so, Sandy found ten combinations but not all of them were different. She repeated two counting units but she was not fully aware of this. This was mainly in part because she did not follow any systematic strategy to keep track of the repeated counting units.

15.5 Discussion

The three young children participating in this experience, at first, used the same strategy to combine shirts and pants in order to find out the different combinations of outfits for the doll. All the children started by pairing up pants with shirts and left one shirt out. They stopped the combinations when they did not have more pants left to combine with the shirts. Similar results were found by Piaget and Inhelder (1975) and later on by English (1991, 1993) who stated that young children initially tend to approach combinatorial problems using very simple counting schemes and empirical approaches.

Valery, as well as Eileen, initially found the same number of outfits by combining shirts with pants using the same rudimentary strategy. They paired the elements of one set with the elements of the other set until they ran out of elements from the smaller set. However, when they were asked what to do with the shirt left aside, the answers were very different. The question the researcher asked did not have any effect on Valery's actions and her task ended there, whereas the same question allowed Eileen to explore other options slightly modifying her strategy and consequently the results. Eileen got four counting units more than in her initial attempt. This is a very interesting result because it shows that the same researcher strategy had different effects on children's actions. These differences in children's performance might be attributed to the different resources children come with to the interview (influence of family, schooling, or culture). Children before being interviewed have previous knowledge that cannot be separated from their essence and constitutes what they are and what they do. In this chapter, knowledge is conceived in a sense similar to Radford: "knowledge [...] is considered to be constituted of forms of human action that have become historically and culturally synthesized" (2016, p. 199). Despite this interesting hypothesis, this experience does not offer sufficient data to support this claim. This is just a hypothesis that could be explored in future studies. What is important to highlight is that the researcher's intervention was essential for one of these children. The researcher's question challenged the child to go further in her combination strategies, and even though she did not use a systematic approach, she was able to find four more outfits. This could be explained using the zone of proximal development in which the learner is able to do something unaided but their capacities are potentiated with the help of an adult or a teacher. In other words, "children 'appropriate' knowledge and skills from more expert members of their society" (Fernández, Wegerif, Mercer, & Rojas-Drummond, 2015, p. 55) and "the child develops through participating in the solution of problems with more experienced members of his or her cultural group" (p. 55).

Sandy's initial strategy was very similar to Valery's and Eileen's strategies. Sandy paired up bottoms with tops and she stopped when she ran out of bottoms. It gives the impression that children see an implicit one-to-one correspondence between the shirts' set and the pants' set, and those single elements (without their respective pair) that cause difficulties in such a correspondence are just left out. The three children in this experience, initially, did not consider interchanging the tops to create more counting units. Apparently, young children's enumerative combinatorial counting strategies are very concrete, probably resembling the same counting strategy they use when counting a simple list of discrete elements as it has been mentioned by Shin and Steffe (2009). In enumerative combinatorial counting situations, the counting units are beyond concrete. The child needs to create those new combinatorial counting units, which usually are a challenge for young children. It is worth noticing Sandy's strategy transformation at the end of the interview. Although she was not able to generate all the new counting units from the combinatorial counting situation, she was able to increase the number of combinations compared to her efforts in her first attempt. This increment in the number of counting units was due to the researcher's intervention through the use of stimulating questions. The researcher did not ask leading questions but those asked made the child either think twice about her decisions or conceive the situation from a different approach. This interaction with a more experienced individual contributes to child development and knowledge in the sense stated by Fernández and colleagues:

> the child develops through participating in the solution of problems with more experienced members of his or her cultural group. […] the development of the child towards more able ways of participation in society is carried out through a process of 'guided participation,' which may or may not include explicit teaching. (2015, p. 55)

That young child–teacher interaction could be oriented, taking into account some aspects of the combinatorial counting. According to the level of the child, the teacher might monitor that there are not elements left out in the new counting units; that there are clear intentions for combining all the elements from one collection with all the elements from the other collection; that there is an explicit use of tools to organize the combinatorial counting units like lists, draws, tables, flow diagrams; that there are clear indications of approaches to keep track of the possibilities to avoid repetitions of the combinatorial counting units. In all these situations, the teacher might ask probing questions or explanations. That does not mean that the child will be successful but at least will be challenged without being explicitly taught. Those "interactions give to each child the opportunity to participate in activities and goals that would be very difficult for them to achieve alone" (Fernández et al., 2015, p. 56).

In terms of the contributions for developing multiplicative reasoning in young children, there is a need to incorporate a wide variety of multiplicative situations in instruction. Most multiplicative situations proposed in the school for young children when learning multiplication have the form of repeated addition, direct proportionality, or rectangular arrangements (arrays). However, on very few occasions, are multiplicative situations that resemble the Cartesian product—like the one discussed in this chapter—used in elementary school to orient the work with multiplication. Some authors have stated that in the proportionality situations or rectangular arrangements, the conception of multiplication as a repeated addition is clear; however, this repeated addition is not as clear in situations that require combinatorial counting to reach a solution (Itzcovich et al., 2009). As a result, in order to contribute to the development of multiplicative reasoning early in elementary education, teachers need to propose a variety of situations in which young children could explore different ways to approach multiplication. Multiplicative reasoning cannot be fully developed using primarily (or exclusively) direct proportionality situations that are very straightforward for most young children. Children's multiplicative reasoning needs to be challenged with multiplicative situations that require deep exploration like Cartesian product tasks. This statement holds firm, taking into account the fact that the literature has shown that young children with no instruction in multiplication are able to solve direct proportion multiplicative situations using their previous knowledge (English, 1991; Park & Nunes, 2001). This suggests that teachers and schools have to do something else. Teachers and schooling must challenge young children to go beyond what they can do using their own resources.

In terms of the familiarity, young children have with the tasks, it is crucial to mention that most combinatorial tasks used in research are too formal and abstract for young children to connect with their daily life. In the experience reported here, having a familiar situation was essential for young children to understand and engage in the solution of the task. Proposing tasks that have some familiarity for young children could potentially activate children's informal knowledge to build new knowledge. In this regard, there are scholars who state that children's abilities are better revealed when the proposed tasks are motivating and meaningful (Falk, et al. 2012).

Providing young children with manipulative to support the exploration of the task was vital to keep track of their approaches to get a solution. By using the manipulative provided, the young children were able to visualize, explore, and model different strategies, and the researcher was able to figure out and follow young children's reasoning while they explored the task.

15.6 Conclusions and Implications

This experience shows that young children's combinatorial reasoning could be stimulated from the moment children begin to work with multiplication. It is not necessary to wait until formal combinatorial instruction that usually takes places in secondary education since the formation of the ideas of probability depends on the evolution of combinatorial counting. Teachers could expose young children to exploring and solving different formats of multiplicative situations, focusing not only on those that follow the structure of either direct proportionality or rectangular arrangements but also on those that follow the structure of Cartesian product. Frequently, in elementary school education, the tasks for developing multiplicative reasoning are based on straightforward strategies without the possibility of exploration. Young children need to be challenged with interesting and familiar situations to enhance their capacities.

In this experience, young children sense of fashion came out in the interviews. Young children wanted to combine tops with bottoms attending to the proper coordination of color. This aspect rises up two contradictory reflections. First, the silhouettes used to help young children visualize, explore, model, and solve the combinatorial task seemed to distract children from creating all the combinatorial units. The researcher could be tempted to simplify the material or the task by taking out the context to warrant young children do not get distracted with fashion issues. However, this could take us back to the criticism received in Piagetian tasks that were too unfamiliar for students. Second, the fashion issue is intrinsic to the task proposed in this experience. Most situations students find in their daily life are not clear and cut. Generally, they incarnate the characteristics of a particular context that in some cases could be considered potential distractions. The researcher could keep the task as it is but emphasizing the probing questions on the creation of the counting units more than in the fashion aspect of it. Either decision the researcher makes will leave something crucial out. This experience also reveals that young children's implicit knowledge can be strengthened by creating hands-on tasks that allow them to deal with combinatorial counting situations early on in schooling in a playful, attractive, and familiar way. Since young children are still concrete thinkers, the use of manipulative is always a welcome support in the modeling and exploration of combinatorial situations. To carry out a simple counting activity, the units to be counted are tangible to the child. However, the combinatorial counting activity requires the child to create the new counting units, which is not a simple task. To help young children with this challenge, teachers could complement the combinatorial task with attractive manipulative that help them in the exploration and modeling.

The results from this experience show that whereas young children explore combinatorial tasks, teachers' questions are essential to focus children's attention and to challenge their reasoning. Teachers' questions could have different purposes: close questions (How many ways do you have to dress the doll?), probing questions (How did you do it? Can you show me those outfits? How many outfits do you have then?), or challenging questions that require the children go beyond their actual state (Do you have any other way to dress the doll? What are you going to do with this piece [*the top left aside*]? What do you think could happen if we do this [*Pairing up one of the pants with the shirt that has been left alone*]?). This young child–teacher interaction is fundamental for child development. After all, learning is the result of interaction with more experienced members of the cultural group.

In this experience, young children did not use structured strategies to find all the different counting units; however, this does not mean that young children were not ready to engage in combinatorial reasoning. The fragments of young child–adult interaction shown here had the intention to illustrate different ways to challenge young children, but literature has revealed previously that young children could develop efficient and sophisticated strategies with the potential to generate all the possible counting units (English, 1991).

Combinatorial reasoning, although developed slowly, can be favored by simple enumeration combinatorial counting activities. Teacher should promote combinatorial tasks early on in schooling to encourage reflection and problem-solving skills that contribute to the development of combinatorial reasoning. Combinatorial tasks, when accompanied with challenging questions from more experienced members of the cultural group (teachers, researchers, parents), could help young children to confront their primary intuitions and polish their reasoning.

Acknowledgements This research was supported by Universidad de Antioquia Research Committee—CODI and Colciencias, Grant Number CT 438-2017.

References

Batanero, C. (2013). La comprensión de la probabilidad en los niños: ¿qué podemos aprender de la investigación? [Understanding Probability in Children: What Can We Learn From Research?]. In J. A. Fernandes, P. F. Correia, M. H. Martinho, & F. Viseu (Eds.), *Atas do III Encontro de Probabilidades e Estatística na Escola*. Braga: Centro de Investigação em Educação da Universidade do Minho.

Batanero, C., Navarro-Pelayo, V., & Godino, J. D. (1997). Effect of the implicit combinatorial model on combinatorial reasoning in secundary school pupils. *Educational Studies in Mathematics, 32*, 181–199.

Bosch, M. (2012). Apuntes teóricos sobre el pensamiento matemático y multiplicativo en los primeros niveles [Theoretical notes on mathematical and multiplicative thinking in the first levels]. *Edma 0-6: Educación Matemática en la Infancia, 1*(1), 15–37.

Cañadas, M. C., & Figueiras, L. (2010). Razonamiento y estrategias en la transición a la generalización en un problema de combinatoria [Reasoning and strategies in the transition to generalization in a combinatorial problem]. *PNA, 4*(2), 73–86.

English, L. D. (1991). Young children's combinatoric strategies. *Educational Studies in Mathematics, 22*(5), 451–474.

English, L. D. (1993). Children's strategies for solving two and three dimensional combinatorial problems. *Journal for Research in Mathematics Education, 24*(3), 255–273.

English, L. D. (2005). Combinatorics and the development of children's combinatorial reasoning. In G. A. Jones (Ed.), *Exploring probability in school: Challenges for teaching and learning* (pp. 121–141). New York: Springer.

Falk, R., Yudilevich-Assouline, P., & Elstein, A. (2012). Children's concept of probability as inferred from their binary choices—Revisited. *Educational Studies in Mathematics, 81*(2), 207–233.

Fernández Millán, E. (2013). Razonamiento Combinatorio y el currículo español [Combinatorial reasoning and the curriculum in Spain]. In J. M. Contreras, G. R. Cañadas, M. M. Gea, & P. Arteaga (Eds.), *Actas de las Jornadas Virtuales en Didáctica de la Estadística, Probabilidad y Combinatoria* (pp. 539–545). Granada, Spain: Departamento de Matemáticas de la Universidad de Granada.

Fernández, M., Wegerif, R., Mercer, N., & Rojas-Drummond, S. (2015). Re-conceptualizing "scaffolding" and the zone of proximal development in the context of symmetrical collaborative learning. *Journal of Classroom Interaction, 50*(1), 54–72.

Fischbein, E., & Grossman, A. (1997). Schemata and intuitions in combinatorial reasoning. *Educational Studies in Mathematics, 34*(1), 27–47.

Fuentes, S., & Roa, R. (2014). Deducción del principio multiplicativo. Una actividad exploratoria en alumnos de 1° de E.S.O [Deduction of the multiplication principle. An exploratory activity in 1st grade students from Compulsory Secundary Education]. *XV Congreso de Enseñanza y Aprendizaje de las Matemáticas- CEAM.* Baeza, Spain.

Itzcovich, H., Ressia de Moreno, B., Novembre, A., & Becerril, M. M. (2009). *La matemática escolar: Las prácticas de enseñanza en el aula [School Mathematics: Teaching practices in the classroom].* Buenos Aires: Aique Educación.

Lockwood, E. (2011). Student connections among counting problems: An exploration using actor-oriented transfer. *Educational Studies in Mathemathics, 78,* 307–322.

Park, J.-H., & Nunes, T. (2001). The development of the concept of multiplication. *Cognitive Development, 16,* 763–773.

Pessoa, C., & Borba, R. (2012). Do young children notice what combinatorial situations require? In T. Y. Tso, *36th Conference of the International Group for the Psychology of Mathematics Education* (p. 261). Taipei, Taiwan: PME.

Piaget, J., & Inhelder, B. (1975). *The origin of the idea of chance in children.* (L. Leake, P. Burrell, & H. D. Fishbein, Trans.). New York: W.W. Norton & Company.

Radford, L. (2016). The theory of objectification and its place among sociocultural research in mathematics education. *International Journal for Research in Mathematics Education (RIPEM), 6*(2), 187–206.

Roa, R., Batanero, C., & Godino, J. (2001). Dificultad de los problemas combinatorios en estudiantes con preparación matemática avanzada [Difficulty of combinatorial problems in students with advanced mathematical preparation]. *Números: Revista de Didáctica de las Matemáticas, 47,* 33–47.

Shin, J., & Steffe, L. (2009). Seventh graders' use of additive and multiplicative reasoning for enumerative combinatorial problems. In: *31st annual meeting of the North American Chapter of the International Group for the Psychology of Mathematics Education* (pp. 170–177). Atlanta, GA: Georgia State University.

Steffe, L. P. (1983). Children's algorithms as schemes. *Educational Studies in Mathematics, 14*(2), 109–125.

Steffe, L. P. (1994). Children's multiplying schemes. In G. Harel & J. Confrey (Eds.), *The development of multiplicative reasoning in the learning of mathematics* (pp. 3–40). Albany: State University of New York Press.

Ura, S. K., Stein-Barana, A. C., & Munhoz, D. P. (2011). Fashion, paper dolls and multiplicatives. *Mathematics Teaching, 221,* 32–33.

Zapata-Cardona, L., & González Gómez, D. (2017). Imágenes de los profesores sobre la estadística y su enseñanza [Teachers' Images about Statistics and its Teaching]. *Educación Matemática, 29*(1), 61–89.

Part V
Teaching Statistics and Probability: Modelling

Chapter 16
Integrating Games into the Early Statistics Classroom: Teachers' Professional Development on Game-Enhanced Learning

Maria Meletiou-Mavrotheris, Efi Paparistodemou and Loucas Tsouccas

Abstract Digital games hold a lot of promise as tools for improving statistics instruction in the early school years. The research discussed in this chapter aimed at equipping a group of in-service primary teachers with the knowledge, skills, and practical experience required to effectively integrate digital games within early statistics education. The study took place within an in-service teachers' professional development program focused on game-enhanced mathematics teaching and learning. The program was designed based on the technological pedagogical and content knowledge (TPACK) framework and was attended by six educators teaching at the lower primary school level (Grades 1–3; ages 6–9). Participants experimented with different ways in constructivist gaming environments could get integrated into the early primary mathematics curriculum to motivate young children and to help them internalize important concepts, including key ideas related to data analysis and probability. They also developed and delivered instructional episodes integrating the use of digital games in different areas of the early mathematics curriculum. This chapter discusses the impact of the study on teachers' perceptions regarding game-based statistical learning and on their competence in selecting, evaluating, and productively utilizing digital games as an instructional tool in the early years of schooling.

Keywords Digital games · Game-based learning · Statistics education · Early statistical reasoning · Professional development · TPACK

M. Meletiou-Mavrotheris (✉) · L. Tsouccas
European University Cyprus, Nicosia, Cyprus
e-mail: m.mavrotheris@euc.ac.cy

L. Tsouccas
e-mail: louevge@gmail.com

E. Paparistodemou
Cyprus Pedagogical Institute, Nicosia, Cyprus
e-mail: Paparistodemou.e@cyearn.pi.ac.cy

A. Leavy et al. (eds.), *Statistics in Early Childhood and Primary Education*,
Early Mathematics Learning and Development,
https://doi.org/10.1007/978-981-13-1044-7_16

16.1 Introduction

Although important for people of all ages, play is essential for young children. Most contemporary theories of learning emphasize its vital role in children's development and its contribution to their cognitive, physical, social, and emotional well-being (Ginsburg, 2007). Through active play, young children can learn in joyful and conceptually rich ways. They can express their feelings and ideas, symbolize and test their knowledge of the world, and strengthen their creativity and imagination (Saracho, 2012). Nevertheless, in a fast-changing world driven by rapid technological advancements, gameplay is also changing. The exponential rate of adoption of tablet computers and other smart mobile devices witnessed worldwide in recent years has dramatically increased young children's accessibility and use of electronic devices for gaming purposes. Research indicates that children as young as two can easily adapt to the intuitive interface of touch-enabled devices and use them with much greater independence compared to desktop computers (Geist, 2012). The ease of using tablets has resulted in a large percentage of young children becoming frequent users of mobile devices, which they tend to use mainly for playing games, often for enormous amount of time (Common Sense Media, 2013). This increased popularity and proliferation of digital games has led to a widespread interest among educators in how this massively popular youth activity could be brought into the classroom to capture students' interest and facilitate their learning. Responding to this trend, there has been an explosive growth in the number of educational game apps, targeting children available on the market. The existing literature strongly indicates the educational value of games and their potential to serve as a powerful perspective for reforming pedagogy at the early school level (Lowrie & Jorgensen, 2015).

Acknowledging the educational potential of games for transforming statistics instruction in the early years, but also the crucial role of teachers in any effort to bring about change and innovation, the current study focused on providing in-service teacher education on the effective utilization of digital games in the lower primary classroom (ages 6–9). The study took place within a Cypriot teachers' professional development program on the integration of games within the early mathematics curriculum, which was designed based on the notion of technological pedagogical content knowledge (TPACK) as a conceptual framework (Mishra and Koehler, 2006). The study aimed at providing teachers with the knowledge, skills, confidence, and practical experience required to effectively exploit digital games as a tool for fostering young children's motivation and learning of statistics (as part of the mathematics curriculum). The program's impact on the study participants was examined from three perspectives:

(i) Influence on teachers' attitudes and perceptions regarding game-based teaching and learning;
(ii) Impact on the development of teachers' TPACK regarding the instructional integration of mobile games;
(iii) Level of transfer and adoption of TPACK competencies acquired through the program to actual teaching practice.

16.2 Theoretical Perspective

The expanding use of data for prediction and decision-making in almost all domains of life makes it a priority for mathematics instruction to help students of all ages develop their statistical reasoning. In recent years, leaders in mathematics education have been advocating a much wider and deeper role for statistics in school mathematics (Franklin et al., 2007). It is now widely recognized that the foundations for statistical reasoning should be laid in the earliest years of schooling rather than being reserved for secondary school or university studies. Consequently, the development of students' statistical literacy has become an important goal of mathematics education at the early school level internationally. This broadening of the mathematics curriculum to encompass statistical literacy, reasoning, and thinking has put considerable demands on teachers (Hannigan, Gill, & Leavy, 2013). In particular, it has been challenging for teachers to design lessons with engaging contexts and a focus on conceptual aspects of statistics and to pose critical questions. As the research literature indicates, many teachers tend to have weak knowledge of the statistical concepts and to focus instruction on the procedural aspects of statistics rather than on conceptual understanding (Watson, 2001).

Recognizing the need for fundamental changes to the instructional practices employed in the mathematics classroom to teach statistical and probabilistic concepts, researchers have in recent years been experimenting with new models of teaching that are focused on inquiry-based, technology-enhanced instruction and on statistical problem-solving (e.g., Meletiou-Mavrotheris & Paparistodemou, 2015). One promising approach lately explored is the potential for digital games to transform statistics instruction. Although—unlike the numerous studies investigating the instructional use of computer simulations, animations, and dynamic software—there only few published studies on the use of games for teaching statistics, the general thrust of the evidence in the existing literature is positive (Boyle et al., 2014). Most of the conducted studies report that employing games has a positive effect on students' motivation and learning of statistical concepts (e.g., Asbell-Clarke et al., 2012; Gresalfi & Barab, 2011).

Findings of the statistics education literature on game-enhanced learning concur with the general educational literature which suggests that, when suitably designed, digital educational games have many potential benefits for teaching and learning at all levels, including the preschool and early school years (Manessis, 2014). One of games' foremost qualities is the capacity to motivate and immerse players (Felicia, 2009). It has been shown that educational games captivate children's attention, contributing to their increased motivation and engagement with learning (see Ke, 2008). However, their greatest strength as a medium, according to a meta-analysis on the impact of games on learning conducted by Clark, Tanner-Smith, and Killingsworth (2014), involves their affordances for supporting higher-order cognitive, intrapersonal, and interpersonal learning objectives. Through the introduction of open-ended, challenging tasks that are meaningful for young children and facilitate their interest in exploration, properly designed games can help focus instruction on conceptual understanding and problem-solving rather than on recipes and formal derivations (Koh, Kin, Wadhwa, & Lim, 2012). Using games, children can collaboratively engage in exploration of virtual worlds and in authentic problem-solving activities and eventually become reflective and self-directed learners (Van Eck, Guy, Young, Winger, & Brewster, 2015). This supports the development of important competencies essential in modern society such as logical and strategic thinking, planning, multi-tasking, self-monitoring, communication, negotiation, group decision-making, pattern recognition, accuracy, speed of calculation, and data-handling (Miller & Robertson, 2010). At the same time, games can match challenges to children's skill level, providing them with immediate feedback about the correctness of their strategies and thought processes, while at the same time enabling teachers to observe students' problem-solving strategies in action and to assess their progress (Koh et al., 2012). Thus, placing a focus on game-enhanced learning offers a powerful perspective for transforming statistics instruction at the early school level and providing children with the tactile and dispositional skills required to meet the needs of a global, information-driven society.

While digital educational games provide a range of potential benefits for mathematics and statistics teaching and learning, not all the available games are designed to promote optimal development among children. Most of the available educational games tend to be a drill and practice and to support mainly procedural fluency rather than high-level thinking. Nonetheless, some exceptional exemplars do exist that can help create constructive and meaningful learning experiences. Larkin (2015), for example, reported on the findings of a long-term research project that comprehensively reviewed mathematical apps to determine their usefulness for primary school students. He found that although the majority of apps provide little more than edutainment, a core group of apps were highly effective in supporting children in their development of higher-order mathematical thinking and learning.

The successful deployment of digital games in the early statistic classroom is highly dependent upon the knowledge, attitudes, and experiences of teachers. Implementing game-based instruction can be a challenge for teachers, requiring skills not necessarily addressed in current teacher education practices. Teachers need to be proactive, choosing high-quality educational games, supporting and scaffolding pupils, and providing appropriate feedback. The research literature indicates that despite having a fundamental understanding of the importance of playtime based on their teacher education, and positive attitudes toward the adoption of games in instruction, the majority of both pre-service and in-service primary teachers have not been educated on the structure and benefits of well-designed educational games, and thus lack the vision and the personal experience of what game-enhanced teaching could look like. They tend to view games as instructional tools to be mainly used for motivational purposes or for reinforcing already acquired concepts (Williamson, 2009). Takeuchi & Vaala (2014), who surveyed 694 K–8 teachers across the USA, found that while around 75% of the teachers reported using digital games in their classrooms, the vast majority used "drill-and-practice"-type games focused on lower-level knowledge and skills. Consequently, the provision of high-quality professional development on the educational applications of games is of paramount importance to their effective integration in classroom settings.

16.3 Methodology

16.3.1 Conceptual Framework

The TPACK conceptual framework guided the program's design and implementation. TPACK is a powerful and influential framework, proposed by Mishra and Koehler (2006), which emphasizes the importance of teachers developing integrated and interdependent understanding of three primary forms of knowledge: technology, pedagogy, and content. In this study, the adoption of TPACK served a twofold purpose: (i) a guiding theory for designing the program so as to create professional development opportunities that would better prepare teachers to effectively integrate digital games in early mathematics and statistics and (ii) a conceptual blueprint for investigating the impact of the program on participants' professional growth in the use of games in early statistics.

16.3.2 Research Design: Scope and Context of Study

A case study design was employed. The case studied consisted of the six primary school teachers (3 males, 3 females) who participated in the program. With one exception, all participants were expert teachers with several years of instructional experience (7–19 years).

Following the TPACK model and action research procedures, the study was designed and carried out in three phases: (i) Phase I: Familiarization with Game-Based Learning; (ii) Phase II: Lesson Planning; (iii) Phase III: Lesson Implementation and Reflection. Each of the three phases, described next in more detail, supported teachers in strengthening the connections among their technological, pedagogical, and content knowledge. At the same time, various forms of data were collected and analyzed in order to track changes in teachers' TPACK regarding game-enhanced statistics learning in the early years.

Phase I—Familiarization with Game-Based Learning

During Phase I (6 h duration), teachers were offered a critical introduction into the potential and challenges of using serious games in early mathematics and statistics instruction. They experienced some of the ways in which purposefully selected games, blended with carefully constructed learning experiences, could help improve children's attitudes toward the subjects, while at the same time advancing their mathematical and statistical thinking and problem-solving skills. The unit also familiarized participants with the design principles for constructivist gaming environments (Munoz-Rosario & Widmeyer, 2009) and promoted the development of their skills in properly evaluating and selecting games with pedagogically sound design features.

Participants worked in various individual or group game-based activities to explore a variety of mathematical concepts included in the lower primary school curriculum, including topics related to data analysis and probability. They experimented with a broad range of serious games, from adventure games to puzzle games. There were also discussions focusing on children's learning and what is required to involve them in learning about mathematics and statistics through use of educational games. These discussions provided the venue for examining the affordances and limitations of educational games and for identifying design considerations that promote the incorporation of educational games in educationally powerful ways. A number of assigned key readings (e.g., Gee, 2007) served as a backdrop for these discussions. In-service teachers were also introduced to research literature (e.g., Pivec & Pivec, 2009) on effective instructional strategies that could be used to facilitate learning with educational games (e.g., debriefing, collaborative gameplaying, supplementing of games with other instructional methods).

A characteristic example of the type of activities included in the program is the *Evaluation of Educational Games Task*. In this activity, teachers worked individually to evaluate six freely available mathematics education game apps. They were instructed to do the following:

1. Use an evaluation rubric to assess each game.
2. Write an evaluation report for each game.

Phase II—Lesson Planning

In Phase II, teachers' TPACK was enhanced through their engagement in lesson planning. They selected a topic from the national mathematics curriculum for Grades 1–3 and developed a lesson plan and accompanying teaching material aligned with the learning objectives specified in the curriculum, which incorporated the use of digital games. They were instructed to integrate into their lessons educational games that adhered to important principles associated well-designed educational games (Gee, 2007) such as facilitation of an authentic learning experience, active participation, collaboration, and promotion of higher-order thinking skills. The lesson plans were shared with the researchers for comments and suggestions and were revised based upon received feedback.

Phase III—Lesson Implementation and Reflection

Next, the participants implemented the lesson plans in their classroom, with the support of the research team. Once the classroom research was completed, teachers prepared and submitted a reflection paper, where they shared their observations on students' reactions during the lesson, noting what went well and what difficulties they faced and making suggestions for improvement.

16.3.3 Instruments, Data Collection, and Analysis Procedure

Multiple forms of data were collected to document changes in teachers' perceptions and attitudes, and in their TPACK of game-enhanced learning as a result of participating in the professional development program:

(i) *Teacher pre-survey*: This open-ended survey administered before the start of the program gathered baseline information about teachers' use of and attitudes toward ICT in general and digital games in particular, as tools in daily life and in the classroom.
(ii) *Individual interviews*: Upon completion of Phase I, the researchers conducted semi-structured interviews with each of the participants to trace possible shifts in their attitudes and perceptions regarding game-enhanced mathematics and statistics learning.
(iii) *Observations and artifacts collected during Phases I & II*: In-service teachers' submitted work (game evaluation reports, lesson plans, etc.), researchers' observations and field notes.
(iv) *Observations and artifacts collected during Phase III*: Given the study's focus on statistical reasoning, data at this phase were mainly collected from the class of the only teacher whose lesson plan had focused on concepts related to statistical data analysis (the rest prepared and implemented lesson plans focusing on other areas of the mathematics curriculum). This intervention took place in a Grade 2 (ages 7–8) classroom with 18 students and lasted for 80 min (two teaching periods). Researchers were present, observing closely and videotaping the lesson, keeping field notes, and collecting student work samples. Qualitative data were also obtained from the reflection papers written by the teachers at the end of Phase III.

For the purpose of analysis, we did not use an analytical framework with predetermined categories to assess how teachers' perceptions and TPACK developed after going through the intervention due to the lack of well-established frameworks and methodological insights for studying game-enhanced statistics education in the context of in-service teacher education. What we did instead was to identify, through careful reviewing of the transcripts, reports, and other data collected during the study, recurring themes or patterns in the data. To increase the reliability of the findings, the activities were analyzed and categorized by all three researchers. Inter-rater discrepancies were resolved through discussion.

16.4 Results

Our research findings illustrate the usefulness of TPACK as a means of studying and facilitating teachers' professional growth in the use of games in early statistics education. They indicate that our TPACK-guided professional development program had a positive impact on all three perspectives of the participants' experiences examined: (i) attitudes and perceptions regarding game-enhanced learning; (ii) TPACK competency for using digital games; and (iii) level of transfer and adoption of acquired TPACK to actual teaching practice.

16.4.1 Changes in Attitudes and Perceptions Regarding Game-Enhanced Learning

In the pre-survey completed at the program outset, although teachers indicated high level of familiarity with technology and very positive attitudes toward its instructional use, they acknowledged that ICT tools were not adequately used in their classrooms. They made extensive use of technology, but this was limited mainly to Word Processing, PowerPoint, Internet browsing, and drill-and-practice-type applets. Teachers uniformly reported low use of technologies promoting more engaging, interactive, student-centered pedagogical approaches, such as simulations, virtual worlds, and electronic voting systems. In mathematics, their students used technology mainly to perform routine calculations, practice skills and procedures, and check answers. They rarely or never used technology to solve complex problems, discover mathematics principles and concepts, process and analyze data, produce graphical representations, or develop mathematics models. As far as digital games are concerned, all teachers agreed that they are worthy of consideration in the classroom and reported already frequently using them in their mathematics classes. However, they had very limited TPACK regarding games as educational tools, viewing them primarily as a useful aid for making instruction more joyful and efficient. The most commonly cited reasons behind their use of games in mathematics were for increasing students' motivation and engagement and for practicing and/or evaluating acquired skills. Thus, teachers lacked understanding of digital games' true potential for transforming the nature of teaching and learning and of how to implement game-based mathematics instruction.

Findings from the interviews at the end of Phase I suggest that the professional development program was quite successful in helping teachers move beyond their restricted views of digital games as educational tools. They came to realize games' true potential for supporting learning of different areas of the early mathematics curriculum, including data analysis and probability, in educationally powerful ways. As they all admitted during the interviews, their past exposure to digital mathematics education games had been limited to drill-and-practice ones. However, their participation in the program gave them the opportunity to be exposed to challenging, complex, and scaffolded (Gee, 2007) games designed to help students build higher-order mathematical and statistical problem-solving skills. This exposure to high-quality digital games helped teachers develop a much more sophisticated view regarding the benefits of gaming:

> I got familiarized with unknown to me games. I saw how the introduction of such games in the classroom could take learning to a higher level, by increasing student initiative and independence.
>
> I realized that games can offer much more than drill-and-practice. The can have an added value in terms of teaching challenging mathematical concepts, promoting children's creativity and analytical skills, supporting differentiated instruction, and enriching student assessment.

Unlike the pre-survey stage, teachers' focus was not on the playfulness of games, but on the fact that their instructional integration offers an effective learning context that refocuses instruction toward a more student-centered instructional experience and promotes the construction of powerful mathematical and statistical knowledge and skills. Using the insights gained from their experimentation with various types of digital games, and their classroom research, teachers listed several advantages that can make appropriately designed games a more meaningful, engaging, and effective learning experience for young students:

> Digital games reinforce student independence, ingenuity, creativity, personalized learning, as well as collaborative learning, and limit the role of the teacher to that of a mentor.
>
> Children get the situation into their own hands and are actively involved in different problem solving scenarios, which are often non-standard and thus force them to think past what they have been taught in class, to think "outside the box".
>
> Good educational games have varying levels of difficulty. So, the game naturally provides feedback on student learning to both teacher and student, since a player needs to achieve the objectives of the previous level to go to the next level. The teacher monitors the progress of each student, and through the game, the learning objectives are redefined for each learner.

16.4.2 Changes in Participants' TPACK Competency for Using Digital Games

Teachers' professional development extended their thinking on how young students learn with digital games and supported the development of their TPACK. Through familiarization with the design principles for constructivist gaming environments, experimentation with a range of game apps, feedback from each other, and reflection, teachers gained better understanding of how to implement game-based mathematics and statistics instruction in the early years. They also improved their ability to assess the educative power of different games, to properly identify their advantages and disadvantages.

Participants' endeavors with the *Evaluation of Educational Games Task* are indicative of the program's effectiveness in helping them acquire the necessary skills for effectively assessing the educational potential and suitability of different digital games. In this task, participants worked individually to compare and contrast six different freely available mathematics education game apps. They used the *Educational Video Game Evaluation Rubric* (http://educators.brainpop.com/wp-content/uploads/2015/04/Game-Rubric-Editable-2015-1.pdf) to appraise the games and then prepared an evaluation report for each selected game. For the purposes of this study, we restricted our analysis to teachers' evaluation rubrics and reports in relation to the following two game apps that have content supporting the development of early statistical reasoning:

- **CarTally**, a mobile game app available on Android and iOS platforms, introduces kindergarten and early elementary school children (Grades K–2; ages 4–8) to the basic mathematical and scientific concepts of observation, classification, quantification and data analysis. Children join Maddie the Giraffe on a journey from the country road to the big city. To move along, they need to identify passing cars and classify them based on color. As they do so, they generate a dataset of the results, which is displayed graphically (as a bar chart). To move forward, children need to answer some questions about what they observe in the graph (e.g., *How many blue cars are there…? Are there more yellow cars than green cars…?*). As they advance through a series of four progressive sorting levels, children are introduced to more complex data analysis concepts, where alternative strategies for visualizing and evaluating data are presented.
- **The Electric Company Prankster Planet**, available on Android and iOS platforms as well as online, is based on the Emmy Award-winning PBS KIDS TV series The Electric Company, and it targets children aged 6–10. It features eight unique quests with math curriculum woven throughout that children have to complete to save Earth from the Reverse-a-ball machines of Prankster character Francine, that are scrambling up all the words on Earth and are causing a lot of confusion. Children complete a series of data collection, representation, and analysis challenges in order to shut down all eight machines hidden in the jungles, cities, junkyards, and underground world of Prankster Planet. The app features side-scrolling play and exploration in a 2D platformer world, an avatar creator with many customization options, a reward system to encourage repeat play, and the option of collaborative play through group activities.

We chose to focus our analysis of students' responses on *CarTally and Prankster Planet* because we wanted to investigate how teachers would compare and contrast two game apps of varied educational value. Specifically, we selected *CarTally* as a typical example of a "drill-and-practice" type of game app, and *Prankster Planet* as a good example of a high-quality game app that includes the elements of collaboration and competition and promotes authenticity of learning, higher-order thinking, and statistical problem-solving.

Table 16.1 shows the median score of teachers' ratings of each game for each of the criteria in the *Educational Video Game Evaluation Rubric* (on **a 1–5 scale with 5 being the best**).

Analysis of teachers' evaluation reports showed that they all based their assessment of the two games on important technical and pedagogical considerations, in accord with the scholarly discussions of positive implementation of games in learning situations and our own ranking of the two games. They justified their preference for *Prankster Planet* by explaining that were attracted by compelling technical features, such as the user-friendly interface and well-designed graphics. They also recognized

Table 16.1 Teachers' median ratings of *Prankster Planet* and *CarTally*

Evaluation feature	Prankster Planet median score	CarTally median score
Presentation of content		
Accurate	4.0	3.0
Interactive	4.0	2.0
Gameplay		
Compelling objectives	4.0	2.0
Integrated content and gameplay	5.0	2.0
Embedded assessment	5.0	2.0
Pedagogy		
Adaptive instruction and feedback	4.0	2.0
Amount of instructions	5.0	1.0
Interface	5.0	3.0
Multimedia		
Audio	5.0	3.0
Artwork	4.0	3.0
Narrative and theme	4.0	1.0

that the pedagogical potentials of *Prankster Planet* are mediated by many factors, including the rich narrative and theme, which establishes an engaging context that promotes authenticity of learning and provides reasons for the student to play the game. They argued that the capturing game scenario, in combination with the reward system, engages and immerses young children in the learning process, while the provision of avatars enables them to identify with the story and the main characters and keeps their interest high. More importantly, teachers recognized that *Prankster Planet* provides a challenging environment that promotes inquiry learning and statistical problem-solving, thus reflecting the emphasis of the program on the importance of selecting games that promote learners' higher-order, critical thinking skills (e.g., Williamson, 2009). Other positive characteristics identified in the evaluation of *Prankster Planet*, again in agreement with the literature, are the fact that the game offers varied difficulty levels for the player to choose from. Teachers also stressed that the app is an excellent example of a social game that can support collaboration and group work and/or competition among players, an important aspect of digital games also discussed in the literature (e.g., Felicia, 2009).

The reasons the teachers gave for their more negative evaluation of *CarTally* were also based on important pedagogical principles. They noted that although *CarTally* does have a user-friendly and well-designed interface and supports differentiated instruction through the provision of multiple levels, it is not a game they would introduce in their classroom to promote higher-order statistical thinking, but would rather "*use it at the end of a unit for drill-and-practice purposes.*" They gave several reasons to justify their more negative evaluation of the game: (i) not based on a really "convincing" scenario; (ii) not adequately challenging and engaging; (iii) does not provide proper feedback; (iv) finding the right answer can be based on random factors; (v) can only be played individually; thus, there is no opportunity for collaboration and/or competition among children.

16.4.3 *Transfer and Adoption of TPACK Competencies to Teaching Practice*

During Phases II and III, participants transferred the knowledge acquired during Phase I into lesson planning and implementation. This was valuable experience that helped them further develop professionally in relation to game integration into the mathematics curriculum. Their hands-on teaching experiences and sharing with peers helped them develop many pedagogical ideas and apply in practice effective instructional strategies for successful instructional integration of games. Analysis of the lesson plans submitted by all six teachers suggests positive gains in their ability to effectively select and integrate digital games within the mathematics curriculum. Everyone prepared lessons that incorporated the use of high-level educational games, which fitted well with their targeted grade level and curricular topic. All lesson plans also included appropriate pre- and post-game activities.

Next, we turn our attention to the only teaching intervention that focused on statistics. As mentioned in the Methodology section, the intervention took place in a Grade 2 mathematics classroom. It targeted the following indicators of achievement included in the national mathematics curriculum for Grade 2: (i) Collect information and data in the environment and present them in an organized way; (ii) record, organize, and present data in tables and graphical representations (pictogram, bar graph, pie chart); (iii) represent the same data in multiple ways. The lesson is briefly described next, followed by some reflection.

Lesson Overview

The lesson started with the teacher informing the class that the headmistress needed information about children's preferences on afternoon activities, to plan for the upcoming school year. Therefore, she would like each class to collect data regarding their preferences, organize them, and present them to her in way that would help her come up with a schedule of activities that would satisfy as many children as possible.

To tackle the task, the class first decided to do a census where each student selected their favorite afternoon activity among five different options. Then, they recorded their preferences on a tally table and counted the total number of students selecting each afternoon activity. This was followed by class discussion on how to best present the survey results to the headmistress. During this discussion, children were introduced to bar graphs, the process of their construction, and the ways in which this graphical representation can help us to present and interpret the results of a survey and to make decisions.

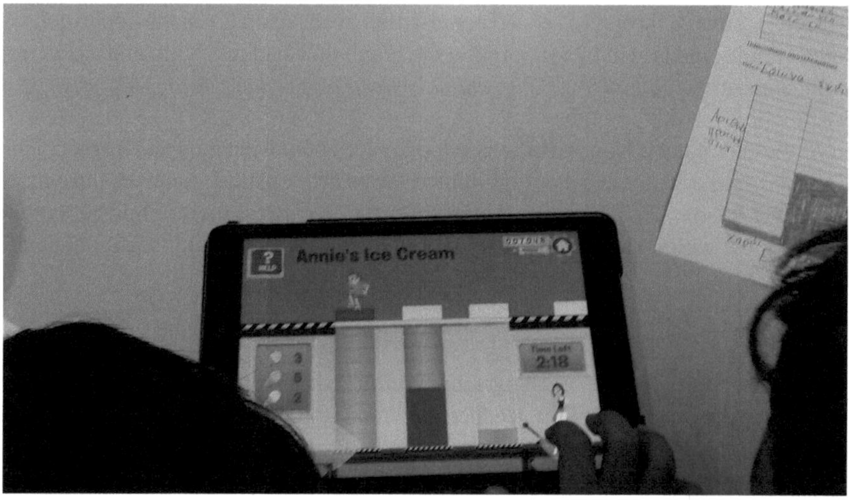

Fig. 16.1 A pair of students interpreting a bar graph in *Prankster Planet*

Next, children played in pairs the digital game *Prankster Planet* on their iPads. As explained earlier, in this game players take the role of a character they construct using the avatar creator and they try to prevent the wicked Francine from reversing the order of letters in all the words on Earth, and thus causing great inconvenience on the planet. Children were asked to go through all eight increasingly difficult missions of the game. Within each mission, they responded to various questions where they had to interpret a bar chart or a pie chart, or to construct such graphs, in order to collect as many points as possible and to achieve their objective (see Fig. 16.1).

Throughout the activity, the teacher went around the class and offered personalized assistance to each pair of students. Although children were fully engaged with the game and quickly learned its mechanics (possibly due to their high degree of familiarity with digital games as the vast majority owned a tablet at home), they needed some help from their teacher in understanding the posed questions, which were in English.

Finally, children completed an exercise adopted from their mathematics textbook. They were presented with a table showing the number of hours of sleep per day of 12 different mammals. First, they worked in pairs to complete a task in the game-platform Kahoot! where they had to respond to various questions related to the information included in the table. The task was set up as a contest, where each correct answer gave different points to the pair, depending on their response time. For each question, the right answer appeared on the whiteboard, as well as a bar chart showing the children that answered it correctly, and the score of each pair. After completing the task, the children worked individually to construct a "bar graph" of the data in their notebook (a simple histogram in reality), with each bar corresponding to an interval of hours of sleep (i.e., animals sleeping between 0 and 5 h, between 6 and

10 h, etc.). Finally, through a second contest, they were again given the opportunity to collect points and to win by interpreting the graph they had just built, and answering various questions (e.g., *How many animals sleep more than 5 h and less than 16 h per day?).*

At the end of the lesson, a discussion took place about what they had learned. During this discussion, the teacher highlighted the main statistical concepts introduced during the lesson. Children were also encouraged to express their opinion about the format of the lesson and whether they would like to again use similar digital games in class.

Reflection on the Lesson

As the observation of the teaching intervention, but also the quality of the lesson plan and tasks prepared by the teacher indicated, the specific educator did indeed acquire the necessary TPACK for teaching this topic and similar statistical topics using tablets and game apps like *Prankster Planet* and *Kahoot!* It can be argued that this teacher knew how to exploit the specific technologies to their full potential. He understood how to represent the specific statistical concepts through use of the technological tools and used appropriate pedagogical techniques in a constructive way. He was also aware of the difficulties/obstacles that students usually face when learning these concepts, and of the ways in which the integration of games could help to overcome these difficulties.

The teacher selected appropriate game apps—*Prankster Planet* for introducing the new statistical concepts to children and *Kahoot!* for reinforcing these concepts in a playful manner. He properly exploited the games to organize teaching in a constructive, learner-centered way, so that his young students would have the opportunity to work together in constructing statistical concepts and processes. Specifically, in several activities, children collaborated to build joint understanding of the new concepts they had just encountered. For example, to respond to some of the *Prankster Planet* questions, students worked together to understand and interpret pie charts, which had never been mentioned in class before (see Fig. 16.2).

Moreover, the process of assessing their learning process was transferred from the teacher to the students themselves. Instead of standing in front of the classroom, the teacher moved around to the different groups and offered assistance whenever necessary.

As expected, the game-based nature of the lesson was well received by the students and increased their morale and motivation. In the discussion that took place at the end of the intervention, children expressed their enthusiasm about the games they engaged with during the lesson, "*because they gave [them] the chance to play and learn at the same time.*" Although expressing a preference for "*the game where [they] had to prevent the Lady from changing the letters*" (i.e., *Prankster Planet*) which they found to be "*more adventurous,*" they also liked the game on *Kahoot!* because it was set up as a contest. The children stressed that they would love all their lessons to have a similar format since games enable them "*to learn more things about mathematics and to have a joyful time in class.*"

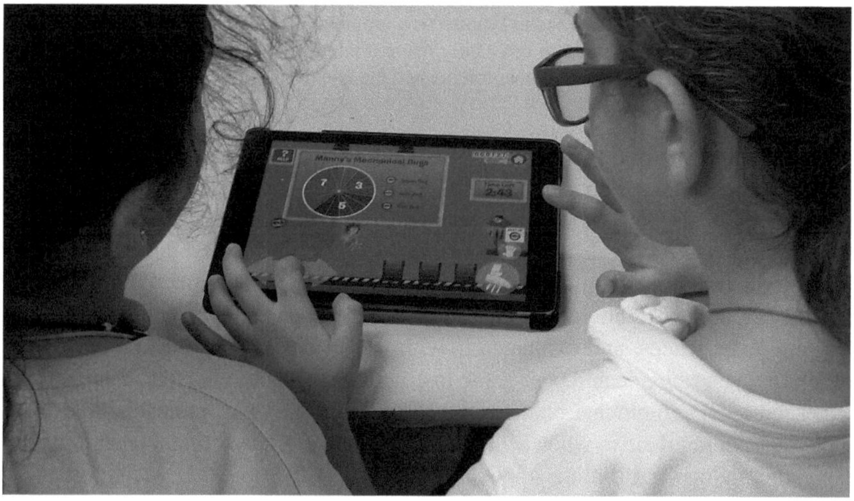

Fig. 16.2 A pair of students interpreting a pie chart in *Prankster Planet*

In the reflection paper he wrote after the teaching intervention's completion, the teacher also noted that he was very impressed by the fact that the lesson ended up being so successful:

> The children got excited with the games, were constantly active, there was rivalry between the teams, but also collaboration within each team, and this had a positive impact on children's motivation, but also on the attainment of the instructional objectives. Very important and noteworthy is also the fact that even the supposedly "weakest" students had strong interest and active participation during the lesson and gave several thoughtful answers, with no trace of fear of making a mistake. These are qualities I do not usually experience in lessons where I do not use digital games.

According to the teacher, the use of the games led to an effortless involvement of all children and contributed substantially to the achievement of the learning objectives, but also to ensuring fruitful cooperation among learners and the collaborative construction of knowledge in a creative and enjoyable way. Lastly, use of technology worked exactly as he had anticipated, providing the opportunity to introduce in the classroom activities with added value, that could not otherwise be implemented.

While the classroom experimentation further strengthened the teacher's belief that appropriate use of game apps can help create motivational and more conducive to learning environments, it also helped him to build more realistic expectations about what games' instructional integration might entail in practice. He recognized that games are not a panacea and that their incorporation into the curriculum does not guarantee improved learning. He mentioned various challenges and drawbacks to digital games' incorporation in the mathematics classroom, including time constraints, difficulties in locating high-level games, the risk of the learning objectives being neglected for the sake of playfulness, and language issues for non-native English speakers:

> The fact that Prankster Planet is in English was prohibitive factor to its best possible use. While students played, I had to constantly move around to translate things to them. The mismatch between children's mother tongue and the language of the game made my work hard and tiring.

The teacher stressed the key role of educators not only in choosing appropriate digital games with the goals of gameplay being closely aligned with instructional objectives, but also in "*coordinating classroom activities appropriately so as to keep children focused on the achievement of the learning objectives,*" and in facilitating learning by providing continued support and scaffolding. He pointed out that games should not dominate class time, but should be used as part of carefully planned learning experiences, and explained how, in his teaching experimentation, the use of alternative pre-game and post-game instructional activities led to a fuller learning experience. His comments concur with the literature, which indicates that digital games are more effective when acting as adjuncts to more traditional teaching methods rather than as stand-alone applications (Gee, 2007).

16.5 Conclusions and Implications for Teaching and Research

Digital games present some exciting opportunities for a transformative shift in early statistics instruction. However, their actual success as an instructional tool will ultimately depend upon the abilities of teachers to take full advantage of their affordances. Findings from the teacher pre-survey in the current study corroborate with the research literature, which indicates that the majority of teachers do have positive attitudes toward the educational adoption of games but lack appreciation of their true potential (e.g., Koh et al., 2012). Similar to other researchers (e.g., Van Eck et al., 2015), we also found that the teachers in our study had limited TPACK of game-enhanced learning, lacking the vision and personal experience of what game-enhanced teaching could look like, and tending to view games as instructional tools to be used for motivational or drill-and-practice purposes.

In accord with the research literature (e.g., Niess et al., 2009), our research has illustrated the usefulness of the TPACK framework as a means of both studying and facilitating teachers' professional growth. Although there was no pre-post test assessment to formally track changes in participants' TPACK, the results do suggest that the in-service teacher education format was successful in achieving all three main objectives, as these derive from the study research questions. Firstly, there are strong indications in the collected data that the program was quite successful in helping teachers move beyond their restricted views of digital games as educational tools. Secondly, it helped to convey to teachers technological and pedagogical knowledge regarding the teaching of specific mathematical and statistical concepts with games. Thirdly, it improved teachers' confidence and ability to transfer and adopt the TPACK competencies acquired through the program to actual teaching practice.

There are important implications of this study for the design or revamping of pre-service or in-service teacher education curricula and programs on digital game integration. The study design and outcomes shed light on what effective teacher education in game-enhanced learning might entail in helping statistics teachers learn about, adopt, and integrate games (but also other technological tools) into their teaching. Insights from the study suggest that utilizing a conceptually based theoretical framework about the relationship between technology and teaching like TPACK can enrich teachers' professional development. Findings also indicate that the development of teachers' TPACK necessitates the provision of opportunities for both theoretical and experiential learning of technology-based pedagogical approaches to mathematics education. Concurring with prior research (e.g., Serradó, Meletiou-Mavrotheris, & Paparistodemou, 2014), this study provides evidence that teachers' involvement in professional development activities such as lesson design and field experience (e.g., conduct of action research, classroom teaching, classroom observation) can support them in developing their teaching competencies with ICT and understanding of TPACK in ways transferable into their own practice.

Although this case study has provided some useful insights, the presented results are only suggestive and warrant further research. There are several limitations to the study, emerging primarily from its exploratory nature, which constrain the interpretation of the research results. A serious drawback is the limited generalizability of the research findings. The qualitative methodology used to research the case, the small scale of the study, and its limited geographical nature means that generalizations to cases that are not very similar should be done cautiously. Thus, the study needs to be repeated with more cohorts of in-service teachers, both within and outside Cyprus. Future iterations of the study ought to employ more rigorous research methods and procedures to investigate the impact of the professional development on in-service teachers' TPACK competencies and skills on game-enhanced early statistics instruction (e.g., use of control groups, collection of pre- and post-data on the actual impact of teachers' developed TPACK on children's motivation, and higher-level statistics learning). This approach could lead to the development of generalized principles and models of professional development that can help foster the expertise of mathematics teachers in incorporating games into early statistics instruction.

References

Asbell-Clarke, J., Edwards, T., Rowe, E., Larsen, J., Sylvan, E., & Hewitt, J. (2012). Martian boneyards: Scientific inquiry in an MMO game. *International Journal of Game-based Learning, 2*(1), 52–76.

Boyle, E. A., MacArthur, E., Connolly, T. M., Hainey, T., Manea, M., Kärki, A., et al. (2014). A narrative literature review of games, animations and simulations to teach research methods and statistics. *Computers & Education, 74,* 1–14.

Common Sense Media. (2013). *Zero to eight: Children's media use in America 2013.* Retrieved from http://www.commonsensemedia.org/research/zero-to-eight-childrens-media-use-in-america-2013.

Clark, D. B., Tanner-Smith, E. E., & Killingsworth, S. (2014). *Digital games, design, and learning: A systematic review and meta-analysis*. Menlo Park, CA: SRI International.

Felicia, P. (2009). *Digital games in schools: A handbook for teachers*. Brussels, Belgium: European Schoolnet.

Franklin, C., Kader, G., Mewborn, D., Moreno, J., Peck, R., Perry, M., et al. (2007). *Guidelines for assessment and instruction in statistics education (GAISE) report: A PreK–12 curriculum framework*. Alexandria, VA: American Statistical Association.

Gee, J. P. (2007). *What video games have to teach us about learning and literacy*. New York, NY: Palgrave Macmillan.

Geist, E. A. (2012). A qualitative examination of two-year-olds interaction with tablet based interactive technology. *Journal of Instructional Psychology, 39*(1), 26–35.

Ginsburg, K. R. (2007). The importance of play in promoting healthy child development and maintaining strong parent–child bonds. *Pediatrics, 119*(1), 182–191.

Gresalfi, M., & Barab, S. (2011). Learning for a reason: Supporting forms of engagement by designing tasks and orchestrating environments. *Theory into Practice, 50*(4), 300–310.

Hannigan, A., Gill, O., & Leavy, A. M. (2013). An investigation of prospective secondary mathematics teachers' conceptual knowledge of and attitudes towards statistics. *Journal of Mathematics Teacher Education, 16*(6), 427–449.

Ke, F. (2008). Computer games application within alternative classroom goal structures: Cognitive, metacognitive, and affective evaluation. *Educational Technology Research and Development, 56*(5–6), 539–556.

Koh, E., Kin, Y. G., Wadhwa, B., & Lim, J. (2012). Teacher perceptions of games in Singapore schools. *Simulation Gaming February, 43*(1), 51–66.

Larkin, K. (2015). An App! An App! My kingdom for an App: An 18-month quest to determine whether Apps support mathematical knowledge building. In T. Lowrie & R. Jorgensen (Eds.), *Digital games and mathematics learning: Potential, promises and pitfalls* (pp. 251–276). New York: Springer.

Lowrie, T., & Jorgensen, R. (2015). *Digital games and mathematics learning: Potential, promises and pitfalls*. New York: Springer.

Manessis, D. (2014). The importance of future kindergarten teachers' beliefs about the usefulness of games-based learning. *International Journal of Game-Based Learning, 4*(1), 78–90.

Meletiou-Mavrotheris, M., & Paparistodemou, E. (2015). Developing young learners' reasoning about samples and sampling in the context of informal inferences. *Educational Studies in Mathematics, 88*(3), 385–404.

Miller, D. J., & Roberstson, D. P. (2010). Using a games-console in the primary classroom: Effects of 'Brain Training' programme on computation and self-esteem. *British Journal of Educational Technology, 41*(2), 242–255.

Mishra, P., & Koehler, M. J. (2006). Technological pedagogical content knowledge: A framework for teacher knowledge. *Teachers College Record, 108*(6), 1017–1054.

Munoz-Rosario, R. A., & Widmeyer, G. R. (2009). An exploratory review of design principles in constructivist gaming learning environments. *Journal of Information Systems Education, 20*(3), 289–300.

Niess, M. L., Ronau, R. N., Shafer, K. G., Driskell, S. O., Harper, S. R., & Johnston, C. (2009). Mathematics teacher TPACK standards and development model. *Contemporary Issues in Technology and Teacher Education, 9*(1), 4–24.

Pivec, M., & Pivec, P. (2009). What do we know from research about the use of games in education? In P. Wastiau, C. Kearney, & W. Van den Berghe (Eds.), *How are digital games used in school* (pp. 123–156). Brussels, Belgium: European Schoolnet.

Saracho, O. N. (2012). *An integrated play-based curriculum for young children*. New York: Routledge.

Serradó, B. A., Meletiou-Mavrotheris, M., & Paparistodemou, E. (2014). EarlyStatistics: A course for developing teachers' statistics technological and pedagogical content. *Le Revue Statistique et Enseignement, 5*(1), 5–29.

Takeuchi, L. M., & Vaala, S. (2014). *Level up learning: A national survey on teaching with digital games*. New York: The Joan Ganz Cooney Center at Sesame Workshop.

Van Eck, R., Guy, M., Young, T., Winger, A., & Brewster, S. (2015). Project NEO: A video game to promote STEM competency for preservice elementary teachers. *Journal of Teaching, Knowledge, and Learning, 20*(3), 277–297.

Watson, J. M. (2001). Profiling teachers' competence and confidence to teach particular mathematics topics: The case of chance and data. *Journal of Mathematics Teacher Education, 4*(4), 305–337.

Williamson, B. (2009). *Computer games, schools, and young people: A report for educators on using games for learning*. Bristol, UK: Futurelab.

Chapter 17
Young Children's Statistical Literacy in Modelling with Data and Chance

Lyn D. English

Abstract This chapter reports on eight-year-old children's responses to data and chance investigations designed to foster their statistical literacy. Explored through the lens of modelling with data, statistical literacy involves a number of processes common to both statistics and probability, with culmination in models from which conclusions and inferences can be drawn. Specifically, consideration is given to how the students identified variation, made informal inferences, created representations, and interpreted their models that displayed the outcomes from their investigations.

17.1 Defining Statistical Literacy

Literacy has had numerous interpretations across disciplines over the years. As diSessa (2018) points out, we need to know more about what forms of literacy can exist, how they evolve over the years, and how they are shaped by sociocultural factors. The representational forms in which literacies are embedded likewise require greater attention. While textual literacy has had extensive coverage (e.g. Kalantzis & Cope, 2012), literacies pertaining to statistics and probability have not been as prolific, especially in the early years, and have been open to debate.

Several definitions of statistical literacy have appeared in the past decade or so (e.g. Ben-Zvi & Garfield, 2004; Gal, 2002; Watson, 2006). In contrast, the notion of probabilistic literacy has received limited attention, frequently remaining implicit in definitions of statistical literacy. Watson's definition (e.g. English & Watson, 2013, 2016) encompasses both statistics and probability, both of which must be targeted in developing fundamental understandings for dealing with an increasingly complex world. Watson considers statistical literacy to be the "meeting point" of statistics and probability with the everyday world, in which encounters involve unrehearsed contexts, chance phenomena, and spontaneous decision making (Watson, 2006, p. 11).

L. D. English (✉)
Faculty of Education, Queensland University of Technology, Brisbane, Australia
e-mail: l.english@qut.edu.au

A. Leavy et al. (eds.), *Statistics in Early Childhood and Primary Education*, Early Mathematics Learning and Development,
https://doi.org/10.1007/978-981-13-1044-7_17

Table 17.1 Foundational literature for early statistics and probability

Knowledge Components	Chance constructs: possible, impossible, certain, uncertain, random, likely, unlikely, equally likely, most/least likely Statistical and chance constructs: variation, expectation, prediction, data distributions, informal measures of centre (mode, median, mean)
Processes	Designing and conducting investigations with chance and data: Mathematical, statistical, and chance processes including measuring, comparing, interpreting, representing, modelling, making decisions, identifying uncertainty, justifying, communicating Making informal inferences: predicting, recognizing uncertainty and variability in drawing conclusions beyond the data, generalizing
Dispositions	Critical awareness, appreciation of uncertainty, flexibility, seeking connections

Watson's definition is a multifaceted one and brings into question the "literature" being targeted. As diSessa (2018) rightly argues, "a literacy needs a literature". By proposing this argument, diSessa emphasizes the need to "get to civilizations' expanse of deep and powerful ideas" (p. 7). What then is the literature of statistical and probabilistic literacy, that is, the core constructs that define the domain? While not delving into controversies on whether statistics and mathematics are discrete domains (see Chap. 9 for a discussion on this issue), I consider the literature for early statistical literacy as encompassing key elements of mathematics, statistics, and chance phenomena and constructs. The role of dispositions in any such literacy also cannot be underestimated (Gal, 2005). Table 17.1 presents what I consider to be the foundational literature for early statistical literacy, including probability.

In this chapter, I describe two investigations implemented in third grade, one focusing primarily on statistics and the other on probability. Examples of children's responses fostered in the early school years are provided. The nature of this learning is addressed through the lens of modelling.

17.2 Developing Statistical Literacy Through Modelling

Models and modelling have been variously interpreted in the literature, as reported by English, Arleback, and Mousoulides (2016). Modelling with statistics and probability can entail a number of processes, including: interpreting and understanding a problem and its context; identifying, posing, and refining questions; collecting and organizing data; recognizing variation; creating models (through representing and re-representing statistical and probabilistic outcomes); and drawing conclusions and informal inferences from models generated (cf. Makar, Bakker, & Ben-Zvi, 2011; Lehrer & English, 2018).

Models are typically conveyed as systems of representation in which structuring and displaying data are essential (Lehrer & Schauble, 2007). Hestenes' (2010) expresses this succinctly, namely "A *model* is a *representation of structure* in a given *system*", with *systems* defined as "a set of related objects", where the structure of a system is the set of relations among the objects (p. 17). Such objects can be actual or imaginary, concrete or cognitive, simple or complex. Young children create the structure of a model; such structure is neither inherent nor given to them, in contrast to many other mathematical situations they meet in class. Furthermore, research has revealed how young learners can create more than one representation to display the same data (e.g. English, 2010; Lehrer & Schauble, 2000), revealing what diSessa (2004) refers to as "met representational competence". Such competence reflects children's "native capacities" (diSessa, 2004, p. 294) to create and recreate their own forms of representation, with such competence appearing to exist before instruction and develop independently of it. Moreover, metarepresentational competence can help children's development of conceptual competence. Further elaboration on metarepresentational competence appears in Chap. 14, where the authors illustrate how preschool students demonstrated features of invention and learning as they created their own representations.

Although the importance of modelling with data and chance has been highlighted in the literature (e.g. English, 2013; Lehrer & English, 2018), young children are often not given credit for their ability to create and work with models, particularly when it comes to probability. Research (e.g. English, 2013; Fielding-Wells, 2014; Lehrer & Schauble, 2005) has revealed; however, that primary school students can engage in modelling involving data and chance, and furthermore, can apply their statistical learning to representing and modelling chance events (English & Watson, 2016).

Young learners' ability to create a range of representations, including those that extend beyond traditionally or expected accepted formats, is underestimated and needs greater recognition. In particular, the explicit consideration of variation in relation to organizing and structuring data, whether the variation is generated in chance experiments or statistical investigations, has not been a key feature of research in the primary years.

17.3 Variation

Variation (and expectation, addressed next) is foundational in developing both probability and statistics in the early years (English & Watson, 2016; Lehrer, 2011). In simple terms, variation is "the quality of an entity (a variable) to vary, including variation due to uncertainty" (Makar & Confrey, 2005, p. 28). Although there is considerable research on older students'/adults' awareness of variation there is less so on how this understanding can be developed with young students. This is a major concern especially when older students frequently apply statistical techniques without understanding or appreciating why, when, or how these are applied sensibly to a range of contexts (Garfield & Ben-Zvi, 2008).

Explicit consideration of variation in relation to representations has not been a key feature of research in the primary years. Yet a major foundational component of young students' statistical literacy is being able to interpret the meaning, within a given context, of a distribution that displays variation, clusters, modes, and unexpected values. Such distributions are not just confined to data generated from children's statistical investigations. As they undertake chance experiments, young children need to become aware of variation in the outcomes, which often are contradictory to their expectations. When children represent their findings from these experiments, they can see and appreciate such variation.

Opportunities for children to represent, in their own way, the outcomes of early chance experiments have remained largely ignored. As is the case for statistical investigations, it is important that children identify and justify the sources of variation that they encounter (but usually don't anticipate) in chance experiments. Their typical experiences tend to involve chance events for well-defined sample spaces in which equally likely outcomes are assumed (e.g. those of a die or coin; Jones, Langrall, & Mooney, 2007). Greater insights are needed into how young students deal with variation in working with statistics and probability, and how such variation confirms or refutes their initial expectations.

17.4 Informal Inference

Informal inference, including expectation and prediction, is the process of using the evidence provided by data to answer questions beyond the data, while acknowledging the uncertainty in reaching a conclusion (Makar, 2016; Makar & Rubin, 2009). Although expectation is usually referred to in connection with drawing informal inferences from the models generated, it can be present, either implicitly or explicitly, throughout the modelling process. As Makar and Rubin (2009) pointed out:

> Focusing on investigating phenomena entails understanding the statistical investigation cycle *as a process of making inferences*. That is, it is not the data in front of us that is of greatest interest, but the more general characteristics and processes that created the data. This process is indeed inferential. (p. 84)

The extent to which any conclusion from an investigation can be reached with some degree of certainty depends on creating a balance between variation and expectation (Watson, 2006). Konold and Pollatsek's (2002) well-known metaphor of signal in noise, with noise indicating variation and signal the expectation (or central tendency) that can be extracted from the noise, nicely conveys these foundational constructs. Unfortunately, a focus on informal inference in the primary school has been limited despite the fact, like adults, young children make predictions in their everyday lives (Doerr, Delmas, & Makar, 2017; Makar & Rubin, 2009). Learning to appreciate variation and its relationship to expectation/prediction needs to begin early with appropriate hands-on experiences and student/teacher questioning.

In the remainder of this chapter, I first describe two investigations, one dealing with statistics and the other with probability. Of particular interest in both activities were students' identification of variation, their informal inferential reasoning, the representations they created, and how they interpreted their models. Findings related to these aspects are presented.

17.5 Investigations in Modelling with Data and Chance

17.5.1 Background

The two investigations, *Manufacturing Licorice*, and *What is the Chance of That?* were implemented during the first year of a four-year longitudinal study being conducted across grades three through six in two Australian capital cities (with collaborators, Jane Watson and Noleine Fitzallen). The first investigation was implemented in both cities, while the second in just the author's city. Examples of children's responses are drawn from data in the author's city only (one third-grade classroom, mean age of 8.8 years). The first investigation, *Manufacturing Licorice*, was implemented in the first half of the third-grade year, with the second, *What is the Chance of That?* in the second half. For the implementation of each investigation, "focus groups" were selected in consultation with the class teacher and comprised three students of mixed achievement levels. Detailed lesson plans for both investigations were prepared for the teacher, as was a workbook for each student. The students worked in small groups of mostly three members, although recorded their own responses to the investigations in their workbooks.

17.5.2 Design and Analysis

A design-based approach was adopted (Cobb, Jackson, & Dunlap, 2016), with such an approach catering for complex classroom situations that contain many variables and real-world constraints. A design-based approach supports learning and informs future learning experiences based on feedback, and facilitates contributions to both theory and practice.

Data collection included videotaping of three focus groups as they worked the investigations, as well as all class discussions, which were subsequently transcribed for analysis. The data reported in this chapter are drawn from the students' workbooks, together with the recorded and transcribed group work and whole-class discussions. In conjunction with an experienced research assistant, content analysis (Patton, 2002) was applied in initially identifying and coding the data recorded in the students' workbooks. A further round of refined coding was undertaken to ensure meaningfulness and accuracy. For example, in coding the students' responses for "What does the shape of the class plot tell you about the variation in the licorice sticks made by the class?", we refined our coding thus: "*code 2*: "student must refer to change or comparison, such as 'there is more on 10 and less on 7'; 'there are a lot of people between 8 and 16'"; *code 1*: "student refers to a single characteristic (no reference to variation), such as 'there is a lot on 10; there is a lot on 13'"; and *code 0*: "the student gave no response, an idiosyncratic response, or one that was out of context" such as, 'there are numbers, *g*, and sticky notes'; they are both like long rectangles'".

Iterative refinement cycles for videotape analyses of conceptual change (Lesh & Lehrer, 2000) were applied in reviewing the transcribed focus group and whole-class discussions to gain greater insights into the development of the students' learning.

17.5.3 Investigation Implementation

Manufacturing Licorice. In this investigation, students experienced the "creation of variation" as they compared the masses of "licorice sticks" they made by hand (using Play-Doh) with those made using a Play-Doh extruder kit ("factory made"; adapted from Watson, Skalicky, Fitzallen, & Wright, 2009). Students chose their own forms of representation in displaying their models for the two forms of licorice production and identified, compared, and explained the features of their data distributions. Following a number of introductory experiences (e.g. exploring the manufacturing processes and roles of engineers in the creation of licorice and other such products), the students discussed questions pertaining to quality control and the overall manufacturing process. In small groups, the students then undertook the two investigations involving handmade licorice sticks and those made by the Play-Doh extruder.

For each licorice manufacturing method, the students identified, measured, compared, and recorded attributes of the sticks including their mass, and compared their findings with their group members. Their group data on the masses were then collated and compared within the group. Each group member then created her own representation of the collated group data. No direct guidance was provided for developing these representations. However, if a student's representation was incomplete or unclear we would remind them to check that their creation could be interpreted clearly. Subsequent class sharing and interpreting of the resultant group models for each method enabled students to identify the range and "typical" masses displayed in each group model. Each method of licorice manufacture ended with all group data being collated and displayed as a class plot (bar graph). Figure 17.2 illustrates the plots created as each child placed a post-it-note, on which they had recorded one of their licorice stick masses, in the appropriate position on the axis drawn on the class whiteboard. The students explored and discussed the data distribution revealed in the whole-class model for each licorice manufacturing method, with inferences drawn regarding the two methods and the predicted masses if further sticks were made.

What is the Chance of That? In this investigation, the children created their own chance experiments and independently developed core probability understandings. Introductory experiences included reading a chance storybook ["Probably Pistachio" (Murphy, 2001)], which helped children appreciate that one cannot predict with certainty the outcomes of a probabilistic situation. On reading the book, the children recorded examples of events that would be certain to happen for them the following weekend, could possibly happen, and would be impossible to occur. The next introductory component engaged the children in playing a "bingo" game where the notions of randomness and variation in chance events were experienced. On playing the game, the children responded to questions including, "Did everyone have an equal chance of winning?"; "Was it certain that someone would win?"; "Was it possible for two people to win?"; and "Do you think some numbers are more likely to roll out than others?" They were to justify their responses.

In the main investigation, the children were presented the scenario of a company seeking their help in designing a game ("What is the chance of that?"). The children were to help the company determine the chances of selecting various coloured counters from a "mystery bag". A container of 36 counters comprising nine of each of four different colours was presented to each group of children. Each group was to select only 12 of the 36 coloured counters to place in their group's mystery bag, using their choice of numbers of each colour but ensuring there was at least one of every colour in the bag (the numbers of each colour did not have to be the same). The remaining counters were returned to the container. Each group member recorded on a table the number of each coloured counter they chose for their mystery bag of 12 counters. Prior to selecting counters, each child was to *predict* and record what coloured counter she would draw from the bag if she only had one chance and was not permitted to look. Once all students in the group had made their prediction, each child recorded the predictions of the other group members.

Next, the children took turns in selecting one counter without looking, returning the counter to the bag, and recording the outcome of each group member's selection. On completion, they responded to questions (and justified their responses) on their initial predictions and any variation in the data they observed. The questions included: "Why did you predict you would select that colour?"; "Did you select the coloured counter that you predicted?"; "Describe any variation you can see in your table of data"; and "If you were to repeat the counter selection over and over, how might the data in your table change?"

The following questions were then posed regarding the children's chances of selecting counters from their bag (responses were again to be justified): "Does each counter have an equal chance of being selected?"; "Is there a coloured counter that has the greatest chance of being selected?"; "Is there a coloured counter that has the least chance of being selected?"; and "Would it be possible to select a purple counter from your bag?" (no purple counters were included in the containers of counters).

The last component of the investigation engaged the children in creating two representations of the chances of selecting the different coloured counters from their group's mystery bag. The first representational form linked the children's learning with their introduction to fractions earlier in the year (e.g. "2 chances *out of* 12 chances;" $\frac{2}{12}$). The second form was primarily a statistical representation of the children's own choice. On creation of their models, the children were to explain how their model displayed the chances of selecting their coloured counters. Following this, the children were invited to represent their chances in a different way (re-representing) and subsequently compare their two models to determine which they considered conveyed their "chance story" more effectively, and why.

In reporting a sample of findings from both investigations, consideration is given to their shared features of identifying variation, drawing informal inferences, creating representations, and interpreting models generated.

17.6 Sample of Findings

17.6.1 Identifying Variation

Manufacturing Licorice. The majority of students were able to identify variation and justify their responses. Over 80% of students ($N = 23$) could detect the variation in the mass of handmade licorice sticks and all ($N = 24$) could do so for the "factory made" ones. Likewise, the students had few difficulties in giving an appropriate initial reason for this variation in the former (87%, $N = 23$), with explanations including reference to some sticks being "fatter" or "too thin" or "thicker". A considerable number of students (71%, $N = 24$) could explain that the Play-Doh extruder was more accurate in producing sticks of a consistent mass (e.g., "Because it's a machine like, the machine makes them all about the same size and when you're doing them with your hands you can't really tell if they're going to be the same size or not").

What is the Chance of That? As for the previous investigation, the students had little difficulty in identifying variation in their predictions and outcomes of counter selection. All students except one ($N = 24$) were able to explain the variation in the table displaying their predictions and outcomes. The students' responses varied from "We all got the same colour which was yellow so there is no variation", to "There wasn't much variation because we had 4 reds and 3 of (each) of the other colours", and "Yes (there was variation) because we picked out different colours and predicted different colours". One student referred to variation in the student names and predictions ("On the table the names are different and the predictions are different").

Students' overall ability to identify variation in both investigations provided a foundation for drawing informal inferences, where one has to acknowledge variation in the data, and hence the uncertainty with which any conclusions can be drawn (cf. Makar, Bakker, & Ben-Zvi, 2011; Lehrer & English, 2018). As noted next, students' reference to chance in drawing conclusions from the *Manufacturing Licorice* investigation suggests they were linking their understanding of statistics and probability in developing statistical literacy.

17.6.2 Drawing Inferences

On both investigations, students were asked to make informal inferences from the data created and/or the models generated.

Manufacturing Licorice. In this investigation, students drew inferences from the whole-class models constructed from the group data collected for each manufacturing method. The scenario was posed: "If you made one more piece of licorice, what do you think (predict) its mass might be? How did you decide?" Students were readily able to respond to the first part of this question, with 88% (N =24) identifying an appropriate mass range for the handmade and 96% for the equipment-made sticks. The majority of students could also offer appropriate reasons for each decision, referring to either their own data or the whole-class data. Their reasons included, "I think because most of mine were around ten and mine were both exactly 1 cm wide and 8 cm long"; "because it is about the average"; and "I decided because 13 g is the typical mass of sticks in the class".

As part of a follow-up class discussion, the students were also asked, "If another student came into our class and made some licorice, what do you think hers would be (mass of licorice stick)?" In their responses, the students frequently referred to chance and uncertainty when explaining what the mass of a licorice stick made by a new student might be. For example, one student explained that, "It might be 13 because most people got … 13 so maybe that's the typical number". Another explained, "I think maybe 12, because if she came in, there's a chance, because the Fun Factory makes all of them um pretty similar and, … but I decided on that [13 g] because I think there's a more likely chance that she would [make that mass] because it won't always be bigger, she might get it a little smaller than some".

The teacher asked a further question, namely, "Would you expect, say, if we did it again next week and we used the same Play-Doh, and we used the same Fun Factory, would you expect the same plots (i.e. the same class plots of the two licorice-making methods)?" Alesha expressed the opinion, "I think they might be different because like we could do something, we may have like cut it a bit further or because it's really hard to get everything exact, so it won't always be exact". Monica agreed, "…maybe or maybe not, I sort of agree … you actually don't know because … when you made three of them like last week they weren't all the same mass, they weren't all 15 or they weren't all 13…"

What is the Chance of That? As mentioned earlier in this section, children were asked to predict the colours they might select from their bag of counters and give reasons for their answer. The students were also asked if their predictions would guarantee the outcomes. The students were readily able to justify their predictions based on the proportions of counters in their bags, with 75% (N =24) offering reasons such as "I predicted green because there were 4 green counters and all the other colours were less". One student simply referred to a random selection: "My prediction was yellow. I chose it because I randomly chose", while 21% offered a general reason unrelated to chance notions or an irrelevant response, such as, "I thought I would get green because I practised it in random and I got green and because it is one of my favourite colours". Another student also referred to predicting her favourite colour, while another explained, "I predicted blue because when the counters were in a pile blue was on top so when we put it in [the bag] it would still be on top".

Almost all students were able to give appropriate reasons for why their predictions would not guarantee the outcomes, with comments such as: "No, because we have an equal chance of getting each colour because there are three of each colour", and "I can't be certain that I will always pick a green counter but it may be likely that I will pick a green or blue".

One student wrote, "I guarantee I will get *a* colour. I don't guarantee that I will get silver" [there were no silver counters].

17.6.3 Creating Representations and Interpreting Resultant Models

Manufacturing Licorice. Perhaps not surprising, given the typical nature of early data experiences in their curriculum, the children mostly created bar graphs to display their licorice-making results. Two students in two groups, however, used a three-way table (Fig. 17.1), with one student using both tallies and a three-way table to represent her data. As can be seen in Fig. 17.1, the former student also indicated the frequencies of some of the masses.

Although the children favoured bar graphs, they differed in their approaches to organizing and structuring their data. For example, many students (78%, $N = 23$) structured their data according to each group member's results (e.g. Monica, Kate, Sarah), while some (13%) ordered the data differently, such as from the "biggest licorice" to "second biggest", to "second smallest", to "smallest licorice" as illustrated in Fig. 17.2. One student displayed each member's heaviest licorice stick only.

On collating the group results to form a class plot for each licorice-making method (see Fig. 17.3), the students were to describe the data distributions of each model. Sixty-two per cent of the students ($N = 24$) could identify one feature of the model for the handmade licorice, (e.g. "very, very lumpy"; "zig-zag"), while 33% of these students could identify multiple characteristics (e.g. "lots of spaces and humps and sections and a lot at the start"). In contrast, all students except one were able to describe the class model developed for the second method, with 79% ($N = 24$) identifying multiple features.

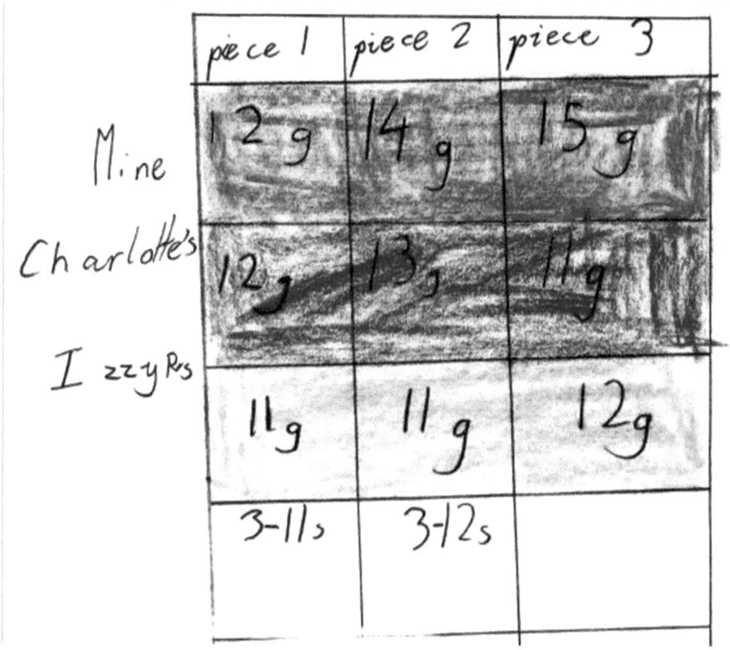

Fig. 17.1 Three-way table displaying one group member's masses

The students' comparisons of the two class plots further suggested their development of statistical literacy as they experimented with the two licorice-making methods (e.g. "handmade was squished together but factory made are apart; factory made looks like a bed to me but handmade looks like boxes in a storage room; handmade are more horizontal but factory made is more vertical. The typical number for handmade was 11 g but in factory made it was 13 g").

What is the Chance of That? In contrast to the *Manufacturing Licorice* representations, the students created a range of ways to display their chances of selecting the counters in their bags. Furthermore, their inscriptions and explanations indicated a linking of their understanding of statistics, probability, and rational number. Table 17.2 displays the forms of representation the children produced for their first and second representations.

It is interesting to note the prevalence of children's use of circle graphs, even though they had not been taught these formally nor had they been introduced to fraction representations using this format (similar findings regarding children's independent use of circle graphs were observed in earlier studies, e.g. English, 2014). Bar graphs were also popular but less so than in the previous investigation. The students increased substantially their use of circle graphs for their second representations; although not observed, it could be that some children had learned about the use of circle graphs from their peers in the first representation. As indicated in Table 17.2, fewer students created bar graphs for their second representation.

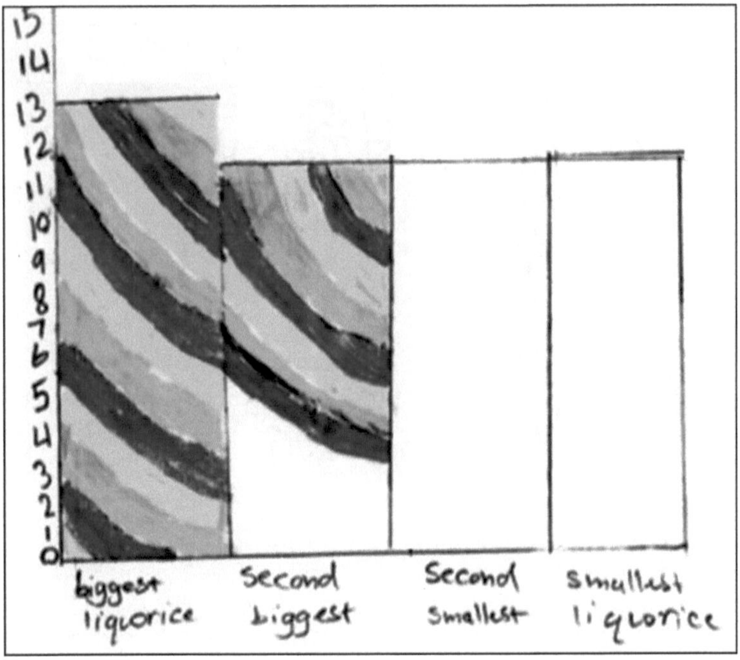

Fig. 17.2 Graph displaying ordering of licorice masses from largest to smallest

Fig. 17.3 Class plots for each licorice-making method

Of particular interest in the students' representations were their inscriptions and accompanying written text. For their first representation, 71% of the students annotated their creations, while 83% did so on their second representation. Their annotations comprised various approaches to documenting the chances of selecting the different colours, as illustrated in Figs. 17.4 and 17.5. Millicent and Greta's represen-

Table 17.2 Students' representations for the chance activity

Representation type	First representation (%)	Second representation (%)
Circle graph	38	63
Bar graph	42	21
Grid	12	4
Picture graph	4	4
Illustration	4	4
Text only	0	4

$N = 24$

tations are chosen as examples of ways in which students linked their understanding of chance, fraction, and statistics.

Millicent (Fig. 17.4) explained that her resultant models "tell you that red is 3 out of 12, blue is 4 out of 12, green is 3 out of 12, and yellow is 2 out of 12 and together they add to twelve". It is interesting that Millicent preferred her bar graph to the circle graph, explaining: "A bar graph because it does the total too, as well as everything else it needs to".

For Millicent, the circle graph did not appear as effective for displaying the total possible outcomes. A desire to display the total possible outcomes is an interesting feature of many of the models created, suggesting further development of statistical literacy. That is, without prompting, the children were able to create models that indicated a linking of chance, fraction, and statistical understandings.

This connected understanding is also evident in Fig. 17.5, where Greta used a range of annotations displaying the chances and likelihood of selecting the different coloured counters. Although Greta did not accurately display the total possible outcomes on her bar graph, she nevertheless indicated the likelihood of each colour being selected. Greta also preferred her bar graph, apparently because of her textual annotations ("The plot because it had it in words").

17.7 Discussion and Concluding Points

This chapter has examined two investigations that revealed 8-year-olds' statistical literacy in modelling with data and chance. Children's responses to both activities were explored in terms of how they identified variation, made informal inferences, created representations, and interpreted their resultant models. Given that the two investigations were the children's first exposure to modelling with data, their responses suggest they were developing important foundational components of statistical literacy. The children could readily identify variation in the data of both investigations, and furthermore, could explain why such variation occurred. They recognized why there was reduced variation in the factory made licorice sticks and understood how varia-

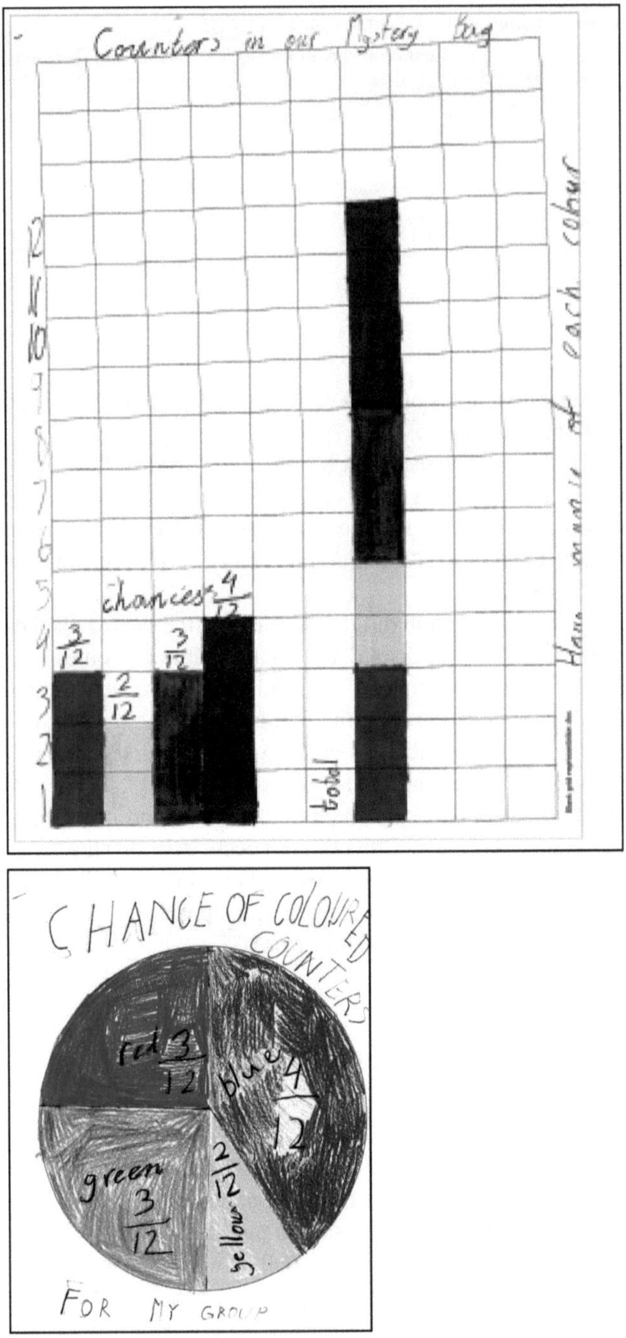

Fig. 17.4 Millicent's annotations

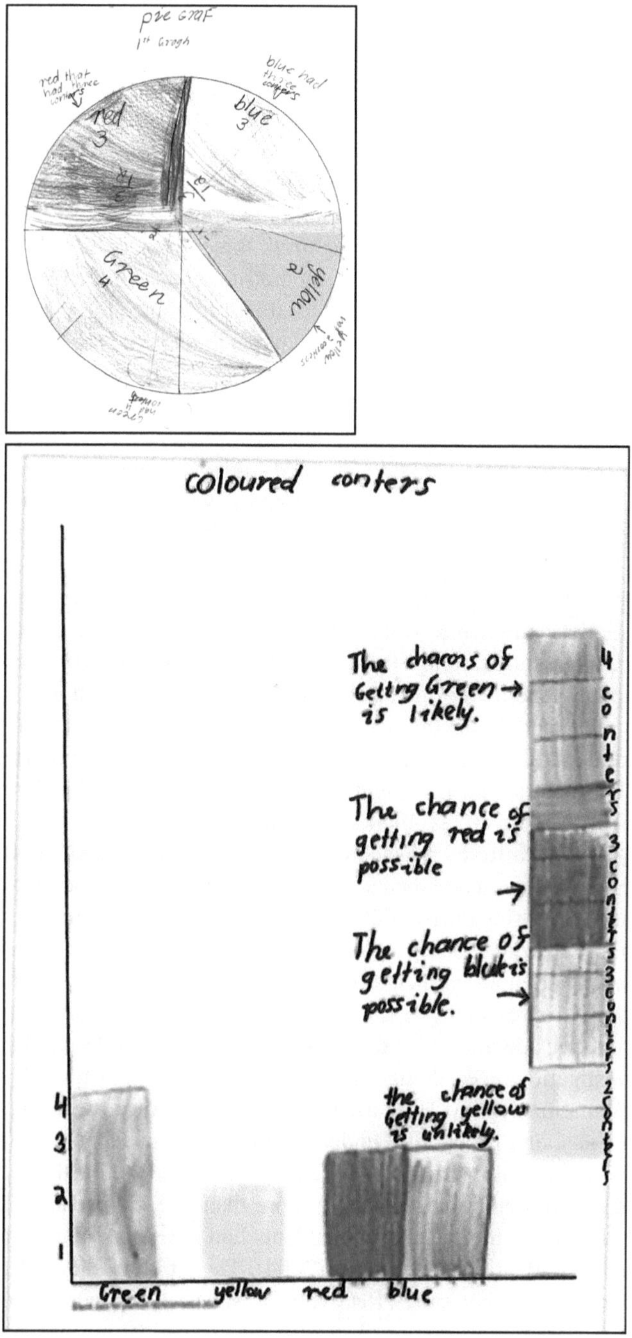

Fig. 17.5 Greta's annotations

tion in their predictions and outcomes of the chance investigation was due largely to the different proportions of their coloured counters.

Using their understanding of variation as a foundation, the students were able to draw informal inferences regarding licorice stick masses that might be produced in future licorice-making activities. Drawing on their understanding of *typical* and *average*, as well as their recognition of other factors that could generate variation, the children displayed a degree of uncertainty in drawing conclusions about future stick masses. For the probability investigation, three quarters of the children made predictions based on their proportions of coloured counters, while a few other students referred to the random nature of selection, or their preference for a particular colour. Nevertheless, the children's "intuitive ideas" or personal beliefs or perceptions about probability (Hawkins & Kapadia, 1984, p. 349) appeared rarely in this particular investigation. In contrast to common activities with equally likely outcomes (e.g. rolling a die), where an "equiprobability" bias can be present (e.g. Khazanov, 2008), the nature of this chance investigation enabled students to appreciate the variation in outcomes possible. Furthermore, students' control over their initial counter selection appeared to facilitate their understanding that predictions and outcomes can vary, and that the former does not guarantee the latter. An appreciation of the relationship between variation and expectation is critical in students' development of formal probability models (English & Watson, 2016).

One of the interesting findings from the children's representations was the greater variety of models generated from the chance investigation, in contrast to the common use of bar graph models for the licorice-making. Despite the preferred use of bar graphs, the students displayed different approaches to organizing and structuring their data in the licorice-making investigation, and furthermore, could identify the data distributions of the whole-class models for the two methods. Their identification of how the distributions differed reflects Konold and Pollatsek's (2002) notion of signal in noise, where the handmade data showed more "noise" than the factory made. In describing and comparing the two data distributions, students identified their features in terms of familiar contexts (e.g. a "bed" and "boxes in a storage room") as well as in terms of statistical notions such as data clusters and typical mass values.

The students' representations for the models produced from the chance investigation varied considerably. There was a greater use of circle graphs than in the previous investigation, even though the children had not been formally introduced to this representational form. Their use of inscriptions, indicating a linking of mathematics, statistics, and probability understandings, was unexpected as, again, they had not received formal instruction in creating such models. Furthermore, with all but a couple of children able to generate more than one representational model to display the same data, it appeared the students had developed the metarepresentational competence identified by diSessa (2004). The children's identification of the representational model that more effectively conveyed the chance outcomes further indicated their conceptual linking of chance and statistical notions. For example, for many children, their desired models needed to indicate clearly the total possible outcomes, which was taken into consideration in model generation. As illustrated in Figs. 17.4 and 17.5, children frequently included an additional bar to display the total outcomes or to document the chances of selecting each colour. Such an inclusion was unexpected and suggests these children had developed a solid grasp of chance outcomes expressed in fraction form. Those students who preferred the circle model gave reasons such as "...because you can easily tell that red and blue are even, and yellow and green are even, and red and blue have more counters", "...because it tells you about how many counters you have and it shows the chance like most likely, least likely and equal", and "I think pie graph because it explains the chance in two ways: in fractions and in chance out of twelve".

Children's responses to both investigations highlight the learning affordances generated when students actually create their own data, experience variation "in action", make predictions based on their findings (rather than someone's else's), and generate their own models to convey their investigative "story". More opportunities that capitalize on, and advance, young children's learning potential in early statistics and probability are clearly warranted, especially when research is revealing the enhanced mathematical skills of today's beginning school students in contrast to previous years (Bassok & Latham, 2017).

Acknowledgments This study was supported by funding from the Australian Research Council (ARC; DP150100120). Views expressed in this paper are those of the author and not the ARC. Collaborators (Jane Watson and Noleine Fitzallen) are acknowledged for their creation of the statistics investigation, while senior research assistant, Jo Macri's contribution to data collection and recording for both investigations is also gratefully acknowledged.

References

Bassok, D., & Latham, S. (2017). Kids today: The rise of children's academic skills at kindergarten entry. *Educational Researcher, 46*(1), 7–20.

Ben-Zvi, D., & Garfield, J. (2004). Statistical literacy, reasoning, and thinking: Goals, definitions, and challenges. In D. Ben-Zvi & J. Garfield (Eds.), *The challenge of developing statistical literacy, reasoning and thinking* (pp. 3–25). Dordrecht, The Netherlands: Kluwer.

Cobb, P., Jackson, K., & Dunlap, C. (2016). Design research: An analysis and critique. In L. D. English & D. Kirshner (Eds.), *Handbook of international research in mathematics education* (3rd ed., pp. 481–503). New York, NY: Routledge.

diSessa, A. A. (2004). Meta-representation: Native competence and targets for instruction. *Cognition and Instruction, 22*(3), 293–331.

diSessa, A. A. (2018). Computational literacy and "The Big Picture" concerning computers in mathematics education. *Mathematical Thinking and Learning*, *20*(1), 3–31.

Doerr, H. M., Delmas, R., & Makar, K. (2017). A modeling approach to the development of students' informal inferential reasoning. *Statistics Education Research Journal, 16*(1). http://iase-web.org/Publications.php?p=SERJ.

English, L. D. (2010). Young children's early modelling with data. *Mathematics Education Research Journal, 22*(2), 24–47.

English, L. D. (2013). Reconceptualising statistical learning in the early years. In L. D. English & J. Mulligan (Eds.), *Reconceptualising early mathematics learning* (pp. 67–82). Dordrecht: Springer.

English, L. D. (2014). Statistics at Play. *Teaching Children Mathematics, 21*(1), 36–45).

English, L., & Watson, J. (2013). Beginning inference in fourth grade: Exploring variation in measurement. In V. Steinle, L. Ball, & C. Bardini (Eds.), Mathematics education: Yesterday, today, and tomorrow. In *Proceedings of the 36th Annual Conference of the Mathematics Education Research Group of Australasia* (Vol. 1, pp. 274–281). Melbourne, Victoria: Mathematics Education Research Group of Australasia. Retrieved from http://www.merga.net.au/documents/English_et_al_MERGA36–2013.pdf.

English, L. D., & Watson, J. M. (2016). Development of probabilistic understanding in fourth grade. *Journal for Research in Mathematics Education, 47*(1), 27–61.

Fielding-Wells, J. (2014). Where's your evidence? Challenging young students' equiprobability bias through argumentation. In K. Makar, B. de Sousa, & R. Gould (Eds.), *Sustainability in statistics education. Proceedings of the Ninth International Conference on Teaching Statistics* (ICOTS9, July, 2014), Flagstaff, Arizona, USA. Voorburg, The Netherlands: International Statistical Institute. iase-web.org [© 2014 ISI/IASE].

Gal, I. (2002). Adults' statistical literacy: Meanings, components, responsibilities. *International Statistical Review, 70*(1), 1–25. https://doi.org/10.1111/j.1751-5823.2002.tb00336.x.

Gal, I. (2005). Towards "probability literacy" for all citizens: Building blocks and instructional dilemmas. In G. A. Jones (Ed.), *Exploring probability in school: Challenges for teaching and learning* (pp. 39–63). New York, NY: Springer. https://doi.org/10.1007/0-387-24530-8_3.

Garfield, J., & Ben-Zvi, D. (2008). *Developing students' statistical reasoning: Connecting research and teaching*. Dordrecht: Springer.

Hawkins, A. S., & Kapadia, R. (1984). Children's conceptions of probability: A psychological and pedagogical review. *Educational Studies in Mathematics, 15*(4), 349–347.

Hestenes, D. (2010). Modeling theory for math and science education. In R. Lesh, P. Galbraith, C. Hines, & A. Hurford (Eds.), *Modeling students' mathematical competencies*. New York: Springer.

Jones, G. A., Langrall, C. W., & Mooney, E. S. (2007). Research in probability: Responding to classroom realities. In F. K. Lester Jr. (Ed.), *Second handbook of research on mathematics teaching and learning* (Vol. 2, pp. 909–955). Charlotte, NC: Information Age.

Kalantzis, M., & Cope, B. (2012). *Literacies*. New York: Cambridge University Press.

Khazanov, L. (2008). Addressing students' misconceptions about probability during the first years of college. *Mathematics and Computer Education, 42*(3), 180–192.

Konold, C., & Pollatsek, A. (2002). Data analysis as the search for signals in noisy processes. *Journal for Research in Mathematics Education, 33*(4), 259–289. https://doi.org/10.2307/749741.

Lehrer, R. (2011, April). *Learning to reason about variability and chance by inventing measures and models*. Paper presented at the annual meeting of the National Association for Research in Science Teaching, Orlando, FL.

Lehrer, R. &, English, L. D. (2018). Modeling variability. In Ben-Zvi, D., Garfield, J., & Makar, K. (Eds.). *International handbook of research in statistics education.* (pp. 229–260). Dordrecht: Springer.

Lesh, R., & Lehrer, R. (2000). Iterative refinement cycles for videotape analyses of conceptual change. In R. Lesh & A. Kelly (Eds.). *Research design in mathematics and science education.* (pp. 665–708). Hillsdale, NJ: Erlbaum.

Lehrer, R., & Schauble, L. (2000). Inventing data structures for representational purposes: Elementary grade students' classification models. *Mathematical Thinking and Learning, 2*, 49–72.

Lehrer, R., & Schauble, L. (2005). Developing modeling and argument in the elementary grades. In T. A. Romberg, T. P. Carpenter, & F. Dremock (Eds.), *Understanding mathematics and science matters* (pp. 29–53). Mahwah, NJ: Erlbaum.

Lehrer, R., & Schauble, L. (2007). Contrasting emerging conceptions of distribution in contexts of error and natural variation. In M. C. Lovett & P. Shah (Eds.), *Thinking with data* (pp. 149–176). Mahwah, NJ: Erlbaum.

Makar, K. (2016). Developing young children's emergent inferential practices in statistics. *Mathematical Thinking and Learning, 18*(1), 1–24.

Makar, K., & Confrey, J. (2005). "Variation-talk": Articulating meaning in statistics. *Statistics Education Research Journal, 4*(1), 27–54. http://www.stat.auckland.ac.nz/serj.

Makar, K., & Rubin, A. (2009). A framework for thinking about informal statistical inference. *Statistics Education Research Journal, 8*(1), 82–105.

Makar, K., Bakker, A., & Ben-Zvi, D. (2011). The reasoning behind informal statistical inference. *Mathematical Thinking and Learning, 13*(1–2), 152–173.

Murphy, S. J. (2001). *Probably Pistachio*. St. Louis, MO: Turtleback Books.

Patton, M. (2002). Qualitative research and evaluation methods (3rd ed.). Thousand Oaks: Sage.

Watson, J. M. (2006). *Statistical literacy at school: Growth and goals*. Mahwah, NJ: Lawrence Erlbaum.

Watson, J., Skalicky, J., Fitzallen, N., & Wright, S. (2009). Licorice production and manufacturing: All-sorts of practical applications for statistics. *Australian primary Mathematics Curriculum, 14*(3), 4–13.

Makar, K., & Confrey, J. (2005). "Variation-talk": Articulating meaning in statistics. *Statistics Education Research Journal, 4*(1), 27–54. http://www.stat.auckland.ac.nz/serj.
Makar, K., & Rubin, A. (2009). A framework for thinking about informal statistical inference. *Statistics Education Research Journal, 8*(1), 82–105.
Makar, K., Bakker, A., & Ben-Zvi, D. (2011). The reasoning behind informal statistical inference. [illegible]
Murphy, S. J. [illegible]
[illegible]
Watson, J. M. (2006). *Statistical literacy at school: Growth and goals*. Mahwah, NJ: Lawrence Erlbaum.
[illegible]

Makar, K., & Rubin, A. (2009). A framework for thinking about informal statistical inference. *Statistics Education Research Journal, 8*(1), 82–105. Retrieved from http://iase-web.org/documents/SERJ/SERJ8(1)_Makar_Rubin.pdf.

Moore, D. S. (1990). Uncertainty. In L. A. Steen (Ed.), *On the shoulders of giants: New approaches to numeracy* (pp. 95–137). Washington, DC: National Academy Press.

[illegible] *[illegible] Standards*. Reston, VA: Author.

[illegible] (2004). Teaching [illegible] the [illegible] approach. *[illegible] Journal of [illegible] Education*, [illegible], 133–152.

[illegible] (2004). [illegible] Children [illegible] [illegible]

[illegible]

Shaughnessy, J. M. (2007). [illegible] and [illegible] and [illegible] (Eds.), [illegible] of the [illegible] of the Mathematics Education Research Group of Australasia, (Vol. [illegible], pp. [illegible]). [illegible]: MERGA.

[illegible] (Ed.), [illegible] Mathematics [illegible] Reston, VA: National Council of Teachers of Mathematics.

Shaughnessy, J. M., Watson, J., Moritz, J., & Reading, C. (1999). [illegible] (Ed.), [illegible]

[illegible]

[illegible]

[illegible]

[illegible]

Watson, J. M., & [illegible] (2008). [illegible] interpreting and predicting from bar graphs. [illegible]

Watson, J. M., & [illegible] [illegible]

[illegible] Dordrecht, The Netherlands: Springer.
Makar, K. (2016). Exposing young children to activities that develop emergent inferential practices in statistics. In [illegible]
[illegible]

Neuman, W. L. ... [illegible] ... Boston: Allyn and Bacon.
Nucci, L. P. (2001). *Education in the moral domain*. New York: Cambridge University Press.
Pa... [illegible] ... M... [illegible] ... (2018). ... [illegible] ... children in informal ... [illegible]
[illegible]

Makar, K., & Confrey, J. (2005). "Variation-talk": Articulating meaning in statistics. *Statistics Education Research Journal, 4*(1), 27–54. http://www.stat.auckland.ac.nz/serj.
Makar, K., & Rubin, A. (2009). A framework for thinking about informal statistical inference. *Statistics Education Research Journal, 8*(1), 82–105.
Makar, K., Bakker, A., & Ben-Zvi, D. (2011). The reasoning behind informal statistical inference. *Mathematical Thinking and Learning* [illegible]
Murphy, [illegible] St. Louis, MO: [illegible] books.
[illegible]
Watson, J. M. [illegible] *Statistical literacy at school: Growth and goals*. Mahwah, NJ: Lawrence Erlbaum.
[illegible]